Wojciech Penczek, Agata Półrola

Advances in Verification of Time Petri Nets and Timed Automata

Studies in Computational Intelligence, Volume 20

Editor-in-chief
Prof. Janusz Kacprzyk
Systems Research Institute
Polish Academy of Sciences
ul. Newelska 6
01-447 Warsaw
Poland
E-mail: kacprzyk@ibspan.waw.pl

Further volumes of this series can be found on our homepage: springer.com

Vol. 5. Da Ruan, Guoqing Chen, Etienne E. Kerre, Geert Wets (Eds.)
Intelligent Data Mining, 2005
ISBN 3-540-26256-3

Vol. 6. Tsau Young Lin, Setsuo Ohsuga, Churn-Jung Liau, Xiaohua Hu, Shusaku Tsumoto (Eds.)
Foundations of Data Mining and Knowledge Discovery, 2005
ISBN 3-540-26257-1

Vol. 7. Bruno Apolloni, Ashish Ghosh, Ferda Alpaslan, Lakhmi C. Jain, Srikanta Patnaik (Eds.)
Machine Learning and Robot Perception, 2005
ISBN 3-540-26549-X

Vol. 8. Srikanta Patnaik, Lakhmi C. Jain, Spyros G. Tzafestas, Germano Resconi, Amit Konar (Eds.)
Innovations in Robot Mobility and Control, 2005
ISBN 3-540-26892-8

Vol. 9. Tsau Young Lin, Setsuo Ohsuga, Churn-Jung Liau, Xiaohua Hu (Eds.)
Foundations and Novel Approaches in Data Mining, 2005
ISBN 3-540-28315-3

Vol. 10. Andrzej P. Wierzbicki, Yoshiteru Nakamori
Creative Space, 2005
ISBN 3-540-28458-3

Vol. 11. Antoni Ligęza
Logical Foundations for Rule-Based Systems, 2006
ISBN 3-540-29117-2

Vol. 12. Jonathan Lawry
Modelling and Reasoning with Vague Concepts, 2006
ISBN 0-387-29056-7

Vol. 13. Nadia Nedjah, Ajith Abraham, Luiza de Macedo Mourelle (Eds.)
Genetic Systems Programming, 2006
ISBN 3-540-29849-5

Vol. 14. Spiros Sirmakessis (Ed.)
Adaptive and Personalized Semantic Web, 2006
ISBN 3-540-30605-6

Vol. 15. Lei Zhi Chen, Sing Kiong Nguang, Xiao Dong Chen
Modelling and Optimization of Biotechnological Processes, 2006
ISBN 3-540-30634-X

Vol. 16. Yaochu Jin (Ed.)
Multi-Objective Machine Learning, 2006
ISBN 3-540-30676-5

Vol. 17. Te-Ming Huang, Vojislav Kecman, Ivica Kopriva
Kernel Based Algorithms for Mining Huge Data Sets, 2006
ISBN 3-540-31681-7

Vol. 18. Chang Wook Ahn
Advances in Evolutionary Algorithms, 2006
ISBN 3-540-31758-9

Vol. 19. Ajita Ichalkaranje, Nikhil Ichalkaranje, Lakhmi C. Jain (Eds.)
Intelligent Paradigms for Assistive and Preventive Healthcare, 2006
ISBN 3-540-31762-7

Vol. 20. Wojciech Penczek, Agata Półrola
Advances in Verification of Time Petri Nets and Timed Automata, 2006
ISBN 3-540-32869-6

Wojciech Penczek
Agata Półrola

Advances in Verification of Time Petri Nets and Timed Automata

A Temporal Logic Approach

With 124 Figures

Springer

Doc. dr. hab. Wojciech Penczek
Institute of Computer Science
Polish Academy of Sciences
ul. Ordona 21
01-237 Warsaw
Poland
and
Institute of Informatics
Podlasie Academy
ul. Sienkiewicza 51
08-110 Siedlce
Poland
E-mail: penczek@ipipan.waw.pl

Dr. Agata Półrola
Faculty of Mathematics
University of Łódz
ul. Banacha 22
90-238 Łódz
Poland
E-mail: polrola@math.uni.lodz.pl

ISSN print edition: 1860-949X
ISSN electronic edition: 1860-9503
ISBN 978-3-642-06942-0
e-ISBN 978-3-540-32870-4

Springer is a part of Springer Science+Business Media
springer.com

Softcover reprint of the hardcover 1st edition 2006

Cover design: deblik, Berlin

To our families

Introduction

Verification of real-time systems is an important subject of research. This is highly motivated by an increasing demand to verify safety critical systems, i.e., time-dependent distributed systems, failure of which could cause dramatic consequences for both people and hardware.

Temporal logic methods have been used for verification over the last twenty years, proving their usefulness for such an application. Whereas infinite state systems still require deductive proof methods, systems of finite abstract models can be verified using algorithmic approaches. This means that the verification process can be fully automated. One of the most promising sets of techniques for verification is known as *model checking*. Essentially, in this formalism verifying that a property follows from a system specification amounts to checking whether or not a temporal formula is valid on a model representing all the possible computations of the system.

Several models of real-time systems are usually considered in the literature, but timed automata (TA) [10] and time Petri nets (TPNs) [106] belong to the most widely used. For these models, one is, usually, interested in checking reachability or more involved temporal properties that are typically expressed either in a standard temporal logic like LTL and CTL*, or in a timed extension of CTL, called TCTL [7]. Unfortunately, practical applicability of model checking methods is strongly limited by the *state explosion problem*, which makes models grow exponentially in the number of the concurrent processes of a system. For real-time systems, this problem occurs with a particular strength, which follows from infinity of the dense time domain. Therefore, existing verification techniques frequently apply symbolic representations of state spaces using either operations on Difference Bound Matrices [68] or similar structures [32] for representing states of abstract models, or variations of Boolean Decision Diagrams like Clock Decision Diagrams (CDDs) [24, 166], Numeric Decision Diagrams (NDDs) [18], Difference Decision Diagrams (DDDs) [107,108], or Clock Restriction Diagrams (CRDs) [168].

Quite recently, a new approach to verification of real time systems, based on SAT-related algorithms, has been suggested. The reason for this is a dramatic increase in efficiency of SAT-solvers over the last few years. The SAT-based approach can exploit either a sequence of translations starting from a timed system and a timed temporal property, going via (quantified) separation logic to quantified propositional logic and further to propositional logic [20, 110, 140] or a direct translation from a timed system and a timed temporal property to propositional logic [120, 171, 177].

Finite state models for timed systems, preserving properties to be checked, are usually built using the detailed region graph approach or (possibly minimal) abstract models, based on state classes or regions. Algorithms for generating such models have been defined for time Petri nets [32, 35, 73, 102, 111, 117, 163, 174], as well as for timed automata [7, 8, 41, 66, 125, 159].

It seems that in spite of the same underlying timed structure, model checking methods for time Petri nets and timed automata have been developed independently of each other. However, several attempts to combine the two approaches were made, concerning both a structural translation of one model into the other [52, 60, 78, 89, 103, 124, 144] or an adaptation of existing verification techniques [73, 111, 163].

The aim of this monograph is to present a recent progress in the development of two model checking methods, based on either building abstract state spaces or application of SAT-based symbolic techniques. The latter is achieved indirectly for time Petri nets, namely via a translation to timed automata. Our special emphasis is put not only on the verification methods, but also on specification languages and their semantics.

Structure of the book

Chapter 1 of this book introduces Petri nets, discusses their time extensions, and provides a definition of time Petri nets. Our attention is focused on a special kind of TPNs – distributed time Petri nets, which, in a sense, correspond to networks of timed automata introduced in Chapter 2. Two main alternative approaches to the semantics of time Petri nets are considered. The first of them consists in assigning clocks to various components of the net, i.e., the transitions, the places, or the processes, whereas the second exploits the so-called *firing intervals* of the transitions. The chapter ends with comparing the above semantics as well as their dense and discrete versions.

Chapter 2 considers timed automata, which were introduced by Alur and Dill [10]. Timed automata are extensions of finite state automata. We give semantics of timed automata and show how to define their product. Typically, we consider *networks* of timed automata, consisting of several concurrent TA running in parallel and communicating with each other. Concrete models are defined and progressiveness is discussed.

Chapter 3 deals with various structural translations from TPNs to TA. They enable an application of specific verification methods for timed automata to time Petri nets. Several methods of translating time Petri nets to timed automata have been already developed. However, in most cases translations produce automata which extend timed automata. We sketch some of the existing approaches, but focus mainly on the translations that correspond to the semantics of time Petri nets, associating clocks with various components of the nets, like the places, the transitions, or the processes.

Chapter 4 introduces temporal specification languages. We start our presentation with the standard branching time logic CTL*, its extension modal μ-calculus, and then discuss timed temporal logics: TCTL and timed μ-calculus. It is important to mention that we consider two versions of syntax of TCTL interpreted over either weakly or strongly monotonic runs.

Chapter 5 gives model abstraction methods based on state classes approaches for TPNs and on partition refinement for TA. For time Petri nets we discuss different abstract models like state class graphs, geometric region graphs, atomic state class graphs, pseudo-atomic state class graphs, and strong state class graphs. For timed automata we concentrate on detailed region graphs, (pseudo-)bisimulating models, (pseudo-)simulating models, and forward-reachability graphs. The last section of this chapter gives an overview of difference bound matrices (DBMs), which are used for representing states of abstract models.

Chapter 6 deals with model checking methods for CTL. These methods include a standard state labelling algorithm as well as an automata-theoretic approach. Moreover, we show that model checking for TCTL over timed automata can be reduced to model checking for CTL.

Chapter 7 discusses SAT-based verification techniques, like bounded (BMC) and unbounded model checking (UMC). The main idea behind BMC [120,171] consists in translating the model checking problem for an existential fragment of some branching-time temporal logic (like CTL or TCTL) on a fraction of a model into a test of propositional satisfiability, for which refined tools already exist [109]. Unlike BMC, UMC [105,140] deals with unrestricted temporal logics checked on complete models at the price of a decrease in efficiency.

Each chapter of our book is accompanied with pointers to the literature, where descriptions of complementary methods or formalisms can be found.

Acknowledgement. The authors would like to thank the following people for commenting on this monograph: Bernard Berthomieu, Franck Cassez, Piotr Dembiński, Magdalena Kacprzak, Sławomir Lasota, Oded Maler, Olivier H. Roux, Maciej Szreter, Stavros Tripakis, Bożena Woźna, Tomohiro Yoneda and Andrzej Zbrzęzny. All the comments from the above experts have greatly helped to improve the book.

The authors acknowledge support from Ministry of Science and Education (grant No. 3T11C 011 28).

Warsaw, Poland,
December 2005

Wojciech Penczek
Agata Półrola

Contents

Part II Model Generation and Verification

List of Figures

List of Symbols

$\cdot$	concatenation, page 159
$\rightarrow$	successor relation of M, page 69
$\rightarrow_{\Pi}$	successor relation of Π, page 123
$\rightarrow_{\Pi \mathrm{L}}$	successor relation between classes of Π (Lee-Yannakakis algorithm, relates marked classes), page 132
$\rightarrow_{\mathfrak{a}}$	abstract transition relation, page 90
$\rightarrow_{\mathfrak{c}}$	concrete successor relation in $C(T)$, page 89
$\rightarrow_{c}$	transition relation of $C_c(\mathcal{A})$, page 38
$\rightarrow_{d_1}$	discrete successor relation of $\mathcal{A}$, page 40
$\rightarrow_{d_2}$	discrete successor relation of $\mathcal{A}$, page 40
$\rightarrow_{\mathfrak{d}}$	transition relation of $DM_{DRG}(\mathcal{A})$, page 188
$\rightarrow_{\mathfrak{d}_A}$	part of $\rightarrow_{\mathfrak{d}}$ where transitions are labelled with elements of $A \cup \{\tau\}$, page 191
$\rightarrow_{\mathfrak{d}_c}$	transition relation of $DM(\mathcal{A})$, page 185
$\rightarrow_{\mathfrak{d}_y}$	part of $\rightarrow_{\mathfrak{d}}$ where transitions are labelled with a_y, page 191
$\rightarrow_{Fc}$	timed consecution relation in $C_c^F(\mathcal{N})$, page 16
$\rightarrow_{Nc}$	timed consecution relation in $C_c^N(\mathcal{N})$, page 14
$\rightarrow_{Pc}$	timed consecution relation in $C_c^P(\mathcal{N})$, page 13
$\rightarrow_{Tc}$	timed consecution relation of $C_c^T(\mathcal{N})$, page 11
$\rightarrow_{\triangleleft}$	successor relation in $DRG(\mathcal{A})$, page 118
$\rightarrow_{\triangleleft b}$	successor relation in $DRG_b(\mathcal{A})$, page 119
$\rightarrow_{\triangleleft b_A}$	page 173
$\rightarrow_{\triangleleft b_{y_i}}$	page 173
$\vert \rightarrow \vert$	the number of transitions of M, page 70
$\equiv_G$	equivalence of classes in geometric region graph, page 104
$\equiv_S$	equivalence of classes in SCG, page 98
$\models$	page 70
$\models$	satisfaction relation for $\mathcal{C}_{\mathcal{X}+}^{\ominus}$, page 30
$\models$	satisfaction relation for $\mathcal{C}_{\mathcal{X}}^{\ominus}$, page 29
$\preceq$	ordering of bounds, page 146

$\simeq_{\mathcal{C}_{\mathcal{A},\varphi}}$	equivalence of clock valuations, page 116
$\lfloor\delta\rfloor$	integral part of δ, page 116
$\bar{\delta}$	negation of the variable δ, page 208
Δ	discretization step, page 183
Δ	transition function of $\mathcal{A}$, page 160
ϵ	label of adjust transitions in $DM(\mathcal{A})$, page 185
ϵ	empty sequence of transitions (state class methods), page 104
ϵ	empty sequence, page 159
η	sequence of transitions, element of C in some state class methods, page 103
η	node of a tree, page 159
ν	fictitious transition denoting start of $\mathcal{N}$, page 104
$\xi_{i,j}$	path from x_i to x_j in G_φ, page 216
π	path (of a tree), page 159
π	path, page 191
π	path in a CTL* model, page 70
π_i	page 70
$\pi(i)$	page 191
Π	partition of a state space, page 122
Π^{ps}	pseudoclasses of Π, page 127
Π_0	initial partition (minimization algorithms), page 122
$\Pi_k(s)$	set of all the k-paths starting at s, page 191
ρ	run of timed automaton, page 38
σ^F	concrete state in the firing interval semantics of $\mathcal{N}$, page 15
$(\sigma^F)^0$	initial state in $C_c^F(\mathcal{N})$, page 16
σ^N	concrete state in the semantics assigning clocks to the processes of $\mathcal{N}$, page 14
$(\sigma^N)^0$	initial state in $C_c^N(\mathcal{N})$, page 14
σ^P	concrete state in the semantics assigning clocks to the places of $\mathcal{N}$, page 12
$(\sigma^P)^0$	initial state in $C_c^P(\mathcal{N})$, page 13
σ^T	concrete state in the semantics assigning clocks to the transitions of $\mathcal{N}$, page 11
$(\sigma^T)^0$	initial state of $C_c^T(\mathcal{N})$, page 11
Σ^F	set of the concrete states in $C_c^F(\mathcal{N})$, page 16
Σ^N	set of the concrete states in $C_c^N(\mathcal{N})$, page 14
Σ^P	set of the concrete states in $C_c^P(\mathcal{N})$, page 13
Σ^T	set of the concrete states of $C_c^T(\mathcal{N})$, page 11
τ	label for time steps in time-abstracted models, page 94
$\mid\varphi\mid$	the length (or size) of φ, page 69
$[\varphi]_k$	page 205
$[\varphi]_k^0$	page 205
${}_h[\varphi]_k^0$	page 205
$[\![\psi]\!]$	page 207
$\psi[\mathbf{A}]$	page 208

ψ^a_{bool}	page 211
ψ^a_{cons}	page 211
$[\psi]_{M_k}$	conjunct of $[M, \psi]_k$, page 195
$\psi_1 \rightsquigarrow \psi_2$	page 208
$\psi[a \leftarrow b]$	substitution (a is substituted with b in ψ)
a	element of $\mathbb{A}$, page 159
$\underline{a}$	page 203
a_{y_i}	label of $\mathcal{A}_\varphi$, page 172
A	universal quantifier of CTL*, page 67
A	set of the actions of $\mathcal{A}$, page 33
$\mathbf{A}$	assignment, page 219
$\mathcal{A}$	timed automaton, page 33
$\mathcal{A}$	alternating automaton over infinite trees, page 160
$\mathbb{A}$	labels (of a tree), finite alphabet, page 159
A'	set of actions of $\mathcal{A}_\varphi$, page 172
$\mathcal{A}_\varphi$	$\mathcal{A}$ extended to verify a TCTL formula φ, page 172
$A_\mathcal{N}$	page 54, 56, 60
$\mathbb{A}_r$	labels of r, page 161
$\mathcal{A}_{\mathcal{D},\varphi}$	page 164
$\mathcal{A}_{M\times\varphi}$	page 166
$A(a)$	indices of the components containing the action a, page 35
$\mathcal{A}_{i_1} \parallel \ldots \parallel \mathcal{A}_{i_{n_j}}$	product of timed automata, page 35
AA	alternating automata, page 158
ACTL	Universal Computation Tree Logic, page 68
ACTL*	universal CTL*, page 68
AE	condition AE, page 91
$AM_\mathcal{N}$	set of the markings of $\mathcal{N}$, page 8
$AM^n_\mathcal{N}$	set of the markings bounded by n of $\mathcal{N}$, page 8
$AM_\mathcal{P}$	set of the markings of $\mathcal{P}$, page 4
$AM^n_\mathcal{P}$	set of the markings of $\mathcal{P}$ bounded by m, page 4
$b(q)$	set of the b-successors of q, page 123
$b^{-1}(q)$	set of the states for which q is b-successor, page 123
B	set of labels of successor relation $\rightarrow_c$ in $\mathcal{T}$, page 89
$\mathcal{B}^+(Y)$	set of the positive boolean formulas over Y, page 159
$c_{max}(\mathcal{A}, \varphi)$	the largest constant in $\mathcal{C}_\mathcal{A}$ and time intervals in φ of TCTL or in clock constraints in φ of TCTL$_\mathcal{C}$, page 116
$c_{max}(\mathcal{A})$	the largest constant appearing in $\mathcal{C}_\mathcal{A}$, page 116
$\mathcal{C}$	simple cycle, page 214
C	state class
C^0	initial state class
$\mathcal{C}^\ominus$	union of $\mathcal{C}^\ominus_\mathcal{X}$ and $\mathcal{C}^\ominus_{\mathcal{X}+}$, page 30
$\mathcal{C}_\mathcal{A}$	clock constraints appearing in enabling conditions and invariants of $\mathcal{A}$, page 116
$\mathcal{C}_\mathcal{X}$	set of the clock constraints without clock differences (over $\mathcal{X}$), page 29

$\mathcal{C}_{\mathcal{X}}^{\ominus}$	set of the clock constraints over $\mathcal{X}$, page 29
$\mathcal{C}_{\mathcal{X}+}^{\ominus}$	set of the normalised clock constraints, page 30
$C(\mathcal{T})$	concrete state space of a system $\mathcal{T}$, page 89
$C_c(\mathcal{A})$	concrete dense state space of $\mathcal{A}$, page 38
$C_c^F(\mathcal{N})$	concrete state space in the firing interval semantics of $\mathcal{N}$, page 16
$C_c^N(\mathcal{N})$	concrete state space in the semantics assigning clocks to the processes of $\mathcal{N}$, page 14
$C_c^P(\mathcal{N})$	concrete state space in the semantics assigning clocks to the places of $\mathcal{N}$, page 13
$C_{d_r}^R(\mathcal{N})$	concrete (discrete) state space, page 22
$C_c^T(\mathcal{N})$	concrete state space in the semantics assigning clocks to the transitions of $\mathcal{N}$, page 11
$\mathfrak{cc}$	clock constraint, page 29
$[\![\mathfrak{cc}]\!]$	set of the clock valuations satisfying $\mathfrak{cc}$, page 30
$cf(D)$	canonical form of DBM D, page 147
$clock^N$	function returning the time elapsed since the marked place of the process became marked most recently in $\mathcal{N}$, page 14
$clock^P$	function returning the time since a place became marked most recently in $\mathcal{N}$, page 12
$clock^T$	function returning the time elapsed since a transition became enabled most recently in $\mathcal{N}$, page 11
$Closure$	page 151
CNF	conjunctive normal form, page 218
$toCNF(\psi)$	page 219
$cor(w)$	core of the state w, page 91
$\mathrm{cr}(\varphi)$	φ translated to $\mathrm{CTL}^{\mathfrak{r}}_{-\mathrm{X}}$, page 172
CTL	Computation Tree Logic, page 68
CTL*	Computation Tree Logic*, page 67
$\mathrm{CTL}^{\mathfrak{z}}$	page 177
$\mathrm{CTL}^{\mathfrak{r}}_{-\mathrm{X}}$	modified $\mathrm{CTL}_{-\mathrm{X}}$, page 172
$\mathbf{d}_{ij}$	element of difference bound matrix, page 147
$d(\eta)$	degree of η, page 159
$\mathcal{D}$	set of degrees, page 159
$\mathbb{D}$	discretized clock space, page 183
$[\![D]\!]$	zone defined by DBM D, page 147
DBM	difference bound matrix, page 147
$DM(\mathcal{A})$	discretized (concrete) model for $\mathcal{A}$, page 185
$DM_{DRG}(\mathcal{A})$	discretized region graph model for $\mathcal{A}$, page 188
$DM_{DRG_b}(\mathcal{A})$	discretized boundary-distinguishing region graph model for $\mathcal{A}$, page 190
$dom(\mathbf{A})$	domain of $\mathbf{A}$, page 219
dpt	auxiliary function in minimization algorithm for simulating models, page 135
$dpt(w)$	depth of state w, page 92

DRG	detailed region graph, page 118
$DRG(\mathcal{A})$	detailed region graph for $\mathcal{A}$ (weakly monotonic semantics), page 118
$DRG_b(\mathcal{A})$	boundary-distinguishing detailed region graph for $\mathcal{A}$ (DRG for strongly monotonic semantics), page 119
$DZ(n_{\mathcal{X}})$	set of all the detailed zones for $\mathcal{X}$, page 117
$\underline{e}$	page 203
e_δ	page 203
e^{i}_{null}	page 203
E	existential quantifier of CTL*, page 67
$\mathbb{E}$	page 183
E	transition relation of $\mathcal{A}$, page 33
E'	transition relation of $\mathcal{A}_\varphi$, page 172
$E_{\mathcal{N}}$	page 54, 57
EA	condition EA, page 91
ECTL	Existential Computation Tree Logic, page 68
ECTL*	existential CTL*, page 68
EE	condition EE, page 90
EE_1	condition EE_1, page 90
EE_2	condition EE_2, page 90
Eft	the earliest firing time function of $\mathcal{N}$, page 8
$en(m)$	set of the transitions enabled at m in $\mathcal{P}$, page 5
$enc_{PL}(l)$	page 209
$extr(Z)$	extrapolation of zone Z, page 142
$extr_d(Z)$	extrapolation of zone Z (diagonal constraints case), page 144
$f_{\mathcal{A}}(q)$	set of the progressive (weakly monotonic) q-runs of $\mathcal{A}$, page 39
$f^+_{\mathcal{A}}(q)$	set of the (progressive) strongly monotonic q-runs of $\mathcal{A}$, page 39
f_k	function determining the number of k-paths of a submodel sufficient to check a formula, page 194
$f_{\mathcal{N}}(\sigma)$	set of the progressive weakly monotonic σ-runs of $\mathcal{N}$, page 20
$f^+_{\mathcal{N}}(\sigma)$	set of the progressive strongly monotonic σ-runs of $\mathcal{N}$, page 20
F	eventually operator of CTL*, page 67
F	flow function of $\mathcal{P}$, page 3
F	acceptance condition for $\mathcal{A}$, page 160
fi	firing interval in $\mathcal{N}$, page 15
$Fill$	page 151
$frac(\delta)$	fractional part of δ, page 116
$free(l)$	condition to check deadlock-freedom of $\mathcal{A}$, page 45
FV	fixed-point variables of μ-calculus, page 71
G_φ	inequality graph for φ, page 213
G	always operator of CTL*, page 67
$guard_{\mathcal{N}}$	page 54, 56, 60

$H(w, w')$	page 197
HAA	hesitant alternating automata, page 158
$\mathcal{I}$	location invariant of $\mathcal{A}$, page 33
I	set of inequalities, page 96
$\mathcal{I}_{\mathcal{A}}$	page 209
$\mathfrak{I}_m$	page 56, 57, 58
$\mathcal{I}_{\mathcal{N}}$	page 54, 56
$I_s(w)$	page 197
$I_1, \ldots, I_r$	intervals in φ, page 172
$inv_{\mathcal{N}}$	page 56, 60
$Iqs(\mathcal{V})$	set of all the inequalities over $\mathcal{V}$, page 96
$IqSets(\mathcal{V})$	set of all the subsets of $Iqs(\mathcal{V})$, page 96
l^0	initial location of $\mathcal{A}$, page 33
l_ψ	page 219
$\mathfrak{l}_b$	length of a vector of global state variable, page 196
L	locations of $\mathcal{A}$, page 33
L_μ	modal μ-calculus, page 71
$\mathfrak{L}_m$	page 56
$L_{\mathcal{N}}$	page 54, 56
$\mathrm{L}_{-\mathrm{X}}$	logic L without the next-step operator X, page 68
$\mathcal{L}(\mathcal{A})$	language of $\mathcal{A}$, page 163
$L(k, h)$	page 205
$L_j(k, h)$	page 197
Lft	the latest firing time function of $\mathcal{N}$, page 8
$lit()$	page 196
$loop$	page 192
LTL	Linear-Time Temporal Logic, page 68
m	marking of $\mathcal{P}$, page 4
m^0	initial marking of $\mathcal{P}$, page 3
$m[t\rangle$	t enabled at m in $\mathcal{P}$, page 5
M	CTL model, page 69
M_k	k-model, page 193
$[M]_k$	page 205
$\mid M \mid$	size of M, page 70
$[M^{\psi, s^0}]_k$	conjunct of $[M, \psi]_k$, page 195
$[M, \psi]_k$	propositional formula to be tested to check ψ over M_k, page 195
$M_{\mathfrak{a}}(\mathcal{T})$	abstract model for $\mathcal{T}$, page 89
$M_c(\mathcal{A})$	concrete dense model for $\mathcal{A}$, page 183
$M_{\mathfrak{c}}(\mathcal{N})$	concrete model of $\mathcal{N}$, page 23
$M_{\mathfrak{c}}(\mathcal{T})$	concrete model of a given system $\mathcal{T}$, page 89
$M_{DRG_b}(\mathcal{A}_\varphi)$	modified bd-region graph model for $\mathcal{A}_\varphi$, page 173
$M'_{DRG_b}(\mathcal{A}_\varphi)$	extended $M_{DRG_b}(\mathcal{A}_\varphi)$, page 174
$\mathbb{N}$	set of natural numbers, page 3
$\mathfrak{N}$	family of 1-safe sequential TPNs, page 9

$\mathcal{N}$	time Petri net, page 7
$\mathbb{N}_+$	set of positive natural numbers, page 3
$\mathbb{N}^*$	set of all the finite sequences of natural numbers, page 159
$newly\_en(m,t)$	set of the transitions newly enabled after firing t, page 5
$\mathcal{O}$	page 161
ox	page 203
$\wp_b$	page 173
$\wp_{y_i \in I_i}$	new proposition for interval I_i, page 173
$\wp(w)$	page 197
$\mathcal{P}$	Petri net, page 3
P	set of places of $\mathcal{P}$, page 3
P_k	set of all the k-paths of a discretized rg($_b$) model, page 192
P_y	set of all the pairs of states $(s, s[y := 0])$ of a discretized rg($_b$) model, page 192
pAE	condition pAE, page 92
$parent(v_t, C)$	variable for the parent of t in C, page 104
PL	propositional logic, page 63
$Post_\Pi(Y)$	successors of Y in Π, page 123
$post_b(Y, Y')$	page 123
$Pre_\Pi(Y)$	predecessors of Y in Π, page 123
$pre_\mathcal{A}(\psi)$	page 209
$pre_b(Y, Y')$	page 123
$pre_e(\psi)$	page 209
pU	condition pU, page 92
PV	propositional variables of PL, page 64
PV'	extended PV, page 173
PV_φ	set of the new propositions for intervals in φ, page 173
PV_P	set of the propositional variables corresponding to P of $\mathcal{N}$, page 23
$PV(\varphi)$	set of propositions occurring in φ, page 67
$\mathfrak{q}^0$	inital state of $\mathcal{A}$, page 160
q^0	initial state of $C_c(\mathcal{A})$, page 38
$\mathcal{Q}$	states of $\mathcal{A}$, page 160
Q	set of the concrete states of $C_c(\mathcal{A})$, page 38
$\mathbb{Q}_+$	set of positive rational numbers, page 3
$\mathbb{Q}_{0+}$	set of non-negative rational numbers, page 3
QPL	quantified propositional logic, page 65
QSL	quantified separation logic, page 66
r	run of $\mathcal{A}$, page 161
r	number of the intervals in φ, page 172
r_Y	representative of the class Y of Π, page 130
R	release operator of CTL*, page 67
$\mathcal{R}$	bisimulation of concrete state spaces of TPNs, page 17
$\mathbb{R}$	set of rational numbers, page 3
$\mathbb{R}$	set of real numbers, page 3

R	region, page 123
$\mathbb{R}_{+}$	set of positive real numbers, page 3
$\mathbb{R}_{0+}$	set of non-negative real numbers, page 3
$R(n_{\mathcal{X}}, L)$	set of the regions over $\mathcal{X}$ and L, page 123
$\mathcal{R}(w, w')$	page 197
$R_x(w, w')$	page 197
$R \setminus R'$	difference of regions R, R', page 123
$Reach_{\mathcal{A}}$	set of the reachable states of $\mathcal{A}$, page 39
$Reach^{+}_{\mathcal{A}}$	set of the reachable states on the strongly monotonic q^0-runs of $\mathcal{A}$, page 39
$Reach^{d_r}_{\mathcal{A}}$	reachable states on the discrete runs of $\mathcal{A}$, page 40
$Reach^{R}_{\mathcal{N}}$	set of the reachable states of $\mathcal{N}$, page 19
$Reach^{+R}_{\mathcal{N}}$	set of the states reachable on the strongly monotonic runs of $\mathcal{N}$ ($R \in \{F, P, N, T\}$ and corresponds to the semantics considered), page 20
$Reach^{Rd_r}_{\mathcal{N}}$	set of the reachable states on the discrete runs of $\mathcal{N}$, page 21
$Reach(\mathcal{T})$	set of all the states reachable in a given concrete state space $C(\mathcal{T})$ for $\mathcal{T}$, page 89
$Reach(\mathcal{U})$	set of all the reachable elements of the set family $\mathcal{U}$, page 89
$reset_{\mathcal{N}}$	page 54, 57, 60
$restrict$	page 153
$Rlx(C)$	page 113
$RM_{\mathcal{N}}$	set of the reachable markings of $\mathcal{N}$, page 19
$RM^{+}_{\mathcal{N}}$	set of the reachable markings on the strongly monotonic runs of $\mathcal{N}$, page 20
$RM^{k_{\mathcal{P}}}_{\mathcal{P}}$	set of all the reachable markings of $\mathcal{P}$ for the capacity $k_{\mathcal{P}}$, page 5
s^0	initial state of M, page 69
$s[y := 0]$	page 191
S	set of states of M, page 69
$S_{\mathbb{D}}$	set of the states of $DM(\mathcal{A})$, page 185
$\mid S \mid$	the number of states of M, page 70
SC	specification clock variables, page 80
$SF(\varphi)$	set of the subformulas of φ, page 69
SL	separation logic or difference logic, page 65
$sol(I)$	set of the solutions of I, page 96
Sp_s	auxiliary function to split classes in minimization algorithm for simulating models, page 133
$Succ_a$	page 138
$succ(s)$	page 166
$t\bullet$	postset of transition t, page 4
$\bullet t$	preset of transition t, page 4
$\mathfrak{T}$	tree, page 159
$\mathcal{T}$	timed system, page 89
T	set of transitions of $\mathcal{P}$, page 3

T^*	set of all the finite sequences of transitions of T, page 103
$\mathfrak{T}_r$	tree of a run r, page 161
$T\mu$	timed μ-calculus, page 80
TCTL	timed CTL, page 75
$\mathrm{TCTL}_\mathcal{C}$	alternative timed CTL, page 82
U	Until operator of CTL*, page 67
U	condition U, page 91
$\mathbb{U}^{n_x}$	valuations to represent detailed zones in discretized model, page 183
v	clock valuation, page 29
v_t	variable corresponding to transition t (state class approaches), page 97
v_t^j	variable for j-th firing of t (state class methods)
$v(i)$	value of clock x_i in v, page 29
$v(x_i)$	value of clock x_i in v, page 29
$v[X := 0]$	valuation like v where the clocks of X are set to 0, page 29
V	valuation function of PL, page 64
V_φ	function labelling states of $M_{DRG_b}(\mathcal{A}_\varphi)$ with propositions from PV_φ and with $\wp_b$, page 173
V_a	valuation function in $M_\mathfrak{a}(\mathcal{T})$, page 90
$V_\mathcal{A}$	valuation function for $\mathcal{A}$, page 41
$V_\mathfrak{c}$	valuation function for a concrete state space of $\mathcal{A}$, page 41
$V_\mathfrak{C}$	valuation function in $M_\mathfrak{c}(\mathcal{T})$, page 89
$V_\mathfrak{D}$	valuation function of $DM_{DRG}(\mathcal{A})$, page 188
$V_{\mathfrak{D}_c}$	valuation function in $DM(\mathcal{A})$, page 185
$V_\mathcal{N}$	valuation function for $\mathcal{N}$, page 23
V_r	valuation function of $\mathfrak{T}_r$, page 161
$V_\mathfrak{T}$	valuation function (a tree), page 159
V_a'	extended V_a, page 173
$V_{\mathcal{A}_\mathcal{N}}$	page 54, 57, 61
(V, v)	model for (Q)SL, page 66
$V[\wp \leftarrow b]$	substitution in QPL, page 65
$var(I)$	set of the variables appearing in I, page 96
w	abstract state
w	global state variable (vector encoding a state), page 196
w^0	initial state of abstract model, page 90
$w[i]$	state variable, page 196
W	states of abstract model, page 90
WAA	weak alternating automata, page 158
x_0	fictitious clock representing constant 0, page 30
$\mathcal{X}$	set of clocks, page 29
X	next-step operator of CTL*, page 67
$\mathcal{X}'$	set of clocks of $\mathcal{A}_\varphi$, page 172
$\mathcal{X}^+$	set of clocks together with x_0, page 30
R_{y_i}	release operator of $\mathrm{CTL}^{\mathrm{r}}_{-\mathrm{X}}$, page 174

Part I

Specifying Timed Systems and Their Properties

1

Petri Nets with Time

We consider two main models of real-time systems: Petri nets with time and timed automata. First, we define Petri nets, discuss their time extensions, and provide a definition of time Petri nets. Our attention is focused on a special kind of TPNs – distributed time Petri nets, which correspond to networks of timed automata considered in Chap. 2.

The following abbreviations are used in the definitions of both TA and TPNs. Let $\mathbb{R}$ denote the set of real numbers, $\mathbb{Q}$ – the set of rational numbers, $\mathbb{Z}$ – the set of integers, and $\mathbb{N}$ – the set of naturals (including 0). For each $\mathbb{S} \in \{\mathbb{R}, \mathbb{Q}, \mathbb{Z}\}$ by $\mathbb{S}_{0+}$ ($\mathbb{S}_{+}$) we denote a subset of $\mathbb{S}$ consisting of all its non-negative (respectively positive) elements. Moreover, by $\mathbb{N}_{+}$ we mean the set of positive natural numbers. When we deal with elements of $\mathbb{R}_{0+} \cup \{\infty\}$, by the notations "$\leq b$" and "$[a, b]$" we mean "$< b$" and "$[a, b)$" if $a \in \mathbb{R}_{0+}$ and $b = \infty$. We assume also $\infty + a = \infty - a = \infty$.

We start with the standard notion of Petri nets.

Definition 1.1. *A Petri net is a four-element tuple*

$$\mathcal{P} = (P, T, F, m^0),$$

where

- $P = \{p_1, \ldots, p_{n_P}\}$ *is a finite set of* places,
- $T = \{t_1, \ldots, t_{n_T}\}$ *is a finite set of* transitions, *where* $P \cap T = \emptyset$,
- $F : (P \times T) \cup (T \times P) \longrightarrow \mathbb{N}$ *is the* flow function, *and*
- $m^0 : P \longrightarrow \mathbb{N}$ *is the* initial marking *of* $\mathcal{P}$.

Intuitively, Petri nets are directed weighted graphs of two types of nodes: places (representing conditions) and transitions (representing events), whose arcs correspond to these elements in the domain of the flow function, for which the value of this function is positive. The arcs are assigned positive weights according to the values of F.

W. Penczek and A. Półrola: *Petri Nets with Time*, Studies in Computational Intelligence (SCI) **20**, 3–27 (2006)
www.springerlink.com

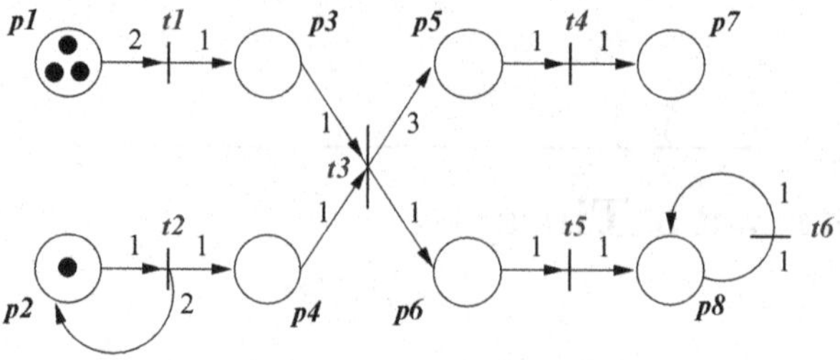

Fig. 1.1. A Petri net

Example 1.2. An example of a Petri net is shown in Fig. 1.1. The set of places of this net is given by $P = \{p_1, \ldots, p_8\}$, the set of transitions by $T = \{t_1, \ldots, t_5\}$, and the initial marking is $m^0(p_1) = 3$, $m^0(p_2) = 1$, and $m^0(p_i) = 0$ for $i = 3, \ldots, 8$. The flow function is defined[1] by

$$F(z) = \begin{cases} 3 & \text{for } z = (t_3, p_5) \\ 2 & \text{for } z \in \{(p_1, t_1), (t_2, p_2)\} \\ 1 & \text{for } z \in \{(t_1, p_3), (p_2, t_2), (t_2, p_4), (p_3, t_3), (p_4, t_3), (t_3, p_6), \\ & \qquad (p_5, t_4), (t_4, p_7), (p_6, t_5), (t_5, p_8), (p_8, t_6), (t_6, p_8)\} \\ 0 & \text{otherwise.} \end{cases}$$

□

For each transition $t \in T$ we define its *preset*

$$\bullet t = \{p \in P \mid F(p, t) > 0\}$$

and its *postset*

$$t\bullet = \{p \in P \mid F(t, p) > 0\}$$

(e.g., for the transition t_2 of the net in Fig. 1.1 we have $\bullet t_2 = \{p_2\}$ and $t_2\bullet = \{p_2, p_4\}$). The elements of $\bullet t$ ($t\bullet$) are called the *input* (*output*, respectively) *places* of the transition t.

In order to simplify some consequent notions, we consider only the nets, for which $\bullet t$ and $t\bullet$ are non-empty for each transition t. We use the following auxiliary notations and definitions:

- A *marking* of $\mathcal{P}$ is any function $m : P \longrightarrow \mathbb{N}$; a place $p \in P$ is called *marked in* m if $m(p) > 0$ (e.g., p_1 and p_2 are marked in m^0 in Fig. 1.1). The value of $m(p)$ is also said to be the number of *tokens* placed in p in the marking m. The set of all the markings of $\mathcal{P}$ is denoted by $AM_{\mathcal{P}}$. Given a number $n \in \mathbb{N}_+ \cup \{\infty\}$, the subset of $AM_{\mathcal{P}}$ consisting of all the markings of $\mathcal{P}$ with $m(p) \leq n$ for each $p \in P$ (i.e., bounded by n) is denoted by $AM^n_{\mathcal{P}}$;

[1] In what follows, we usually identify the notation $F(s, t)$ with $F((s, t))$.

- Given a number $k_{\mathcal{P}} \in \mathbb{N}_+ \cup \{\infty\}$ denoting a maximal capacity of the places of $\mathcal{P}$ (i.e., a bound on the number of tokens in its places). A transition $t \in T$ is (*safely*) *enabled* at a marking m ($m[t\rangle$ for short) if

$$(\forall p \in \bullet t)\ m(p) \geq F(p,t) \text{ and } (\forall p \in t\bullet)\ m(p) - F(p,t) + F(t,p) \leq k_{\mathcal{P}}$$

(e.g., for the net in Fig. 1.1 and $k_{\mathcal{P}} = 3$ the transitions t_1 and t_2 are enabled at m^0). Clearly, we assume that the initial marking m^0 conforms with the maximal capacity $k_{\mathcal{P}}$ of the places. It is worth noticing that in most cases $k_{\mathcal{P}} = \infty$ or $k_{\mathcal{P}} = 1$ is considered;
- A transition $t \in T$ *leads from a marking m to m'* if t is enabled at m, and

$$m'(p) = m(p) - F(p,t) + F(t,p) \quad \text{for each } p \in P$$

(e.g., in the net in Fig. 1.1 for each $k_{\mathcal{P}} \geq 3$ the transition t_2 leads from m^0 to the marking m' given by $m'(p_1) = 3$, $m'(p_2) = 2$, $m'(p_4) = 1$ and $m'(p_i) = 0$ for $i = 3, 5, \ldots, 8$).
The marking m' is denoted by $m[t\rangle$ as well, if this does not lead to misunderstanding;
- By

$$en(m) = \{t \in T \mid m[t\rangle\}$$

we denote the set of the transitions enabled at m (e.g., in Fig. 1.1 for each $k_{\mathcal{P}} \geq 3$, $en(m^0) = \{t_1, t_2\}$);
- For $t \in en(m)$,

$$\begin{aligned} newly\_en(m,t) = \{u \in T \mid u \in en(m[t\rangle) \wedge u \notin en(m') \\ \text{with } m'(p) = m(p) - F(p,t) \text{ for each } p \in P\} \end{aligned}$$

is the set of transitions newly enabled after firing t (e.g., for the net in Fig. 1.1 we have $newly\_en(m^0, t_2) = \{t_2\}$ for $k_{\mathcal{P}} \geq 3$);
- A marking m is *reachable* if there exists a sequence of transitions $t_1, \ldots, t_l \in T$ and a sequence of markings $m_0, \ldots, m_l$ such that $m_0 = m^0$, $m_l = m$, and $t_i \in en(m_{i-1})$, $m_i = m_{i-1}[t_i\rangle$ for each $i \in \{1, \ldots, l\}$ (e.g., in the net in Fig. 1.1 for $k \geq 3$ the marking m' given by $m'(p_1) = 1$, $m'(p_2) = 3$, $m'(p_3) = 1$, $m'(p_4) = 2$ and $m'(p_i) = 0$ for $i = 5, \ldots, 8$ is reachable from m^0 via the sequence of transitions t_2, t_1, t_2). The set of all the reachable markings of $\mathcal{P}$ for a given capacity $k_{\mathcal{P}}$ of the places is denoted by $RM_{\mathcal{P}}^{k_{\mathcal{P}}}$ (notice that $RM_{\mathcal{P}}^{k_{\mathcal{P}}} \subseteq AM_{\mathcal{P}}^{k_{\mathcal{P}}}$);
- A net $\mathcal{P}$ is said to be *bounded* if there is a bound on all its reachable markings assuming unlimited capacity of the places, i.e., there is a bound for each $p \in P$ and $m \in RM_{\mathcal{P}}^{\infty}$ (e.g., the net in Fig. 1.2 is bounded, whereas that in Fig. 1.1 is not since each firing of t_2 adds a token to the place p_2, and therefore there is no bound on the number of the tokens in this place);
- A marking m *concurrently enables* two transitions $t, t' \in T$ if $t \in en(m)$ and $t' \in en(m')$ with $m'(p) = m(p) - F(p,t)$ for each $p \in P$ (e.g., t_1 and t_2 are concurrently enabled in m^0 in the net in Fig. 1.2);

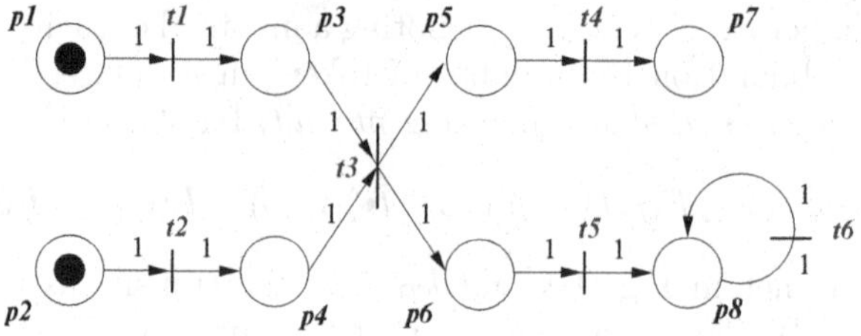

Fig. 1.2. A bounded Petri net

- A net $\mathcal{P}$ is *sequential* if none of its reachable markings concurrently enables two transitions;
- A net $\mathcal{P}$ is *ordinary* if the flow function F maps onto $\{0,1\}$;
- A net $\mathcal{P}$ is *1-safe* if it is ordinary and $m(p) \leq 1$, for each $p \in P$ and each $m \in RM_{\mathcal{P}}^{\infty}$ (an example of such a net is in Fig. 1.2). However, in this book we will use an alternative definition saying that a net $\mathcal{P}$ is *1-safe* if it is ordinary and the capacity of its places is set to 1. Such a definition ensures that for each reachable marking m and each $p \in P$ we have $m(p) \leq 1$ without analysing the reachable markings of the net.

The theory of Petri nets provides a general framework for modelling distributed and concurrent systems. Since for many of them timing dependencies play an important role, a variety of extensions to the main formalism, enabling to reason about temporal properties, has been introduced. In what follows, we present a brief survey of such approaches, based on [46,138].

1.1 Incorporating Time into Petri Nets

Petri nets with timing dependencies can be classified according to the way of specifying timing constraints (these can be timing intervals [106, 165] or single numbers [133]), or elements of the net these constraints are associated with (i.e., places [59], transitions [106,133] or arcs [3,80,147,165]). The next criterion is an interpretation of the timing constraints. When associated with a *transition*, the constraint can be viewed as

- its *firing time* (a transition consumes the input tokens when it becomes enabled, but does not create the output ones until the delay time associated with it has elapsed [133]),
- *holding time* (when the transition fires, the actions of removing and creating tokens are performed instantaneously, but the tokens created are not available to enable new transitions until they have been in their output places for the time specified as the duration time of the transition which created them [1]), or

- *enabling time* (a transition is forced to be enabled for a specified period of time before it can fire, and tokens are removed and created in the same instant [106]).

A time associated with a *place* usually refers to the period the tokens must spend in the place before becoming available to enable a transition [59]. A timing interval on an *input arc* usually expresses the conditions under which tokens can potentially leave the place using this arc [80,147], whereas a timing interval on an *output arc* denotes the time when tokens produced on this arc become available [165].

Nets can be also classified according to *firing rules*:

- the *weak firing rule* means that the time which passes between enabling of the transition and its firing is not determined [161],
- the *strong earliest firing rule* requires the transition to be fired as soon as it is enabled and the appropriate timing conditions are met [80], whereas
- the *strong latest firing rule* means that the transition can be fired in a specified period of time, but no later than after certain time from its enabling, unless it becomes disabled by firing of another one [106].

The best known timed extensions of Petri nets are *timed Petri nets* by Ramchandani [133] and *time Petri nets* by Merlin and Farber [106]. In this book, we focus on the latter in order to provide also verification methods via translations to timed automata.

Timed extensions are known also for high-level Petri nets. One of them are *timed coloured Petri nets* [92], in which the time concept is based on introducing a global clock used to represent the *model time*. Tokens are equipped with *time stamps*, which describe the earliest model times at which they can be used to fire a transition. Stamps are modified according to expressions associated either with transitions, or with their output arcs. Timing intervals can be interpreted as periods of non-activity of tokens, and the transitions are fired according to the strong earliest firing rule.

1.2 Time Petri Nets

Time Petri nets by Merlin and Farber [106] are considered in many papers [32, 33, 35, 37, 39, 43, 73, 78, 102, 103, 111, 117, 124, 127, 130, 163, 174]. In what follows, we introduce the definition and some alternative semantics of time Petri nets, used in the literature.

Definition 1.3. *A time Petri net (TPN, for short) is a six-element tuple*

$$\mathcal{N} = (P, T, F, m^0, Eft, Lft),$$

where

- (P, T, F, m^0) *is a Petri net, and*
- $Eft : T \longrightarrow \mathbb{N}$, $Lft : T \longrightarrow \mathbb{N} \cup \{\infty\}$ *are functions describing respectively the* earliest *and the* latest firing times *of the transitions, where clearly* $Eft(t) \leq Lft(t)$ *for each* $t \in T$.

Example 1.4. An example of a time Petri net is shown in Fig. 1.3. Its underlying Petri net is depicted in Fig. 1.2. The values of the functions Eft and Lft are given by $Eft(t) = 1$ and $Lft(t) = 2$ for $t \in \{t_1, t_3, t_5, t_6\}$, $Eft(t_2) = 0$, $Lft(t_2) = 3$ and $Eft(t_4) = Lft(t_4) = 1$.

□

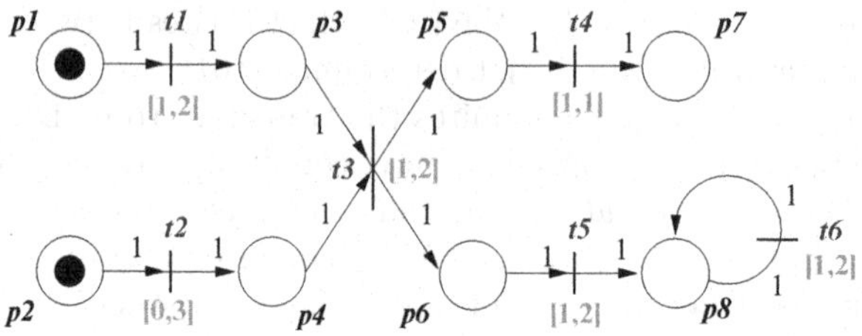

Fig. 1.3. A time Petri net

The earliest and latest firing times of a transition t specify the timing interval in which the transition t can be fired. If the time passed since the transition t has become enabled reaches the value $Lft(t)$, the transition has to be fired, unless disabled by a firing of another transition[2].

A time Petri net $\mathcal{N} = (P, T, F, m^0, Eft, Lft)$ is said to be *sequential* if the net $\mathcal{P}_\mathcal{N} = (P, T, F, m^0)$ is so. Moreover, by a *1-safe* time Petri net we mean a time Petri net whose underlying net is so[3].

Unless otherwise stated, in what follows 1-safe TPNs are considered only. Moreover, by $AM_\mathcal{N}$ we denote the set $AM_{\mathcal{P}_\mathcal{N}}$ of all the markings of the net $\mathcal{P}_\mathcal{N}$, whereas $AM^n_\mathcal{N}$, for $n \in \mathbb{N}_+ \cup \{\infty\}$, denotes the set $AM^n_{\mathcal{P}_\mathcal{N}}$ of all the markings of $\mathcal{P}_\mathcal{N}$ whose values do not exceed n for any $p \in P$.

1.2.1 Distributed Time Petri Nets

In order to benefit from a distributed representation of a system, we define a notion of a *distributed time Petri net*, which is an adaptation of the one from [88], and provide an alternative semantics for these nets in Sect. 1.2.2.

[2] There exists also an alternative approach in which inequalities constraining the time, a transition can be fired at, are allowed to be strict [35]. However, this is not discussed in our book.

[3] Here, we assume a definition which allows to reason about a net without analysing its time-dependent behaviour. An alternative definition is provided on p. 27.

Definition 1.5. *Let* $\mathfrak{I} = \{\mathfrak{i}_1, \ldots, \mathfrak{i}_{n_\mathfrak{I}}\}$ *be a finite ordered set of indices, and let* $\mathfrak{N} = \{N_\mathfrak{i} \mid \mathfrak{i} \in \mathfrak{I}\}$, *where* $N_\mathfrak{i} = (P_\mathfrak{i}, T_\mathfrak{i}, F_\mathfrak{i}, m_\mathfrak{i}^0, Eft_\mathfrak{i}, Lft_\mathfrak{i})$ *be a family of 1-safe, sequential time Petri nets (called* processes*), indexed with* $\mathfrak{I}$, *with the pairwise disjoint sets* $P_\mathfrak{i}$ *of places, and satisfying the condition*

$$(\forall \mathfrak{i}_1, \mathfrak{i}_2 \in \mathfrak{I})(\forall t \in T_{\mathfrak{i}_1} \cap T_{\mathfrak{i}_2})\ (Eft_{\mathfrak{i}_1}(t) = Eft_{\mathfrak{i}_2}(t)\ \wedge\ Lft_{\mathfrak{i}_1}(t) = Lft_{\mathfrak{i}_2}(t)).$$

A distributed time Petri net $\mathcal{N} = (P, T, F, m^0, Eft, Lft)$ *is the union of the* processes $N_\mathfrak{i}$, *i.e.,*

- $P = \bigcup_{\mathfrak{i} \in \mathfrak{I}} P_\mathfrak{i}$,
- $T = \bigcup_{\mathfrak{i} \in \mathfrak{I}} T_\mathfrak{i}$,
- $F = \bigcup_{\mathfrak{i} \in \mathfrak{I}} F_\mathfrak{i}$,
- $m^0 = \bigcup_{\mathfrak{i} \in \mathfrak{I}} m_\mathfrak{i}^0$,
- $Eft = \bigcup_{\mathfrak{i} \in \mathfrak{I}} Eft_\mathfrak{i}$, *and*
- $Lft = \bigcup_{\mathfrak{i} \in \mathfrak{I}} Lft_\mathfrak{i}$.

Notice that the function $Eft_{\mathfrak{i}_1}$ ($Lft_{\mathfrak{i}_1}$) coincides with $Eft_{\mathfrak{i}_2}$ ($Lft_{\mathfrak{i}_2}$, respectively) for the joint transitions of each two processes $\mathfrak{i}_1$ and $\mathfrak{i}_2$.

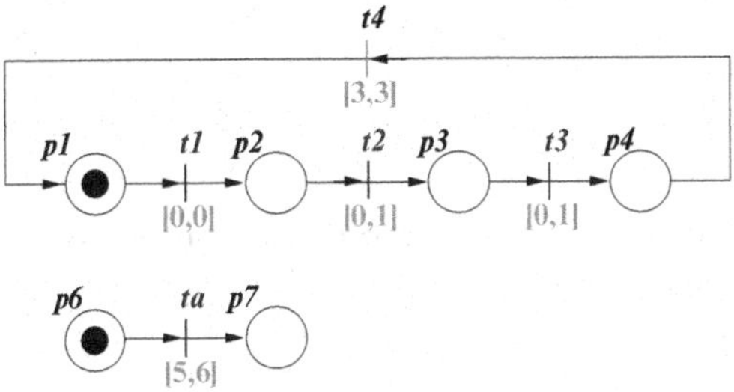

Fig. 1.4. A distributed time Petri net with disjoint processes

Example 1.6. Examples of (distributed) time Petri net are shown[4] in Fig. 1.4, Fig. 1.3 and Fig. 1.5. The net in Fig. 1.4 consists of two disjoint processes with the sets of places $P_1 = \{p_1, p_2, p_3, p_4\}$ and $P_2 = \{p_6, p_7\}$. The net in Fig. 1.3, considered in the previous example, consists of two processes with the sets of places $P_1 = \{p_1, p_3, p_5, p_7\}$ and $P_2 = \{p_2, p_4, p_6, p_8\}$, communicating via the joint transition t_3. The net in Fig. 1.5 (Fischer's mutual exclusion protocol) is composed of three communicating processes with the sets of places: $P_i = \{idle_i, trying_i, waiting_i, critical_i\}$ with $i = 1, 2$, and $P_3 = \{place0, place1, place2\}$. All the transitions of the process 1 are joint with the process 3, and similarly all the transitions of the process 2 are joint with the process 3.

□

[4] From now on, the annotation of the edges with the value of the flow function equal to 1 is omitted.

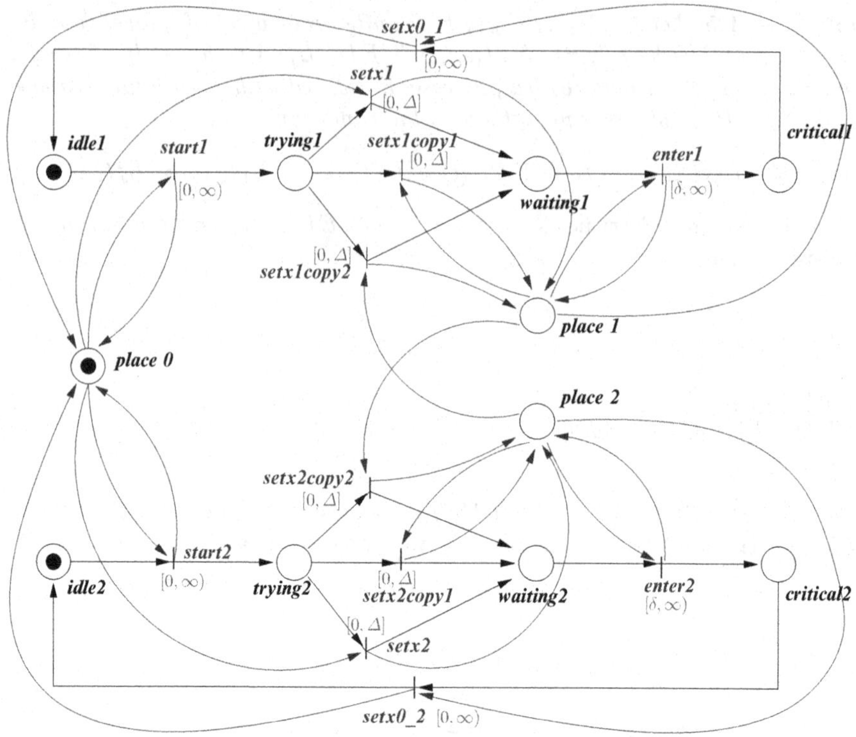

Fig. 1.5. A distributed time Petri net with communicating processes (Fischer's mutual exclusion protocol for $n = 2$)

It is easy to notice that a distributed net is 1-safe. The interpretation of such a net is a collection of sequential, non-deterministic processes with communication capabilities (via joint transitions). In what follows, for simplicity, we consider distributed nets whose all the processes are *state machines* (i.e., for each $N_{\mathfrak{i}}$ and each $t \in T_{\mathfrak{i}}$, $|\bullet t| = |t \bullet| = 1$), which implies that for each $\mathfrak{i} \in \mathfrak{I}$ there is exactly one $p \in P_{\mathfrak{i}}$ with $m^0_{\mathfrak{i}}(p) > 0$ (and, in fact, with $m^0_{\mathfrak{i}}(p) = 1$), and that in any reachable marking m of $\mathcal{N}$ there is exactly one place p of each process with $m(p) = 1$. It is important to mention that a large class of distributed nets can be decomposed to satisfy the above requirement [91].

Two main alternative approaches to the semantics of time Petri nets are typically considered in the literature. The first of them consists in assigning clocks to various components of the net, i.e., the transitions, the places or the processes, whereas the second one exploits the so-called *firing intervals* of the transitions. Below, we provide a brief description of both of them. In the description, we focus on 1-safe TPNs only. Semantics for other kinds of time Petri nets are then discussed in Sect. 1.3.5.

1.2.2 Semantics: The Clock Approach

In this section we consider semantics based on assigning clocks to various components of the net, i.e., the transitions, the places or the processes.

Clocks Assigned to the Transitions

One of the approaches to the concrete (dense) semantics of time Petri nets, widely used in the literature [103, 111, 117, 163, 174], consists in associating clocks with the transitions of the net. A *concrete state* σ^T of $\mathcal{N}$ is then an ordered pair

$$(m, clock^T),$$

where

- m is a marking, and
- $clock^T : T \longrightarrow \mathbb{R}_{0+}$ is a function which for each transition $t \in en(m)$ gives the time elapsed since t became enabled most recently[5].

For $\delta \in \mathbb{R}_{0+}$, by $clock^T + \delta$ we denote the function given by $(clock^T + \delta)(t) = clock^T(t) + \delta$ for all $t \in T$. Moreover, let $(m, clock^T) + \delta$ denote $(m, clock^T + \delta)$. The *(dense) concrete state space* of a time Petri net $\mathcal{N}$ is a transition system

$$C_c^T(\mathcal{N}) = (\Sigma^T, (\sigma^T)^0, \rightarrow_{Tc}),$$

where

- Σ^T is the set of all the concrete states of $\mathcal{N}$,
- $(\sigma^T)^0 = (m^0, clock_0^T)$, with $clock_0^T(t) = 0$ for each $t \in T$, is the initial state, and
- a timed consecution relation $\rightarrow_{Tc} \subseteq \Sigma^T \times (T \cup \mathbb{R}_{0+}) \times \Sigma^T$ is defined by action- and time successors as follows:
  - for $\delta \in \mathbb{R}_{0+}$, $(m, clock^T) \xrightarrow{\delta}_{Tc} (m, clock^T + \delta)$ iff[6]
    - $(clock^T + \delta)(t) \leq Lft(t)$ for all $t \in en(m)$

    (*time successor*),
  - for $t \in T$, $(m, clock^T) \xrightarrow{t}_{Tc} (m_1, clock_1^T)$ iff
    - $t \in en(m)$,
    - $Eft(t) \leq clock^T(t) \leq Lft(t)$,
    - $m_1 = m[t\rangle$, and
    - for all $u \in T$ we have $clock_1^T(u) = 0$ for $u \in newly\_en(m, t)$, and $clock_1^T(u) = clock^T(u)$ otherwise

    (*action successor*).

---

[5] In fact, the enabled transitions are of our interest only. The clocks associated with the transitions that are not enabled can be either assigned a fixed value (e.g., 0), or be undefined, or can "uselessly" count the time. We follow the third of these approaches to comply with the literature [52, 174].

[6] We use the term "iff" as a shorthand for "if and only if".

Intuitively, a time successor does not change the marking m of a concrete state, but it increases the clocks assigned to each transition, provided all of these corresponding to the transitions t enabled at m would not exceed $Lft(t)$. An action successor corresponding to a transition t which is enabled at m can be executed when the clock corresponding to t belongs to the interval $[Eft(t), Lft(t)]$. Then, the marking m is modified, the clocks corresponding to the newly enabled transitions are set to 0, whereas the other clocks remain unchanged.

Example 1.7. Consider the net shown in Fig. 1.3. In the initial state $(\sigma^T)^0$, $m^0(p_1) = m^0(p_2) = 1$ and $m^0(p_i) = 0$ for $i = 3, \ldots, 8$, whereas $clock_0^T(t) = 0$ for all $t \in T$. Passing of two time units results in changing the state into $\sigma_1^T = (m^0, clock_1^T)$, with $clock_1^T(t) = 2$ for all $t \in T$. At the state σ_1^T, firing of both the transitions t_1 and t_2 is possible. Firing of t_1 leads to the state $\sigma_2^T = (m_2, clock_2^T)$, with $m_2(p_2) = m_2(p_3) = 1$ and $m_2(p_i) = 0$ for $i = 1, 4, 5, 6, 7, 8$, and with $clock_2^T(t) = clock_1^T(t)$ for all $t \in T$. Firing of t_2 at the state σ_2^T leads to the state $\sigma_3^T = (m_3, clock_3^T)$, with $m_3(p_3) = m_3(p_4) = 1$ and $m_3(p_i) = 0$ for $i = 1, 2, 5, 6, 7, 8$, and with $clock_3^T(t_3) = 0$ and $clock_3^T(t) = clock_2^T(t)$ for all $t \in T \setminus \{t_3\}$. Then, passing one unit of time results in the state $\sigma_4^T = (m_3, clock_4^T)$, with $clock_4^T(t_3) = 1$ and $clock_4^T(t) = 3$ for all $t \in T \setminus \{t_3\}$. Firing of t_3 at σ_4^T leads to the state $\sigma_5^T = (m_5, clock_5^T)$, with $m_5(p_5) = m_5(p_6) = 1$ and $m_5(p_i) = 0$ for $i = 1, 2, 3, 4, 7, 8$, and with $clock_5^T(t_4) = clock_5^T(t_5) = 0$ and $clock_5^T(t) = clock_4^T(t)$ for all $t \in T \setminus \{t_4, t_5\}$.

□

Clocks Assigned to the Places

Another approach consists in assigning clocks to the places of a net[7]. A *concrete state* σ^P of $\mathcal{N}$ is then an ordered pair

$$(m, clock^P),$$

where

- m is a marking, and
- $clock^P : P \longrightarrow \mathbb{R}_{0+}$ is a function which for each place $p \in P$ gives the time elapsed since p became marked most recently[8].

For $\delta \in \mathbb{R}_{0+}$, by $clock^P + \delta$ we denote the function given by $(clock^P + \delta)(p) = clock^P(p) + \delta$ for all $p \in P$. Moreover, let $(m, clock^P) + \delta$ denote $(m, clock^P + \delta)$. The (*dense*) *concrete state space* of $\mathcal{N}$ is now a transition system

[7] Recall that we consider 1-safe time Petri nets only.

[8] Only the marked places influence the behaviour of the net, so the clocks assigned to the non-marked ones can be treated in various ways, i.e., be assigned a fixed value, be undefined or "uselessly" count the time. Here we follow the third of these approaches, similarly as in the semantics associating clocks with the transitions.

$$C_c^P(\mathcal{N}) = (\Sigma^P, (\sigma^P)^0, \rightarrow_{Pc}),$$

where

- Σ^P is the set of all the concrete states of $\mathcal{N}$,
- $(\sigma^P)^0 = (m^0, clock_0^P)$, with $clock_0^P(p) = 0$ for each $p \in P$, is the initial state, and
- a timed consecution relation $\rightarrow_{Pc} \subseteq \Sigma^P \times (T \cup \mathbb{R}_{0+}) \times \Sigma^P$ is defined by action- and time successors as follows:
  - for $\delta \in \mathbb{R}_{0+}$, $(m, clock^P) \xrightarrow{\delta}_{Pc} (m, clock^P + \delta)$ iff
    - for each $t \in en(m)$ there exists a place $p \in \bullet t$ such that $(clock^P + \delta)(p) \leq Lft(t)$

    (*time successor*),
  - for $t \in T$, $(m, clock^P) \xrightarrow{t}_{Pc} (m_1, clock_1^P)$ iff
    - $t \in en(m)$,
    - for each $p \in \bullet t$ we have $clock^P(p) \geq Eft(t)$,
    - there is $p \in \bullet t$ such that $clock^P(p) \leq Lft(t)$,
    - $m_1 = m[t\rangle$, and
    - for all $p \in P$ we have $clock_1^P(p) = 0$ for $p \in t\bullet$ and $clock_1^P(p) = clock^P(p)$ otherwise

    (*action successor*).

Intuitively, a time successor does not change the marking m of a concrete state, but it increases the value of the clock assigned to each place, provided for each transition t enabled at m at least one of the clocks corresponding to the input places of t would not exceed $Lft(t)$. An action successor corresponding to a transition t enabled at m can be executed when the values of the clocks corresponding to each input place p of t are greater than $Eft(t)$, and at least one of them is smaller than $Lft(t)$. Then, the marking m is modified, the clocks corresponding to the output places of t are set to 0, and the other clocks remain unchanged.

Example 1.8. Consider the net shown in Fig. 1.3. In the initial state $(\sigma^P)^0$, $m^0(p_1) = m^0(p_2) = 1$ and $m^0(p_i) = 0$ for $i = 3, \ldots, 8$, whereas $clock_0^P(p) = 0$ for all $p \in P$. Passing of two time units results in changing the state into $\sigma_1^P = (m^0, clock_1^P)$, with $clock_1^P(p) = 2$ for all $p \in P$. At the state σ_1^P, firing of both the transitions t_1 and t_2 is possible. Firing of t_1 leads to the state $\sigma_2^P = (m_2, clock_2^P)$, with $m_2(p_2) = m_2(p_3) = 1$ and $m_2(p_i) = 0$ for $i = 1, 4, 5, 6, 7, 8$, and with $clock_2^P(p_3) = 0$ and $clock_2^P(p) = clock_1^P(p)$ for all $p \in P \setminus \{p_3\}$. Firing of t_2 at the state σ_2^P leads to the state $\sigma_3^P = (m_3, clock_3^P)$, with $m_3(p_3) = m_3(p_4) = 1$ and $m_3(p_i) = 0$ for $i = 1, 2, 5, 6, 7, 8$, and with $clock_3^P(p_4) = 0$ and $clock_3^P(p) = clock_2^P(p)$ for all $p \in P \setminus \{p_4\}$. Then, passing one unit of time results in the state $\sigma_4^P = (m_3, clock_4^P)$, with $clock_4^P(p_3) = clock_4^P(p_4) = 1$ and $clock_4^P(p) = 3$ for all $p \in P \setminus \{p_3, p_4\}$. Firing of t_3 at σ_4^P leads to the state $\sigma_5^P = (m_5, clock_5^P)$, with $m_5(p_5) = m_5(p_6) = 1$ and $m_5(p_i) = 0$ for $i = 1, 2, 3, 4, 7, 8$, and with $clock_5^P(p_5) = clock_5^P(p_6) = 0$ and $clock_5^P(p) = clock_4^P(p)$ for all $p \in P \setminus \{p_5, p_6\}$. □

Clocks Assigned to the Processes

If $\mathcal{N}$ is a distributed net, another approach to the concrete semantics is possible. In this case, clocks correspond to the processes of the net. This follows from the fact that for the processes which are state machines, in each marking exactly one place of each process is marked, so the clock associated with this process can be considered as associated with the marked place. Let $\mathfrak{I} = \{\mathfrak{i}_1, \ldots, \mathfrak{i}_{n_\mathfrak{I}}\}$ be the set indexing the processes of $\mathcal{N}$. A *concrete state* σ^N of $\mathcal{N}$ is defined as an ordered pair

$$(m, clock^N),$$

where

- m is a marking, and
- $clock^N : \mathfrak{I} \longrightarrow \mathbb{R}_{0+}$ is a function which for each index $\mathfrak{i} \in \mathfrak{I}$ gives the time elapsed since the marked place $p \in P_\mathfrak{i}$ of the process $N_\mathfrak{i}$ of $\mathcal{N}$ became marked most recently.

For $\delta \in \mathbb{R}_{0+}$, by $clock^N + \delta$ we denote the function given by $(clock^N + \delta)(\mathfrak{i}) = clock^N(\mathfrak{i}) + \delta$ for all $\mathfrak{i} \in \mathfrak{I}$. Moreover, let $(m, clock^N) + \delta$ denote $(m, clock^N + \delta)$. The (*dense*) *concrete state space* of $\mathcal{N}$ is now a transition system

$$C_c^N(\mathcal{N}) = (\Sigma^N, (\sigma^N)^0, \rightarrow_{Nc}),$$

where

- Σ^N is the set of all the concrete states of $\mathcal{N}$,
- $(\sigma^N)^0 = (m^0, clock_0^N)$ with $clock_0^N(\mathfrak{i}) = 0$ for each $\mathfrak{i} \in \mathfrak{I}$ is the initial state, and
- a timed consecution relation $\rightarrow_{Nc} \subseteq \Sigma^N \times (T \cup \mathbb{R}_{0+}) \times \Sigma^N$ is defined by action- and time successors as follows:
  - for $\delta \in \mathbb{R}_{0+}$, $(m, clock^N) \xrightarrow{\delta}_{Nc} (m, clock^N + \delta)$ iff
    - for each $t \in en(m)$ there exists $\mathfrak{i} \in \mathfrak{I}$ with $\bullet t \cap P_\mathfrak{i} \neq \emptyset$ such that $(clock^N + \delta)(\mathfrak{i}) \leq Lft(t)$

    (*time successor*),
  - for $t \in T$, $(m, clock^N) \xrightarrow{t}_{Nc} (m_1, clock_1^N)$ iff
    - $t \in en(m)$,
    - for each $\mathfrak{i} \in \mathfrak{I}$ with $\bullet t \cap P_\mathfrak{i} \neq \emptyset$ we have $clock^N(\mathfrak{i}) \geq Eft(t)$,
    - there is $\mathfrak{i} \in \mathfrak{I}$ with $\bullet t \cap P_\mathfrak{i} \neq \emptyset$ such that $clock^N(\mathfrak{i}) \leq Lft(t)$,
    - $m_1 = m[t\rangle$, and
    - for all $\mathfrak{i} \in \mathfrak{I}$ we have $clock_1^N(\mathfrak{i}) = 0$ if $\bullet t \cap P_\mathfrak{i} \neq \emptyset$ and $clock_1^N(\mathfrak{i}) = clock^N(\mathfrak{i})$ otherwise

    (*action successor*).

Intuitively, a time successor does not change the marking m of a concrete state, but it increases the value of the clock assigned to each process, provided for each transition t enabled at m, at least one of the clocks corresponding to a process containing an input place of t would not exceed $Lft(t)$. An action successor, corresponding to a transition t which is enabled at m, can be executed when the values of all the clocks corresponding to the processes containing an input place of t are greater than $Eft(t)$ and at least one of them is smaller than $Lft(t)$. Then, the marking m is modified, the clocks corresponding to the processes containing input places of t are set to 0, and the other clocks remain unchanged. Notice that a process contains an input place of t iff it contains an output place of t, as we are dealing with state machines.

Example 1.9. Again, consider the (distributed) net shown in Fig. 1.3, whose processes are indexed by the set $\mathfrak{I} = \{1,2\}$, and the sets of places of these processes are $P_1 = \{p_1, p_3, p_5, p_7\}$ and $P_2 = \{p_2, p_4, p_6, p_8\}$. In the initial state $(\sigma^N)^0$, $m^0(p_1) = m^0(p_2) = 1$ and $m^0(p_i) = 0$ for $i = 3, \ldots, 8$, whereas $clock_0^N(\mathrm{i}) = 0$ for all $\mathrm{i} \in \mathfrak{I}$. Passing of two time units results in changing the state into $\sigma_1^N = (m^0, clock_1^N)$, with $clock_1^N(\mathrm{i}) = 2$ for all $\mathrm{i} \in \mathfrak{I}$. At the state σ_1^P, firing of both the transitions t_1 and t_2 is possible. Firing of t_1 leads to the state $\sigma_2^N = (m_2, clock_2^N)$, with $m_2(p_2) = m_2(p_3) = 1$ and $m_2(p_i) = 0$ for $i = 1,4,5,6,7,8$, and with $clock_2^N(1) = 0$ and $clock_2^N(2) = clock_1^N(2)$. Firing of t_2 at the state σ_2^N leads to the state $\sigma_3^N = (m_3, clock_3^N)$, with $m_3(p_3) = m_3(p_4) = 1$ and $m_3(p_i) = 0$ for $i = 1,2,5,6,7,8$, and with $clock_3^N(2) = 0$ and $clock_3^N(1) = clock_2^N(1) = 0$. Then, passing one unit of time results in the state $\sigma_4^N = (m_3, clock_4^N)$, with $clock_4^N(1) = clock_4^N(2) = 1$. Firing of t_3 at σ_4^N leads to the state $\sigma_5^N = (m_5, clock_5^N)$, with $m_5(p_5) = m_5(p_6) = 1$ and $m_5(p_i) = 0$ for $i = 1,2,3,4,7,8$, and with $clock_5^N(1) = clock_5^N(2) = 0$.

□

1.2.3 Semantics: Firing Intervals

There is one more approach to the semantics of time Petri nets. Instead of associating clocks with the places, the transitions, or the processes, one can assign to each transition enabled at a given marking a firing interval, i.e., a time interval within which the transition is supposed to fire.

Formally, let $\mathbb{I}$ be the set of all the intervals in $\mathbb{R}_{0+}$, and let $[\,]$ denote the empty interval. For a non-empty interval $I \in \mathbb{I}$, by $lb(I)$ and $ub(I)$ we denote, respectively, the lower and the upper bound of I. In this approach, described in the papers [32, 33, 35], a *concrete state* σ^F of $\mathcal{N}$ is defined as an ordered pair

$$(m, fi),$$

where

- m is a marking, and
- $fi : T \longrightarrow \mathbb{I}$ is a function which for each transition $t \in en(m)$ gives the *firing interval*, i.e., a timing interval in which t is (individually) allowed to fire.

Intuitively, when a transition t becomes enabled, its firing interval is initialised to its *static firing interval* $[Eft(t), Lft(t)]$. The bounds of the interval decrease synchronously while the time passes until t is fired or disabled by firing of another transition. The transition can be fired if the lower bound of its firing interval reaches 0, and must be fired without any additional delay if the upper bound of its firing interval reaches 0.

In what follows, by $fi + \delta$, where $\delta \in \mathbb{R}_{0+}$, we denote the function given by $(fi+\delta)(t) = [\max(0, lb(fi(t)) - \delta), \max(0, ub(fi(t)) - \delta)]$ if $t \in en(m)$, and $(fi + \delta)(t) = [\,]$ otherwise. Moreover, let $(m, fi) + \delta$ denote $(m, fi + \delta)$. The *(dense) concrete state space* of $\mathcal{N}$ is a transition system

$$C_c^F(\mathcal{N}) = (\Sigma^F, (\sigma^F)^0, \rightarrow_{Fc}),$$

where

- Σ^F is the set of all the concrete states of $\mathcal{N}$, whereas
- $(\sigma^F)^0 = (m^0, fi_0)$ is the initial state with $fi_0(t) = [Eft(t), Lft(t)]$ for $t \in en(m^0)$ and $fi_0(t) = [\,]$ otherwise.
- the timed consecution relation $\rightarrow_{Fc} \subseteq \Sigma^F \times (T \cup \mathbb{R}_{0+}) \times \Sigma^F$ is defined by action- and time successors as follows:
  - for $\delta \in \mathbb{R}_{0+}$, $(m, fi) \xrightarrow{\delta}_{Fc} (m, fi + \delta)$ iff
    - $\delta \in [0, \min\{ub(fi(t)) \mid t \in en(m)\}]$

    (*time successor*),
  - for $t \in T$, $(m, fi) \xrightarrow{t}_{Fc} (m_1, fi_1)$ iff
    - $t \in en(m)$,
    - $lb(fi(t)) = 0$,
    - $m_1 = m[t\rangle$ and
    - for all $u \in T$ we have $fi_1(u) = [\,]$ if $u \notin en(m_1)$, $fi_1(u) = [Eft(u), Lft(u)]$ if $u \in newly\_en(m, t)$, and $fi_1(u) = fi(u)$ otherwise

    (*action successor*).

Notice that when an action successor corresponding to t is executed and the marking m_1 is reached, the firing intervals of all the transitions that are not enabled at m_1 are set to $[\,]$, the firing interval of each newly enabled transition u is set to $[Eft(u), Lft(u)]$, whereas the firing intervals of the other transitions remain unchanged.

Example 1.10. Consider the net in Fig. 1.3. In the initial state $(\sigma^F)^0$, $m^0(p_1) = m^0(p_2) = 1$ and $m^0(p_i) = 0$ for $i = 3, \ldots, 8$, whereas $fi_0(t_1) = [1, 2]$, $fi_0(t_2) = [0, 3]$ and $fi_0(t_i) = [\,]$ for $i = 3, 4, 5, 6$. Passing of two time units results in changing the state into $\sigma_1^F = (m^0, fi_1)$, with $fi_1(t_1) = [0, 0]$, $fi_1(t_2) = [0, 1]$ and $fi_1(t_i) = [\,]$ for $i = 3, 4, 5, 6$. At the state σ_1^T, firing of both the transitions t_1 and t_2 is possible, since $lb(fi(t_1)) = 0$ and $lb(fi(t_2)) = 0$. Firing of t_1 leads to the state $\sigma_2^F = (m_2, fi_2)$, with $m_2(p_2) = m_2(p_3) = 1$ and $m_2(p_i) = 0$ for $i = 1, 4, 5, 6, 7, 8$, and with $fi_2(t_2) = fi_1(t_2)$ and $fi_2(t_i) = [\,]$ for $i =$

1, 3, 4, 5, 6. Firing of t_2 at the state σ_2^F leads to the state $\sigma_3^F = (m_3, fi_3)$, with $m_3(p_3) = m_3(p_4) = 1$ and $m_3(p_i) = 0$ for $i = 1, 2, 5, 6, 7, 8$, and with $fi_3(t_3) = [1, 2]$ and $fi_3(t) = [\,]$ for all $t \in T \setminus \{t_3\}$. Then, passing one unit of time results in the state $\sigma_4^F = (m_3, fi_4)$ with $fi_4(t_3) = [0, 1]$ and $fi_4(t) = [\,]$ for all $t \in T \setminus \{t_3\}$. Firing of t_3 at σ_4^F leads to the state $\sigma_5^F = (m_5, fi_5)$, with $m_5(p_5) = m_5(p_6) = 1$ and $m_5(p_i) = 0$ for $i = 1, 2, 3, 4, 7, 8$, and with $fi_5(t_4) = [1, 1]$, $fi_5(t_5) = [1, 2]$ and $fi_5(t) = [\,]$ for all $t \in T \setminus \{t_4, t_5\}$.

□

1.3 Reasoning about Time Petri Nets

In this section we show that there is a clear relationship between the semantics defined above as long as we consider the transitions enabled at a given marking. However, such a connection does not exist for the other transitions. Next, we compare dense and discrete semantics for time Petri nets as well as define concrete models for them.

1.3.1 Comparison of the Semantics

Given a state $\sigma^P = (m, clock^P) \in \Sigma^P$, it is easy to see that it corresponds to a state $\sigma^T = (m, clock^T) \in \Sigma^T$ which satisfies the condition

$$(\forall t \in en(m))\ clock^T(t) = \min\{clock^P(p) \mid p \in \bullet t\}.$$

On the other hand, given $\sigma^N = (m, clock^N) \in \Sigma^N$, the corresponding state $\sigma^P = (m, clock^P) \in \Sigma^P$ satisfies

$$(\forall t \in en(m))\ \min\{clock^P(p) \mid p \in \bullet t\} = \min\{clock^N(\mathrm{i}) \mid \bullet t \cap P_\mathrm{i} \neq \emptyset\}.$$

Notice, however, that it is impossible to establish a relationship for the transitions which are not enabled.

It can be proven that the concrete state spaces $C_c^T(\mathcal{N})$, $C_c^P(\mathcal{N})$ and $C_c^N(\mathcal{N})$ are equivalent w.r.t. branching time temporal properties considered in Chap. 4 [124]. This is obtained by showing that for each pair of the above concrete state spaces there exists a symmetric relation $\mathcal{R}$ (called a *bisimulation*) connecting their states such that the initial states are related by $\mathcal{R}$, and the condition $\sigma\mathcal{R}\sigma'$ implies that if one of the related states has an a-successor[9] for any $a \in T \cup \mathbb{R}_{0+}$, then also its counterpart does, and these successors are related by $\mathcal{R}$. The state spaces for which such a relation exists are called *bisimilar*.

Considering the firing interval approach, it is easy to see that a state $(m, clock^T) \in \Sigma^T$ corresponds to such a state $(m, fi) \in \Sigma^F$ for which

[9] For a set S and a successor relation $\rightarrow \subseteq S \times B \times S$, where B is a set of labels, by an *a-successor* of $s \in S$ we mean each $s' \in S$ such that $s \xrightarrow{a} s'$.

$$fi(t) = [\max(0, Eft(t) - clock^T(t)), \max(Lft(t) - clock^T(t), 0)]$$

for each $t \in en(m)$, and

$$fi(t) = [\,]$$

for $t \notin en(m)$. The bounds of the firing interval for a transition $t \in en(m)$ are, respectively, the minimal and maximal time remaining before firing this transition. Notice, however, that this is not necessarily the one-to-one correspondence. On the one hand, the states $(m, clock^T), (m, clock_1^T) \in \Sigma^T$ satisfying $(\forall t \in en(m))\ clock^T(t) = clock_1^T(t)$ correspond to the same state $\sigma^F \in \Sigma^F$. On the other hand, in the case when $t \in en(m)$ and $Lft(t) = \infty$, for all the states with $clock^T(t) \geq Eft(t)$ the corresponding firing interval is given by $fi(t) = [0, \infty)$. However, it can be proven that the state spaces $C_c^F(\mathcal{N})$ and $C_c^T(\mathcal{N})$ are equivalent w.r.t. branching time temporal properties considered in Sect. 4 bisimilar)[10].

1.3.2 Dense versus Discrete Semantics

So far we have been looking at the four types of concrete state spaces. Depending on the notion of a run we can distinguish between dense and discrete semantics for time Petri nets. Notice that when defining discrete semantics we do not depart from the dense notion of the time domain, but rather combine time steps with action steps calling this a *discrete view*. Alternatively, one could restrict the time domain to integers [127, 129], so that passing of time could be measured only using integer numbers. We do not discuss such an approach in this book.

In what follows we define dense runs in which action- and time successors are not combined. We give an example of such a run for the net $\mathcal{N}$ shown in Fig. 1.3. Next, we define discrete runs by combining time- and action successors. Similarly, we show an example of a discrete run, and compare reachability sets for both the dense and discrete semantics.

Let $C_c^R(\mathcal{N}) = (\Sigma^R, (\sigma^R)^0, \rightarrow_{Rc})$, where $R \in \{T, P, N, F\}$, be one of the above-defined concrete state spaces of a time Petri net $\mathcal{N}$ (recall that T, P and N refer to the semantics where the clocks are assigned respectively to the transitions, places and processes of a distributed net, whereas F corresponds to the firing interval semantics). *Concatenation* of two time steps $\sigma^R \xrightarrow{\delta}_{Rc} \sigma^R + \delta$ and $\sigma^R + \delta \xrightarrow{\delta'}_{Rc} \sigma^R + \delta + \delta'$ is the time step $\sigma^R \xrightarrow{\delta+\delta'}_{Rc} \sigma^R + \delta + \delta'$. Similarly, if $\sigma^R \xrightarrow{\delta}_{Rc} \sigma^R + \delta$, then for any $k \in \mathbb{N}_+$ there exist $\delta_1, \ldots, \delta_k \in \mathbb{R}_{0+}$

[10] It should be noticed that if the semantics is redefined such that for each $(m, clock^T) \in \Sigma^T$ and each $t \notin en(m)$ the value of $clock^T(t)$ is equal to some fixed value, then there is a bijection between the states of Σ^T and Σ^F if the values of Lft are finite only. However, if some of the latest firing times is infinite, there is only a surjection, since possibly many concrete states of Σ^T correspond to the same state of Σ^F [31].

such that $\delta_1+\ldots+\delta_k=\delta$ and $\sigma^R \xrightarrow{\delta_1}_{Rc} \sigma^R+\delta_1 \xrightarrow{\delta_2}_{Rc} \ldots \xrightarrow{\delta_k}_{Rc} \sigma^R+\delta$ (i.e., each time step can be *split* into an arbitrary number of consecutive time successors). A *(dense)* σ_0^R*-run* ρ of $\mathcal{N}$ is a maximal (i.e., non-extendable) sequence

$$\rho^R = \sigma_0^R \xrightarrow{\delta_0}_{Rc} \sigma_0^R + \delta_0 \xrightarrow{t_0}_{Rc} \sigma_1^R \xrightarrow{\delta_1}_{Rc} \sigma_1^R + \delta_1 \xrightarrow{t_1}_{Rc} \sigma_2^R \xrightarrow{\delta_2}_{Rc} \ldots,$$

where $t_i \in T$ and $\delta_i \in \mathbb{R}_{0+}$, for all $i \in \mathbb{N}$ (notice that due to the fact that δ can be equal to 0, two subsequent firings of transitions without any time passing in between are allowed as well, and that consecutive time passings are concatenated). Such runs are called *weakly monotonic* in order to distinguish them from strongly monotonic runs defined later.

Example 1.11. Consider the net $\mathcal{N}$ shown in Fig. 1.3 and its concrete state space $C_c^T(\mathcal{N})$. For simplicity of the description below, let $((m(p_1),\ldots,m(p_8)), (clock^T(t_1),\ldots,clock^T(t_6)))$ denote the concrete state $(m, clock^T) \in \Sigma^T$. One of the possible $(\sigma^T)^0$-runs is

$((1,1,0,0,0,0,0,0),(0,0,0,0,0,0)) \xrightarrow{2}_{Tc} ((1,1,0,0,0,0,0,0),(2,2,2,2,2,2)) \xrightarrow{t_1}_{Tc} ((0,1,1,0,0,0,0,0),(2,2,2,2,2,2)) \xrightarrow{0}_{Tc} ((0,1,1,0,0,0,0,0),(2,2,2,2,2,2)) \xrightarrow{t_2}_{Tc} ((0,0,1,1,0,0,0,0),(2,2,0,2,2,2)) \xrightarrow{1}_{Tc} ((0,0,1,1,0,0,0,0),(3,3,1,3,3,3)) \xrightarrow{t_3}_{Tc} ((0,0,0,0,1,1,0,0),(3,3,1,0,0,3)) \xrightarrow{1}_{Tc} ((0,0,0,0,1,1,0,0),(4,4,2,1,1,4)) \xrightarrow{t_4}_{Tc} ((0,0,0,0,0,1,1,0),(4,4,2,1,1,4)) \xrightarrow{0}_{Tc} ((0,0,0,0,0,1,1,0),(4,4,2,1,1,4)) \xrightarrow{t_5}_{Tc} ((0,0,0,0,0,0,1,1),(4,4,2,1,1,0)) \xrightarrow{1.5}_{Tc} ((0,0,0,0,0,0,1,1),(5.5,5.5,3.5,2.5,2.5,1.5)) \xrightarrow{t_6}_{Tc} ((0,0,0,0,0,0,1,1),(5.5,5.5,3.5,2.5,2.5,0)) \xrightarrow{1.5}_{Tc} \ldots$

□

A state $\sigma^R \in \Sigma^R$ is *reachable* if there exists a $(\sigma^R)^0$-run ρ^R and $i \in \mathbb{N}$ such that $\sigma^R = \sigma_i^R + \delta$ for some $0 \le \delta \le \delta_i$, where $\sigma_i^R + \delta_i$ is an element of ρ^R. The set of all the reachable states of $\mathcal{N}$ is denoted by $Reach_{\mathcal{N}}^R$. A marking m is reachable if there is a state $(m,\cdot) \in Reach_{\mathcal{N}}^R$. The set of all the reachable markings of $\mathcal{N}$ is denoted by $RM_{\mathcal{N}}$ (it is easy to see that this set does not depend on the way the concrete states of the net are defined).

Given a run $\rho^R = \sigma_0^R \xrightarrow{\delta_0}_{Rc} \sigma_0^R + \delta_0 \xrightarrow{t_0}_{Rc} \sigma_1^R \xrightarrow{\delta_1}_{Rc} \sigma_1^R + \delta_1 \xrightarrow{t_1}_{Rc} \ldots$, we say that a run $(\rho^R)' = (\sigma_0^R)' \xrightarrow{\delta_0'}_{Rc} (\sigma_0^R)' + \delta_0' \xrightarrow{t_0'}_{Rc} \ldots$ is a *suffix* of ρ^R if there exists $i \in \mathbb{N}$ and $0 \le \delta \le \delta_i$ such that $(\sigma_0^R)' = \sigma_i^R + \delta$, $\delta_i = \delta + \delta_0'$, and $t_j = t_{i+j}$ and $\delta_{j+1} = \delta_{i+j+1}$ for each $j \in \mathbb{N}$ (notice that a suffix of ρ can start at a state which results from splitting a time step occurring in the run). A run ρ^R is said to be *progressive* iff $\Sigma_{i\in\mathbb{N}}\delta_i$ is unbounded. Note that a run is progressive iff all its suffixes are so.

A time Petri net is called *progressive* if all its runs starting at the initial state are progressive. This means that they are infinite[11] and time divergent. It is possible to formulate sufficient conditions ensuring progressiveness of a net (we discuss this subject in Sect. 1.3.4). In what follows, progressive time Petri nets are considered only.

Independently of the approach to the concrete semantics, for any concrete state σ, the set of all the progressive weakly monotonic σ-runs of $\mathcal{N}$ is denoted by $f_{\mathcal{N}}(\sigma)$. The above notion of a run can be used for interpreting untimed branching time temporal logics [111] or for checking reachability. It can be also applied to interpreting timed languages like Timed Computation Tree Logic (TCTL) or timed μ-calculus. However, for interpreting TCTL it is sometimes more convenient to restrict the runs to contain non-zero time passings only (i.e., $\delta_i > 0$ for all $i \in \mathbb{N}$), which prevents firings of two transitions immediately one after the other, i.e., at the same time. Such runs are called *strongly monotonic*. By $f^+_{\mathcal{N}}(\sigma)$ we denote the set of all the progressive strongly monotonic σ-runs of $\mathcal{N}$. The set of all the states of $\mathcal{N}$ which are reachable on the (progressive[12]) strongly monotonic σ^0-runs is denoted by $Reach^{+R}_{\mathcal{N}}$, where $R \in \{T, P, N, F\}$ refers to the semantics. A marking m is *reachable* (on strongly monotonic runs) if there is a state $(m, \cdot) \in Reach^{+R}_{\mathcal{N}}$. The set of all the reachable markings of $\mathcal{N}$ is denoted by $RM^+_{\mathcal{N}}$ (again, this set does not depend on the definition of the concrete states applied). It is clear that we have $RM^+_{\mathcal{N}} \subseteq RM_{\mathcal{N}}$, but there are nets for which the inclusion is strict (see Example 1.13). The semantics based on weakly (strongly) monotonic runs is called *weakly* (*strongly*, respectively) *monotonic*. The reason for the above distinction will become clear in Sect. 4.3.

Example 1.12. Consider the net $\mathcal{N}$ shown in Fig. 1.3 and its concrete state space $C^T_c(\mathcal{N})$. Notice that the run shown in Example 1.11 is not permitted in the strongly monotonic semantics. Notice, moreover, that the sets of states $Reach^T_{\mathcal{N}}$ and $Reach^{+T}_{\mathcal{N}}$ are different: the state $(m^0[t_2\rangle, clock^T_0)$, which can be obtained only by the immediate firing of t_2 at the state $(\sigma^T)^0$, belongs to $Reach^T_{\mathcal{N}}$, but it does not belong to $Reach^{+T}_{\mathcal{N}}$.

□

Example 1.13. Consider the net shown in Fig. 1.6. The marking m with $m(p_3) = 1$, $m(p_4) = 1$ and $m(p_i) = 0$ for $i = 1, 2$ is reachable when the weakly monotonic semantics is assumed, but is not when the time is strongly monotonic. The sets $RM_{\mathcal{N}}$ and $RM^+_{\mathcal{N}}$ are therefore not equal.

□

---

[11] This can be checked by applying algorithms looking for deadlocks (i.e., states which prevent firing of any transition in the future, see Sect. 1.3.4).

[12] Notice that this requirement is in fact redundant as we assume that $\mathcal{N}$ is progressive.

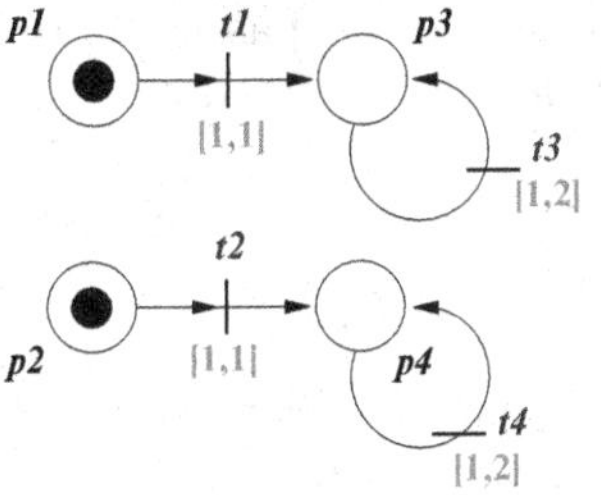

Fig. 1.6. The time Petri net used in Example 1.13

Alternative (Discrete) Semantics

Alternatively, the runs can be defined such that passing some time and firing a transition (and then, possibly, passing some time again) are combined into a single step. Such an approach, called sometimes *discrete*, enables verification of untimed temporal properties as well as reachability checking, and is very popular in the literature [32,33,35,117,174]. Thus, let $C_c^R(\mathcal{N}) = (\Sigma^R, (\sigma^R)^0, \rightarrow_{Rc})$, where $R \in \{T, P, N, F\}$, be a concrete state space of a time Petri net $\mathcal{N}$. We can define the following modifications of the successor relation $\rightarrow_{Rc}$:

- $\rightarrow_{Rd_1} \subseteq \Sigma^R \times T \times \Sigma^R$, where
  $\sigma \xrightarrow{t}_{Rd_1} \sigma'$ iff $(\exists \sigma_1 \in \Sigma^R)(\exists \delta \in \mathbb{R}_{0+})\ \sigma \xrightarrow{\delta}_{Rc} \sigma_1 \xrightarrow{t}_{Rc} \sigma'$;
- $\rightarrow_{Rd_2} \subseteq \Sigma^R \times T \times \Sigma^R$; where
  $\sigma \xrightarrow{t}_{Rd_2} \sigma'$ iff $(\exists \sigma_1, \sigma_2 \in \Sigma^R)(\exists \delta, \delta' \in \mathbb{R}_{0+})\ \sigma \xrightarrow{\delta}_{Rc} \sigma_1 \xrightarrow{t}_{Rc} \sigma_2 \xrightarrow{\delta'}_{Rc} \sigma'$.

The relation $\rightarrow_{Rd_2}$ includes $\rightarrow_{Rd_1}$ (due to the possibility of the δ's to be equal to zero), but usually in the literature $\rightarrow_{Rd_1}$ is used. It is important to notice that the above inclusion does not need to hold if the relations are redefined to contain non-zero time passings only. This follows from the fact that in this case the relation $\rightarrow_{Rd_1}$ can have as its last component only states resulting from firing of a transition, whereas the relation $\rightarrow_{Rd_2}$ – only states resulting from passing some positive time from the states mentioned above.

For each of the above relations we can define a notion of a (*discrete*) σ_0^R*-run* of $\mathcal{N}$, which is a sequence

$$\rho_r^R = \sigma_0^R \xrightarrow{t_0}_{Rd_r} \sigma_1^R \xrightarrow{t_1}_{Rd_r} \sigma_2^R \xrightarrow{t_2}_{Rd_r} \cdots,$$

where $R \in \{T, P, N, F\}$, $r \in \{1, 2\}$ and $t_i \in T$ for all $i \in \mathbb{N}$. A state $\sigma^R \in \Sigma^R$ is *reachable* if there exists a $(\sigma^R)^0$-run ρ_r^R and $i \in \mathbb{N}$ such that $\sigma^R = \sigma_i^R$, where σ_i^R is an element of ρ_r^R. The set of all the reachable states of $\mathcal{N}$ is denoted by $Reach_{\mathcal{N}}^{Rd_r}$ (notice that this set can vary depending on the transition relation). It can be proven that for each $R \in \{T, P, N, F\}$ and for the weakly monotonic semantics we have $Reach_{\mathcal{N}}^{Rd_1} \subseteq Reach_{\mathcal{N}}^{Rd_2} \subseteq Reach_{\mathcal{N}}^R$. Notice, however, that the set of all the reachable markings is always the same and equal to $RM_{\mathcal{N}}$

($RM^+_{\mathcal{N}}$, respectively, if non-zero time passings are allowed only). *Concrete (discrete) state spaces* of $\mathcal{N}$ are now defined as

$$C^R_{d_r}(\mathcal{N}) = (\Sigma^R, (\sigma^R)^0, \rightarrow_{Rd_r}),$$

where $R \in \{T, P, N, F\}$ and $r \in \{1, 2\}$.

Example 1.14. Consider the net $\mathcal{N}$ shown in Fig. 1.3. For simplicity of the presentation, let $((m(p_1), \ldots, m(p_8)), (clock^T(t_1), \ldots, clock^T(t_6)))$ denote the concrete state $(m, clock^T) \in \Sigma^T$. In the case of the concrete state space $C^T_{d_1}(\mathcal{N})$, one of the possible $(\sigma^T)^0$-runs, obtained from the one shown in Example 1.11 by combining together time- and action steps, is

$((1,1,0,0,0,0,0,0),(0,0,0,0,0,0)) \xrightarrow{t_1}_{Td_1} ((0,1,1,0,0,0,0,0),\ (2,2,2,2,2,2)) \xrightarrow{t_2}_{Td_1} ((0,0,1,1,0,0,0,0),(2,2,0,2,2,2)) \xrightarrow{t_3}_{Td_1} ((0,0,0,0,1,1,0,0),(3,3,1,0,0,3)) \xrightarrow{t_4}_{Td_1} ((0,0,0,0,0,1,1,0),(4,4,2,1,1,4)) \xrightarrow{t_5}_{Td_1} ((0,0,0,0,0,0,1,1),(4,4,2,1,1,0)) \xrightarrow{t_6}_{Td_1} ((0,0,0,0,0,0,1,1),(5.5,5.5,3.5,2.5,2.5,0)) \xrightarrow{t_6}_{Td_1} \ldots$

Considering the concrete state space $C^T_{d_2}(\mathcal{N})$, we can obtain, for instance, the $(\sigma^T)^0$-run

$((1,1,0,0,0,0,0,0),(0,0,0,0,0,0)) \xrightarrow{t_1}_{Td_2} ((0,1,1,0,0,0,0,0),(2.5,2.5,2.5,2.5,2.5,2.5)) \xrightarrow{t_2}_{Td_2} ((0,0,1,1,0,0,0,0),(2.5,2.5,0,2.5,2.5,2.5)) \xrightarrow{t_3}_{Td_2} ((0,0,0,0,1,1,0,0),(3.5,3.5,1,0,0,3.5)) \xrightarrow{t_4}_{Td_2} ((0,0,0,0,0,1,1,0),(5,5,2.5,1.5,1.5,5)) \xrightarrow{t_5}_{Td_2} ((0,0,0,0,0,0,1,1),(5,5,2.5,1.5,1.5,0)) \xrightarrow{t_6}_{Td_2} ((0,0,0,0,0,0,1,1),(6.5,6.5,4,3,3,0.5)) \xrightarrow{t_6}_{Td_2} \ldots.$

The first transition of the run corresponds to passing of two units of time, firing t_1 and then passing 0.5 time unit. The second one is an immediate firing of t_2 with no time passing after that. The third corresponds to passing one unit of time and then firing t_3, with no further time delay, whereas the next one consists of passing one time unit, firing t_4 and then passing 0.5 time unit again. The fifth transition is equivalent to firing of t_5 with zero-time delays before and after that, whereas the sixth one consists of passing one unit of time, firing t_6 and then passing 0.5 unit of time again.

Notice that the sets $Reach^{Td_1}_{\mathcal{N}}$ and $Reach^{Td_2}_{\mathcal{N}}$ differ (e.g., the state $(m^0[t_1\rangle, clock^T_1)$ with $clock^1_T(t) = 2.5$ for all $t \in T$, which can be obtained from the initial state by passing two units of time, then firing t_1, and then passing 0.5 unit of time, belongs to $Reach^{Td_2}_{\mathcal{N}}$, but does not belong to $Reach^{Td_1}_{\mathcal{N}}$, since it cannot be obtained from the initial state by firing the transition without any further passage of time). Notice, moreover, that the above sets are different from the set $Reach^T_{\mathcal{N}}$, since the state $(m^0, clock^T_1) \in Reach^T_{\mathcal{N}}$ with $clock^T_1(t) = 0.5$ for each $t \in T$ does belong neither to $Reach^{Td_1}_{\mathcal{N}}$ nor to $Reach^{Td_2}_{\mathcal{N}}$.

□

1.3.3 Concrete Models for TPNs

In order to reason about systems represented by TPNs, we define a set of propositional variables

$$PV_P = \{\wp_p \mid p \in P\},$$

which correspond to the places of $\mathcal{N}$, and a valuation function $V_{\mathcal{N}} : P \to PV_P$ assigning propositions to the places of $\mathcal{N}$, given by

$$V_{\mathcal{N}}(p) = \wp_p \text{ for each } p \in P.$$

Let $C(\mathcal{N}) = (\Sigma, \sigma^0, \to)$, where $C(\mathcal{N}) \in \{C_c^R(\mathcal{N}) \mid R \in \{T, P, N, F\}\} \cup \{C_{d_r}^R(\mathcal{N}) \mid R \in \{T, P, N, F\} \wedge r \in \{1, 2\}\}\}$, be a concrete state space of a time Petri net $\mathcal{N}$. Let $V_{\mathfrak{C}} : \Sigma \longrightarrow 2^{PV}$ be a valuation function such that

$$V_{\mathfrak{C}}((m, \cdot)) = \bigcup_{\{p \in P \mid m(p) > 0\}} V_{\mathcal{N}}(p),$$

i.e., $V_{\mathfrak{C}}$ assigns the propositions corresponding to the places marked in m. The ordered pair

$$M_{\mathfrak{c}}(\mathcal{N}) = (C(\mathcal{N}), V_{\mathfrak{C}})$$

is called a *concrete model* of $\mathcal{N}$ (*based on the state space* $C(\mathcal{N})$). The concrete model is called *dense* (*discrete*) if $C(\mathcal{N})$ is so.

1.3.4 Progressiveness in Time Petri Nets

As it has been already stated (see p. 20), we consider only nets whose all the runs starting at the initial state are progressive. However, for a given $\mathcal{N}$ one cannot say without any prior analysis whether or not $\mathcal{N}$ does satisfy this condition. Below, we discuss this problem more thoroughly.

The first case in which runs of a net are non-progressive can occur if the net contains no loop, and therefore is of finite runs only (consider, for instance, the net like in Fig. 1.3 but without the transition t_6). Such a situation, however, can be easily avoided, since if the system $\mathcal{S}$ modelled by a net $\mathcal{N}_{\mathcal{S}}$ can finish a sequence of its actions in a certain "desired" state, one can add to a place, which becomes marked when that state is reached, a fictitious loop transition with arbitrary non-zero values of the earliest and the latest firing time.

Another case, when a net $\mathcal{N}$ is non-progressive, occurs if it is not *deadlock-free*. A deadlock-free TPN $\mathcal{N}$ is a net whose all the reachable states are not *deadlocks*, where by a *deadlock* we mean such a concrete state of $\mathcal{N}$ which occurs in a finite run starting at the initial state, that no further passage of time at that state can enable firing of any transition. Later, we show that a net, which is deadlock-free in the weakly monotonic semantics, does not need to be so in the strongly monotonic one (see Example 1.16).

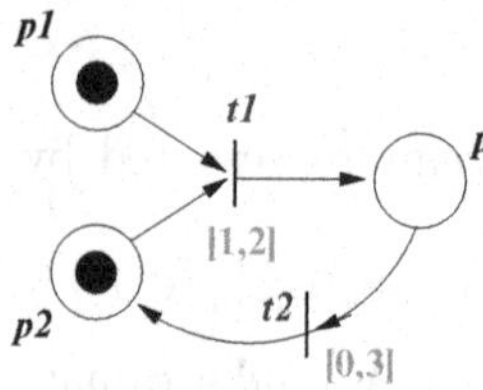

Fig. 1.7. A time Petri net with structural deadlock

A deadlock can be *structural*, i.e., independent on the timing parameters (an example of a net with such a deadlock is shown in Fig. 1.7). Since such a deadlock occurs also in the underlying Petri net $\mathcal{P}_{\mathcal{N}}$, it can be found using an algorithm for untimed nets (see, e.g., [16,21,58,162]). Since the set of all the reachable markings of a given time Petri net $\mathcal{N}$ is a subset of the set of all the reachable markings of its underlying Petri net $\mathcal{P}_{\mathcal{N}}$, and the firings possible in $\mathcal{N}$ in the weakly monotonic semantics are possible also in the underlying untimed net, it is easy to see that if $\mathcal{P}_{\mathcal{N}}$ is deadlock-free, then also $\mathcal{N}$ is so, for time to be weakly monotonic. However, timing conditions can sometimes prevent deadlocks (see below).

Example 1.15. Consider the time Petri net $\mathcal{N}$ shown in Fig. 1.8. The net contains a structural deadlock (if the transition t_2 is fired, no further firings are possible). However, when p_3 becomes marked, the timing intervals of the transitions t_2 and t_3 force t_3 to be fired before t_2, which prevents firing of t_2 and deadlocks.

□

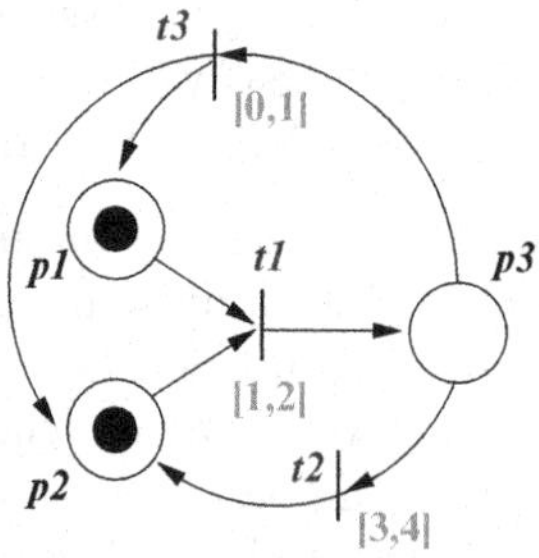

Fig. 1.8. A deadlock-free time Petri net with a structural deadlock

The structural analysis of the untimed net can therefore exclude a deadlock for the weakly monotonic semantics, but to check whether it occurs, or to deal with the strongly monotonic case, another method needs to be applied. Thus, one can also look for deadlocks, for instance, by building a bisimulating model for the net (see, e.g., [32,35] and Sect. 5.1), and checking whether it contains a reachable abstract state with no successors. This method is applicable mainly to the weakly monotonic semantics, for which these models are usually built[13]. However, *detailed region graphs* (see Sect. 5.2.1[14]) can be used for both the (weakly and strongly monotonic) semantics.

Notice also that dealing with the strongly monotonic semantics introduces an additional complexity. First of all, in order to prevent some deadlocks we restrict the nets to contain no transition with both the earliest and the latest firing time equal to zero, since such a transition is never fired according to this semantics. However, in spite of this restriction, deadlocks still can occur, even in the case when the underlying untimed net is deadlock-free. Such a situation can be illustrated by the following example:

Example 1.16. Consider the net in Fig. 1.9. If the transition t_2 fires in time 1, then t_1 should be fired without any additional delay. This, however, is impossible for the strongly monotonic semantics. The corresponding concrete state is thus a deadlock.

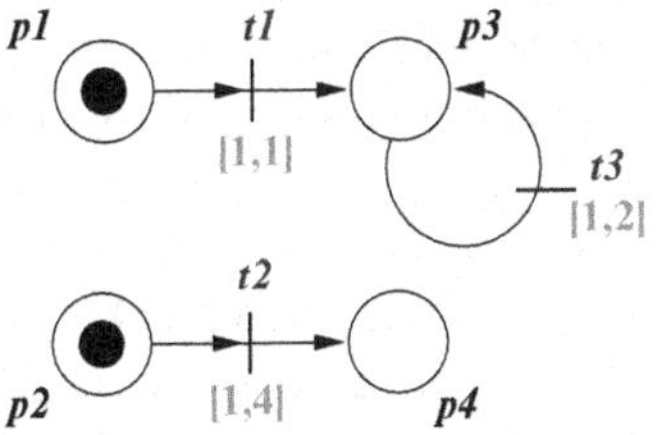

Fig. 1.9. A time Petri net with deadlock (only non-zero time steps allowed)

Another example of a net with a deadlock in the strongly monotonic semantics is shown in Fig. 1.6. Notice that both the states obtained by firing t_1 or t_2 at the initial marking are deadlocks, since no further firings of transitions are possible.

□

Deadlocks of this kind are difficult to predict and analyse. Since their occurrence follows from the specificity of the successor relation of time Petri nets (i.e., from the fact that transitions cannot be disabled by passage of time), and since the class of nets considered (i.e., these without transitions of the

[13] More precisely, the models are generated for a discrete semantics, in which passage of time "inside" the transition relation is not restricted to be non-zero. It is, however, easy to see that these models are sufficient to check the existence of deadlocks for the semantics in which time- and action steps are separated.

[14] The section defines detailed region graphs for timed automata. Such models for TPNs can be built in a similar way [111, 163], or result from translating a net to a timed automaton (see Chap. 3).

earliest and the latest firing times both equal to zero) seems too restricted, the semantics with non-zero time steps is usually not considered in practice. In this book, we provide it only for compatibility with the timed automata approach, and do not consider unless stated otherwise.

Moreover, notice that the absence of deadlocks in the net does not guarantee that all its runs are progressive, which is shown in Example 1.17.

Example 1.17. Consider the net $\mathcal{N}$ shown in Fig. 1.10 and its concrete state space $C_c^T(\mathcal{N})$. For simplicity of the below description, let $((m(p_1), m(p_2)),$

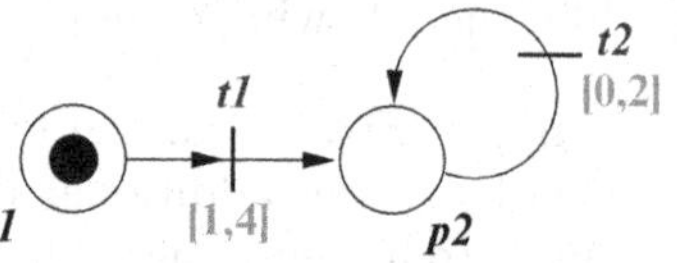

Fig. 1.10. A time Petri net with a non-progressive run

$(clock^T(t_1), clock^T(t_2)))$ denote the concrete state $(m, clock^T) \in \Sigma^T$. Although all the runs of $\mathcal{N}$ are infinite, some of them are not progressive. An example of such a run is

$$\rho = ((1,0),(0,0)) \xrightarrow{1}_{Tc} ((1,0),(1,1)) \xrightarrow{t_1}_{Tc} ((0,1),(1,0)) \xrightarrow{1/2}_{Tc} ((0,1),(1.5, 0.5)) \xrightarrow{t_2}_{Tc} ((0,1),(1.5,0)) \xrightarrow{1/4}_{Tc} ((0,1),(1.75,0)) \xrightarrow{t_2}_{Tc} \ldots$$

It is easy to see that the example applies to both the weakly and strongly monotonic semantics.

□

However, one can formulate a sufficient condition for a TPN to be progressive. To do this, we introduce a notion of a *structural loop.*

Definition 1.18. *A* structural loop *of a time Petri net* $\mathcal{N} = (P, T, F, m^0, Eft, Lft)$ *is a sequence of distinct transitions* $t_1, \ldots, t_n \in T$ *such that firing of* t_i *enables* $t_{i+1 \ mod \ n}$ *for all* $i = 1, \ldots, n$.

Now, progressiveness can be ensured by restricting the nets to satisfy the following two conditions: deadlock-freedom, and containing no structural loop of transitions whose all the earliest firing times are equal to 0. Notice that the second condition means that in each structural loop there is a transition $t \in T$ such that $Eft(t) \geq 1$, and therefore at least one unit of time has to pass in every loop.

1.3.5 The Case of "General" TPNs

In the previous subsections, we considered 1-safe time Petri nets only. However, it is worth noticing that the semantics of the clocks assigned to the transitions, as well as the firing interval semantics, are applicable also to the case of unrestricted TPNs. The corresponding definitions of monotonicity, discrete semantics, runs, reachability, as well as of a set of all the reachable markings (in this case denoted by $RM_{\mathcal{N}}^{k_{\mathcal{N}}}$ to reflect the maximal capacity $k_{\mathcal{N}}$ of the places assumed or by $RM_{\mathcal{N}}^{k_{\mathcal{N}}+}$, respectively, if the time is strongly monotonic), can be easily adapted to this case. But then one should comment on the case of *multiple enabledness of transitions*. Such an enabledness happens when in a given marking a transition is enabled "several times". Formally, this occurs when for some marking m and $t \in en(m)$ we have

$$(\forall p \in \bullet t)\ m(p) \geq 2{\times}F(p,t).$$

One should decide what the firing interval or the value of the clock assigned to t is. Many strategies can be exploited (the "oldest" one, anyone randomly chosen etc.); however, what seems to be most natural is that the transitions which are enabled several times simultaneously are considered as independent occurrences of the same transition.

Consider now unrestricted TPNs, for which one can define the notion of *boundedness* as well. Thus, a time Petri net $\mathcal{N}$ is *bounded* if there is a bound on all the reachable markings of $\mathcal{N}$, i.e., there is a bound on $m(p)$ for each $p \in P$ and $m \in RM_{\mathcal{N}}^{\infty}$ ($m \in RM_{\mathcal{N}}^{\infty+}$, depending on the semantics assumed). Similarly, we can define a time Petri net to be *1-safe* if for each $p \in P$ and each $m \in RM_{\mathcal{N}}^{\infty}$ ($m \in RM_{\mathcal{N}}^{\infty+}$, respectively) we have $m(p) \leq 1$. The methods provided in the book for 1-safe TPNs do not depend on which of the two definitions[15] of these nets is used. However, it is important to notice that reachability and boundedness problems for unrestricted time Petri nets are undecidable [93].

Further Reading

For an introduction to Petri nets the reader is referred to the books by W. Reisig [135] and P. Starke [148]. Surveys on incorporating time into Petri nets can be found in [46, 47]. Several books dealing with Petri nets with time [54, 169] have been published recently. The interval semantics for TPNs is discussed in [32], whereas a description of all the clock-related semantics can be found in [124].

---

[15] The first definition is given in Sect. 1.2.

2

Timed Automata

In this chapter we consider timed automata, which were introduced by Alur and Dill [10]. Timed automata are extensions of finite state automata with constraints on timing behaviour. The underlying finite state automata are augmented with a set of real time variables. We start with formalising the above notions.

2.1 Time Zones

Let $\mathcal{X} = \{x_1, \ldots, x_{n_\mathcal{X}}\}$ be a finite set of real-valued variables, called *clocks*. The set of *clock constraints* over $\mathcal{X}$ is defined by the following grammar:

$$\mathfrak{cc} := true \mid x_i \sim c \mid x_i - x_j \sim c \mid \mathfrak{cc} \wedge \mathfrak{cc},$$

where $x_i, x_j \in \mathcal{X}$, $c \in \mathbb{N}$, and $\sim \in \{\leq, <, =, >, \geq\}$. The constraints of the form $true$, $x_i \sim c$ and $x_i - x_j \sim c$ are called *atomic*. The set of all the clock constraints over $\mathcal{X}$ is denoted by $\mathcal{C}^{\ominus}_{\mathcal{X}}$, whereas its restriction, where inequalities involving differences of clocks are not allowed, is denoted by $\mathcal{C}_{\mathcal{X}}$.

A *clock valuation* on $\mathcal{X}$ is a $n_\mathcal{X}$-tuple $v \in \mathbb{R}^{n_\mathcal{X}}_{0+}$. For simplicity, we assume a fixed ordering on $\mathcal{X}$. The value of the clock x_i in v can be then denoted by $v(x_i)$ or $v(i)$, depending on the context.

- For a valuation v and $\delta \in \mathbb{R}_{0+}$, $v + \delta$ denotes the valuation v' s.t.[1] for all $x \in \mathcal{X}$, $v'(x) = v(x) + \delta$.
- Moreover, for a subset of clocks $X \subseteq \mathcal{X}$, $v[X := 0]$ denotes the valuation v' such that $v'(x) = 0$ for all $x \in X$, and $v'(x) = v(x)$ for all $x \in \mathcal{X} \setminus X$.

The satisfaction relation $\models$ for a clock constraint $\mathfrak{cc} \in \mathcal{C}^{\ominus}_{\mathcal{X}}$ is defined inductively as follows:

- $v \models true$,

[1] The abbreviation "s.t." stands for "such that".

W. Penczek and A. Półrola: *Timed Automata*, Studies in Computational Intelligence (SCI) **20**, 29–49 (2006)
www.springerlink.com

- $v \models (x_i \sim c)$ iff $v(x_i) \sim c$,
- $v \models (x_i - x_j \sim c)$ iff $v(x_i) - v(x_j) \sim c$, and
- $v \models (\mathfrak{cc} \wedge \mathfrak{cc}')$ iff $v \models \mathfrak{cc}$ and $v \models \mathfrak{cc}'$.

Unless otherwise stated, we shall use the clock constraints defined by the grammar above. However, for some applications (to be seen in the further part of the book) it is convenient to use the atomic clock constraints described in a more unified manner. To this aim, we augment the set $\mathcal{X}$ with an additional fictitious clock $x_0 \notin \mathcal{X}$, which represents the constant 0. The set $\mathcal{X} \cup \{x_0\}$ is denoted by $\mathcal{X}^+$. Then, without loss of generality we can assume that the set of all the clock constraints for $\mathcal{X}$ is defined by the grammar

$$\mathfrak{cc} := x_i - x_j \sim c \mid x_i - x_j < \infty \mid x_i - x_j < -\infty \mid \mathfrak{cc} \wedge \mathfrak{cc},$$

where $x_i, x_j \in \mathcal{X}^+$, $c \in \mathbb{Z}$, and $\sim \in \{<, \leq\}$. The idea is to *normalise* every atomic clock constraint of $\mathcal{C}_{\mathcal{X}}^{\ominus}$ in the following steps (see also Example 2.1):

- the constraints of the form $x_i \sim c$ with $x_i \in \mathcal{X}$, $\sim \in \{<, \leq, \geq, >\}$ and $c \in \mathbb{N}$ are expressed by $x_i - x_0 \sim c$,
- these of the form $x_i = c$ or $x_i - x_j = c$ – by the conjunctions $x_i - x_0 \leq c \wedge x_0 - x_i \leq -c$ and $x_i - x_j \leq c \wedge x_j - x_i \leq -c$, respectively,
- *true* is expressed by $x_0 - x_0 < \infty$, and
- all the constraints of the form $x_i - x_j > c$ with $x_i, x_j \in \mathcal{X}^+$ and $c \in \mathbb{N}$ are replaced by $x_j - x_i < -c$, whereas these of the form $x_i - x_j \geq c$ – by $x_j - x_i \leq -c$.

The constraints of the form $x_i - x_j < -\infty$ are introduced to express constraints which are never satisfied. The modified set of the clock constraints, consisting of the above-defined *normalised atomic clock constraints* and their conjunctions, is denoted by $\mathcal{C}_{\mathcal{X}^+}^{\ominus}$. It will be used in Sect. 5.3.

Example 2.1. Consider the clock constraint $x_1 > 2$. Firstly, it is converted to $x_1 - x_0 > 2$, and then to its normal form $x_0 - x_1 < -2$.

□

The satisfaction relation $\models$ for a clock constraint $\mathfrak{cc} \in \mathcal{C}_{\mathcal{X}^+}^{\ominus}$ is defined inductively as follows:

- $v \models (x_0 - x_0 \sim c)$ iff $0 \sim c$,
- $v \models (x_i - x_0 \sim c)$ iff $v(x_i) \sim c$,
- $v \models (x_0 - x_i \sim c)$ iff $-v(x_i) \sim c$,
- the definitions for the other clock constraints follow these given before for the elements of $\mathcal{C}_{\mathcal{X}}^{\ominus}$.

Let $\mathcal{C}^{\ominus} = \mathcal{C}_{\mathcal{X}}^{\ominus} \cup \mathcal{C}_{\mathcal{X}^+}^{\ominus}$. For a constraint $\mathfrak{cc} \in \mathcal{C}^{\ominus}$, let $[\![\mathfrak{cc}]\!]$ denote the set of all the clock valuations satisfying $\mathfrak{cc}$, i.e.,

$$[\![\mathfrak{cc}]\!] = \{v \in \mathbb{R}_{0+}^{n_{\mathcal{X}}} \mid v \models \mathfrak{cc}\}.$$

By a (*time*) *zone* in $\mathbb{R}^{n_{\mathcal{X}}}_{0+}$ we mean each convex polyhedron $Z \subseteq \mathbb{R}^{n_{\mathcal{X}}}_{0+}$ defined by a clock constraint, i.e.,

$$Z = [\![\mathfrak{cc}]\!] \text{ for some } \mathfrak{cc} \in \mathcal{C}^{\ominus}$$

(for simplicity, we identify the zones with the clock constraints which define them). The set of all the zones for $\mathcal{X}$ is denoted by $Z(n_{\mathcal{X}})$.

Example 2.2. The left-hand side of Fig. 2.2 presents some examples of time zones of $Z(2)$.

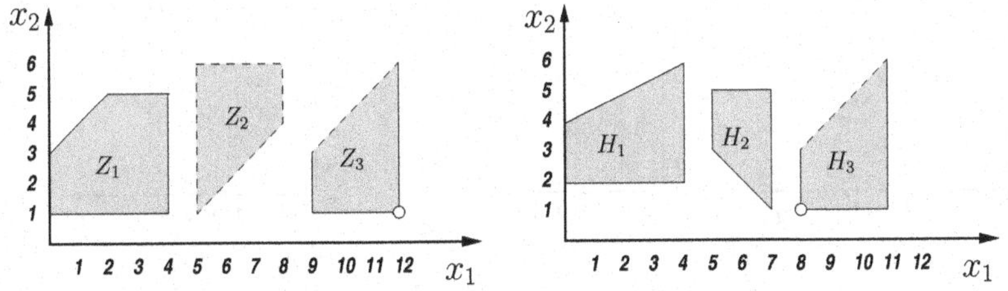

Fig. 2.1. Polyhedra in $\mathbb{R}^2_{0+}$. Only these in the left-hand side (i.e., Z_1, Z_2, Z_3) are time zones of $Z(2)$

The zone Z_1 is given by the clock constraint $x_1 \geq 0 \wedge x_1 \leq 4 \wedge x_2 \geq 1 \wedge x_2 \leq 5 \wedge x_2 - x_1 \leq 3$. Moreover, we have $Z_2 = [\![x_1 > 5 \wedge x_1 < 8 \wedge x_2 < 6 \wedge x_1 - x_2 < 4]\!]$ and $Z_3 = [\![x_1 \geq 9 \wedge x_1 \leq 12 \wedge x_2 \geq 1 \wedge x_1 - x_2 > 6 \wedge x_1 - x_2 < 11]\!]$. On the other hand, polyhedra on the right (denoted by H_1, H_2, H_3) are not time zones, since there is no clock constraint in $\mathcal{C}^{\ominus}$ which can describe any of them.

□

Given $v, v' \in \mathbb{R}^{n_{\mathcal{X}}}_{0+}$ and $Z, Z' \in Z(n_{\mathcal{X}})$, we use the following operations:

- $v \leq v'$ iff $\exists \delta \in \mathbb{R}_{0+}$ s.t. $v' = v + \delta$;
- $Z \setminus Z'$ is a finite set of disjoint zones[2] s.t. $\{Z'\} \cup (Z \setminus Z')$ is a partition[3] of Z,
- $Z \nearrow := \{v' \in \mathbb{R}^{n_{\mathcal{X}}}_{0+} \mid (\exists v \in Z)\ v \leq v'\}$,

[2] Obviously, an alternative definition of zone difference is also possible, i.e., one could find a finite family of zones $\mathcal{Z} \subseteq 2^{Z(n_{\mathcal{X}})}$ s.t. $Z = \bigcup_{Z_1 \in \mathcal{Z}} Z_1 \cup Z'$, and define $Z \setminus Z' = \bigcup_{Z_1 \in \mathcal{Z}} Z_1$ (i.e., define zone difference as a union of zones). However, we follow the approach of [8]. In our book, this operation is used mainly by the *partitioning algorithms* (see Chap. 5) and such a definition corresponds to what these algorithms produce in practice.

[3] By a *partition* of a set B we mean a family of its disjoint subsets $\mathcal{B}$ such that $\bigcup_{B' \in \mathcal{B}} B' = B$.

- $Z \swarrow := \{v' \in \mathbb{R}_{0+}^{n_{\mathcal{X}}} \mid (\exists v \in Z)\, v' \leq v\}$,
- $Z \Uparrow Z' = \{v \in Z \mid (\exists v' \in Z')\, v \leq v' \ \wedge (\forall v \leq v'' \leq v')\, v'' \in Z \cup Z'\}$,
- $Z[X := 0] = \{v[X := 0] \mid v \in Z\}$,
- $[X := 0]Z = \{v \in \mathbb{R}_{0+}^{n_{\mathcal{X}}} \mid v[X := 0] \in Z\}$.

Notice that the operations $\nearrow$, $\swarrow$, clock reset and its inverse (i.e., $Z[X := 0]$ and $[X := 0]Z$) and the standard intersection preserve zones [8, 159]. A description of the implementation of $Z \setminus Z'$, following [8], is given also in Sect. 5.3. Some examples of the operations are presented in Fig. 2.2.

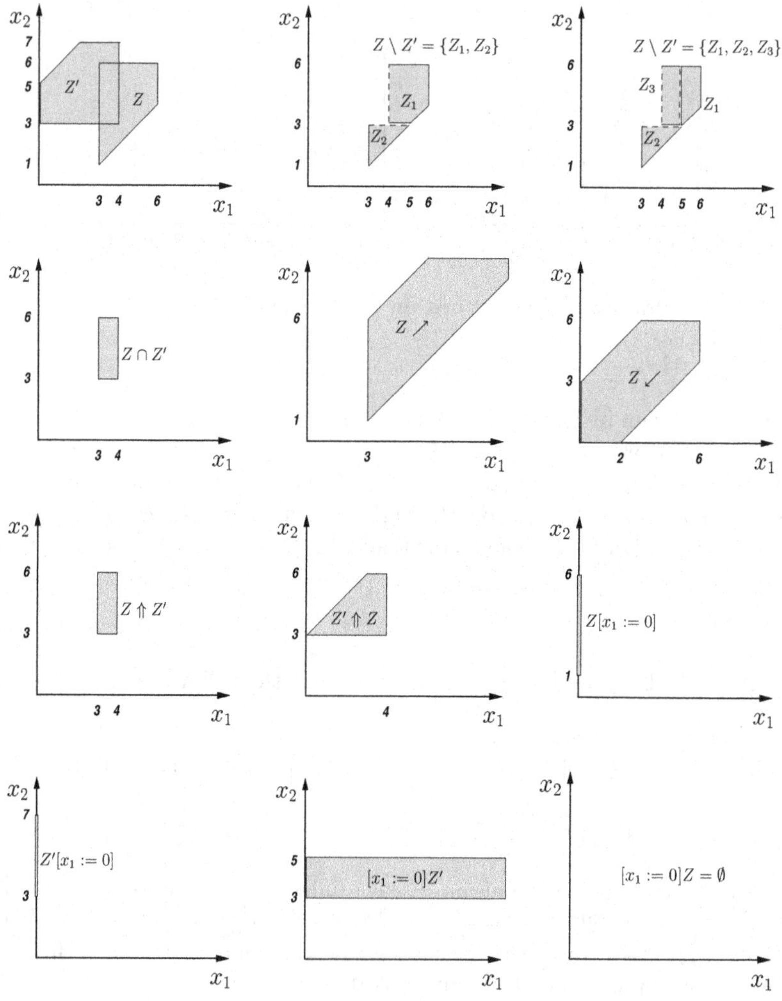

Fig. 2.2. Zones $Z, Z' \in Z(2)$ and examples of operations on them

2.2 Networks of Timed Automata

In this section we define timed automata, give their semantics, and show how to define a product of timed automata. Typically, we consider *networks* of timed automata, consisting of several timed automata running in parallel and communicating with each other.

Definition 2.3. A timed automaton *(TA, for short) is a six-element tuple*

$$\mathcal{A} = (A, L, l^0, E, \mathcal{X}, \mathcal{I}),$$

where

- A *is a finite set of* actions, *where* $A \cap \mathbb{R}_{0+} = \emptyset$,
- L *is a finite set of* locations,
- $l^0 \in L$ *is an* initial location,
- $\mathcal{X}$ *is a finite set of clocks,*
- $E \subseteq L \times A \times \mathcal{C}_{\mathcal{X}}^{\ominus} \times 2^{\mathcal{X}} \times L$ *is a* transition relation,
- $\mathcal{I} : L \longrightarrow \mathcal{C}_{\mathcal{X}}^{\ominus}$ *is a* (location) invariant.

Each element e *of* E *is denoted by* $l \xrightarrow{a,cc,X} l'$, *which represents a transition from the location* l *to the location* l', *executing the action* a, *with the set* $X \subseteq \mathcal{X}$ *of clocks to be reset, and with the clock constraint* $\mathfrak{cc}$ *defining the enabling condition for* e. *The function* $\mathcal{I}$ *assigns to each location* $l \in L$ *a clock constraint defining the conditions under which* $\mathcal{A}$ *can stay in* l.

If the enabling conditions and the values of the location invariant are in the set $\mathcal{C}_{\mathcal{X}}$ only, then the automaton is called *diagonal-free*. Given a transition $e := l \xrightarrow{a,cc,X} l'$, we write *source*$(e)$, *target*$(e)$, *action*$(e)$, *guard*$(e)$ and *reset*(e) for l, l', a, $\mathfrak{cc}$ and X, respectively. The clocks of a timed automaton allow to express the timing properties. An enabling condition constrains the execution of a transition without forcing it to be taken. An invariant condition permits an automaton to stay at the location l only as long as the clock constraint $\mathcal{I}(l)$ is satisfied.

Example 2.4. Figure 2.3 shows a timed automaton $\mathcal{A}$ of four locations, numbered from 0 to 3, one clock x and the set of actions $A = \{approach, in, out, exit\}$. The initial location is coloured. The invariant of the location 0 is *true*[4], whereas for all the other locations is given by $x \leq 500$. The transition relation of $\mathcal{A}$ consists of the elements $0 \xrightarrow{approach,true,\{x\}} 1$, $1 \xrightarrow{in,x\geq 300,\emptyset} 2$, $2 \xrightarrow{out,true,\emptyset} 3$, and $3 \xrightarrow{exit,x\leq 500,\emptyset} 0$. The location 2 is annotated by the proposition *is_inside*, which holds *true* at this location (a valuation of the locations is discussed in Sect. 2.4).

□

[4] In the pictures we omit the invariants and the guards equal to *true* as well as the empty sets of clocks to be reset.

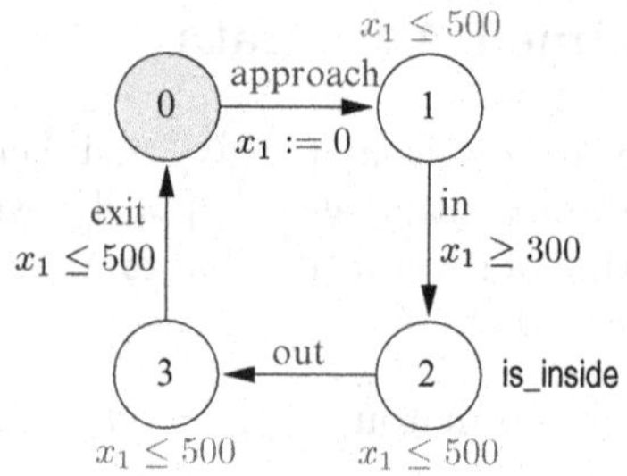

Fig. 2.3. A timed automaton

Real-time systems are usually represented by *networks* (sets) of timed automata. An example of such a system, widely considered in the literature, is the Fischer's mutual exclusion protocol (see below). Another system, modelling an automated railroad crossing (known as the Train–Gate–Controller example) is considered in Example 2.8.

Example 2.5. In Fig. 2.4, a network of TA for Fischer's mutual exclusion protocol with two processes is depicted[5]. The protocol is parameterised by the number of processes involved. In the general case, the network consists of n automata of processes, together with one automaton modelling a global variable V, used to coordinate the processes' access to their critical sections. □

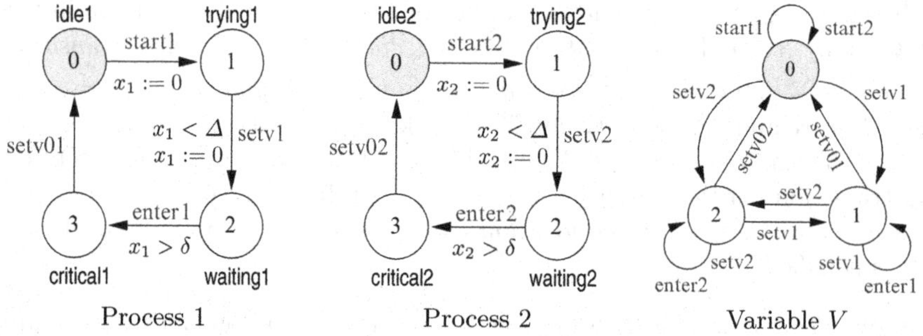

Fig. 2.4. Fischer's mutual exclusion protocol for two processes

A set of timed automata can be composed into a global (*product*) timed automaton as follows: the transitions of the timed automata that do not correspond to a shared action are interleaved, whereas the transitions labelled

[5] Two clocks are denoted by x_1 and x_2, whereas Δ and δ are parameters. The initial locations are numbered by 0.

with a shared action are synchronized. There are many different definitions of a parallel composition. Our definition determines the *multi-way synchronization*, i.e., it requires that each component that contains a communication transition (labelled with a shared action) has to perform this action.

Definition 2.6. *Let* $\mathfrak{I} = \{\mathfrak{i}_1, \ldots, \mathfrak{i}_{n_\mathfrak{I}}\}$ *be a finite ordered set of indices, and* $\mathfrak{A} = \{\mathcal{A}_\mathfrak{i} \mid \mathfrak{i} \in \mathfrak{I}\}$, *where* $\mathcal{A}_\mathfrak{i} = (A_\mathfrak{i}, L_\mathfrak{i}, l^0_\mathfrak{i}, E_\mathfrak{i}, \mathcal{X}_\mathfrak{i}, \mathcal{I}_\mathfrak{i})$, *be a set (network) of timed automata indexed with* $\mathfrak{I}$. *The automata in* $\mathfrak{A}$ *are called* components. *Let* $A(a) = \{\mathfrak{i} \in \mathfrak{I} \mid a \in A_\mathfrak{i}\}$ *be a set of the indices of the components containing the action* $a \in \bigcup_{\mathfrak{i}\in\mathfrak{I}} A_\mathfrak{i}$. A composition *(*product*)* of the timed automata $\mathcal{A}_{\mathfrak{i}_1} \parallel \ldots \parallel \mathcal{A}_{\mathfrak{i}_{n_\mathfrak{I}}}$ *is a timed automaton* $\mathcal{A} = (A, L, l^0, E, \mathcal{X}, \mathcal{I})$, *where*

- $A = \bigcup_{\mathfrak{i}\in\mathfrak{I}} A_\mathfrak{i}$,
- $L = \prod_{\mathfrak{i}\in\mathfrak{I}} L_\mathfrak{i}$,
- $l^0 = (l^0_{\mathfrak{i}_1}, \ldots, l^0_{\mathfrak{i}_{n_\mathfrak{I}}})$,
- $\mathcal{X} = \bigcup_{\mathfrak{i}\in\mathfrak{I}} \mathcal{X}_\mathfrak{i}$,
- $\mathcal{I}((l_{\mathfrak{i}_1}, \ldots, l_{\mathfrak{i}_{n_\mathfrak{I}}})) = \bigwedge_{\mathfrak{i}\in\mathfrak{I}} \mathcal{I}_\mathfrak{i}(l_\mathfrak{i})$,

and the transition relation is given by

- $((l_{\mathfrak{i}_1}, \ldots, l_{\mathfrak{i}_{n_\mathfrak{I}}}), a, \bigwedge_{\mathfrak{i}\in A(a)} \mathfrak{cc}_\mathfrak{i}, \bigcup_{\mathfrak{i}\in A(a)} X_\mathfrak{i}, (l'_{\mathfrak{i}_1}, \ldots, l'_{\mathfrak{i}_{n_\mathfrak{I}}})) \in E \iff$ $(\forall \mathfrak{i} \in A(a))\ (l_\mathfrak{i}, a, \mathfrak{cc}_\mathfrak{i}, X_\mathfrak{i}, l'_\mathfrak{i}) \in E_\mathfrak{i}$ *and* $(\forall \mathfrak{i} \in \mathfrak{I} \setminus A(a))\ l'_\mathfrak{i} = l_\mathfrak{i}$.

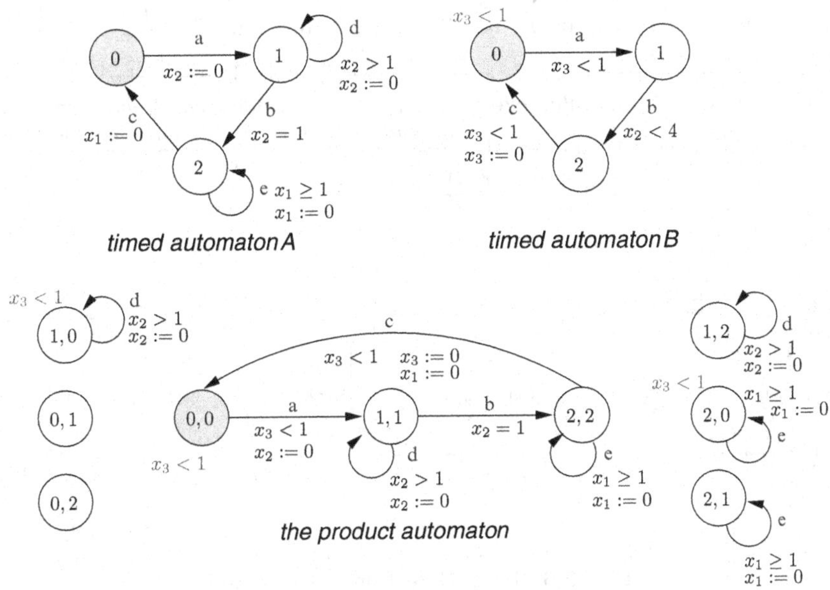

Fig. 2.5. Two timed automata and their composition

Example 2.7. In Fig. 2.5, a network of timed automata, consisting of two automata A and B, and the product automaton $A \parallel B$ of A and B are presented. The initial locations of all the automata are coloured. The automaton A, over two clocks x_1, x_2, can execute the actions a, b, c, d and e with the transitions $0 \xrightarrow{a,true,\{x_2\}} 1$, $1 \xrightarrow{b,x_2=1,\emptyset} 2$, $1 \xrightarrow{d,x_2>1,\{x_2\}} 1$, $2 \xrightarrow{c,true,\{x_1\}} 0$ and $2 \xrightarrow{e,x_1\geq 1,\{x_1\}} 2$. The possible transitions of the automaton B of two clocks x_2, x_3 (notice that the automata have a common clock x_2) are $0 \xrightarrow{a,x_3<1,\emptyset} 1$, $1 \xrightarrow{b,x_2<4,\emptyset} 2$, and $2 \xrightarrow{c,x_3<1,\{x_3\}} 0$. Both the components execute in parallel and synchronize through the actions a, b, c. The locations of the product automaton are given as pairs j_A, j_B whose elements corresponds to the numbers of locations of the components (notice that only three of them are reachable). □

The next example shows the railroad crossing system (known also as a Train–Gate–Controller example), often considered in the literature:

Example 2.8. Figure 2.6 depicts the timed automata of the train, gate and controller, modelling the behaviour of the automatic railroad crossing system. The component Train can execute the actions *approach*, *in*, *out* and *exit* labelling the transitions $0 \xrightarrow{approach,true,\{x_1\}} 1$, $1 \xrightarrow{in,x_1\geq 300,\emptyset} 2$, $2 \xrightarrow{out,true,\emptyset} 3$, $3 \xrightarrow{exit,x_1\leq 500,\emptyset} 0$. Similarly, the actions of the component Gate are *lower*, *down*, *raise* and *up* labelling the transitions $0 \xrightarrow{lower,true,\{x_2\}} 1$, $1 \xrightarrow{down,x_2\leq 100,\emptyset} 2$, $2 \xrightarrow{raise,true,\{x_2\}} 3$, $3 \xrightarrow{up,100\leq x_2\leq 200,\emptyset} 0$. The component Controller executes the actions *approach*, *lower*, *exit* and *raise* labelling the transitions $0 \xrightarrow{approach,true,\{x_3\}} 1$, $1 \xrightarrow{lower,x_3=100,\emptyset} 2$, $2 \xrightarrow{exit,true,\{x_3\}} 3$, $3 \xrightarrow{raise,x_3\leq 100,\emptyset} 0$.

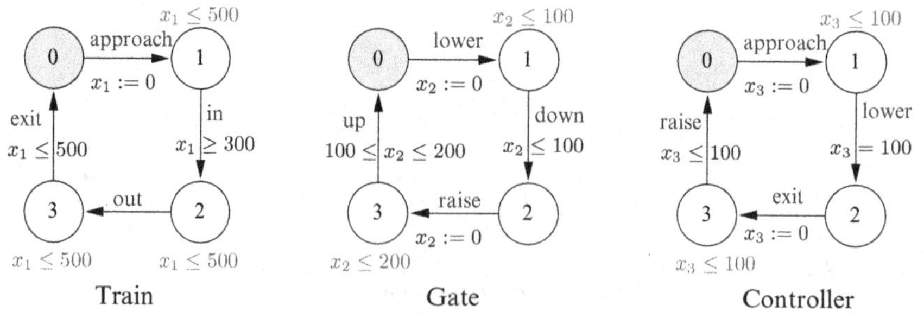

Fig. 2.6. Train–Gate–Controller example

All the components execute in parallel and synchronize through the actions *approach*, *exit*, *lower* and *down*. When Train approaches the crossing, it sends an *approach* signal to Controller and enters the crossing at least 300 seconds later. When Train leaves the crossing, it sends an *exit* signal to Controller. The *exit* signal is sent within 500 seconds after the *approach* signal. Controller sends a signal *lower* to Gate exactly 100 seconds after the *approach* signal and sends a *raise* signal within 100 seconds after *exit*. Gate responds to *lower* by moving *down* within 100 seconds, and responds to *raise* by moving *up* between 100 and 200 seconds.

The product automaton for the above example is depicted in Fig. 2.7.

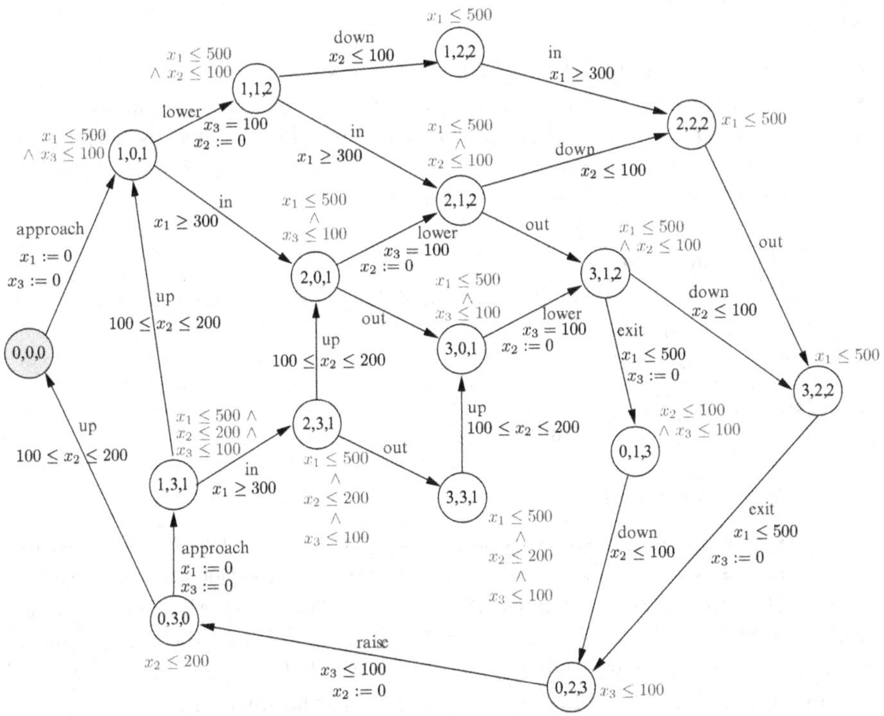

Fig. 2.7. The product automaton for the Train–Gate–Controller example

□

In what follows we assume that all the transitions of $\mathcal{A}$ labelled with the same action reset the same clocks[6].

---

[6] Such an assumption is made in order to efficiently encode the transition relation in the bounded model checking method for timed automata (see Sect. 7.1).

2.3 Semantics of Timed Automata

Let $\mathcal{A} = (A, L, l^0, E, \mathcal{X}, \mathcal{I})$ be a timed automaton. A *concrete state* of $\mathcal{A}$ is defined as an ordered pair

$$(l, v),$$

where $l \in L$ and $v \in \mathbb{R}_{0+}^{n_\mathcal{X}}$ is a valuation. The *concrete (dense) state space* of $\mathcal{A}$ is a transition system

$$C_c(\mathcal{A}) = (Q, q^0, \rightarrow_c),$$

where

- $Q = L \times \mathbb{R}_{0+}^{n_\mathcal{X}}$ is the set of all the *concrete states*,
- $q^0 = (l^0, v^0)$ with $v^0(x) = 0$ for all $x \in \mathcal{X}$ is the *initial state*, and
- $\rightarrow_c \subseteq Q \times (E \cup \mathbb{R}_{0+}) \times Q$ is the *transition relation*, defined by action- and time successors as follows:
  - for $\delta \in \mathbb{R}_{0+}$, $(l, v) \xrightarrow{\delta}_c (l, v + \delta)$ iff
    - $v, v + \delta \in [\![\mathcal{I}(l)]\!]$

    (*time successor*),
  - for $a \in A$, $(l, v) \xrightarrow{a}_c (l', v')$ iff $(\exists \mathfrak{cc} \in \mathcal{C}_\mathcal{X})(\exists X \subseteq \mathcal{X})$ such that
    - $l \xrightarrow{a, \mathfrak{cc}, X} l' \in E$,
    - $v \in [\![\mathfrak{cc}]\!]$,
    - $v \in [\![\mathcal{I}(l)]\!]$,
    - $v' = v[X := 0]$, and
    - $v' \in [\![\mathcal{I}(l')]\!]$

    (*action successor*).

Intuitively, a time successor does not change the location l of a concrete state, but it increases the clocks, provided their values still satisfy the invariant of l. Since the invariants are zones, if v and v' satisfy $\mathcal{I}(l)$, then all the clock valuations between v and v' satisfy $\mathcal{I}(l)$. An action successor corresponding to an action a is executed when the guard $\mathfrak{cc}$ holds for v and the valuation v' obtained after resetting the clocks in X satisfies the invariant of l'.

For $(l, v) \in Q$ and $\delta \in \mathbb{R}_{0+}$, let $(l, v) + \delta$ denote $(l, v + \delta)$. *Concatenation* of two time steps $q \xrightarrow{\delta}_c q + \delta$ and $q + \delta \xrightarrow{\delta'}_c q + \delta + \delta'$ is the time step $q \xrightarrow{\delta + \delta'}_c q + \delta + \delta'$. Similarly, if $q \xrightarrow{\delta}_c q + \delta$, then for any $k \in \mathbb{N}_+$ there exist $\delta_1, \ldots, \delta_k \in \mathbb{R}_{0+}$ such that $\delta_1 + \ldots + \delta_k = \delta$ and $q \xrightarrow{\delta_1}_c q + \delta_1 \xrightarrow{\delta_2}_c \ldots \xrightarrow{\delta_k}_c q + \delta$ (i.e., each time step can be *split* into an arbitrary number of consecutive time successors). A (*dense*) q_0-*run* ρ of $\mathcal{A}$ is a maximal (i.e., non-extendable) sequence

$$\rho = q_0 \xrightarrow{\delta_0}_c q_0 + \delta_0 \xrightarrow{a_0}_c q_1 \xrightarrow{\delta_1}_c q_1 + \delta_1 \xrightarrow{a_1}_c q_2 \xrightarrow{\delta_2}_c \ldots,$$

where $a_i \in A$ and $\delta_i \in \mathbb{R}_{0+}$, for each $i \geq \mathbb{N}$ (notice that due to the fact that δ can be equal to 0 two consecutive transitions can be executed without any time passing in between, and that consecutive time passings are concatenated). Such runs are called *weakly monotonic.*

Example 2.9. Given the product automaton shown in Fig. 2.5, let $((j_A, j_B),(v(x_1), v(x_2), v(x_3)))$ denote the state (l, v) of the automaton, where l is a pair j_A, j_B. One of the possible q^0-runs is

$((0,0),(0,0,0)) \xrightarrow{0.5}_c ((0,0),(0.5,0.5,0.5)) \xrightarrow{a}_c ((1,1),(0.5,0,0.5)) \xrightarrow{1}_c ((1,1),(1.5,1,1.5)) \xrightarrow{b}_c ((2,2),(1.5,1,1.5)) \xrightarrow{0.5}_c ((2,2),(2,1.5,2)) \xrightarrow{e}_c ((2,2),(0,1.5,2)) \xrightarrow{1.5}_c \ldots$

□

A state $q \in Q$ is *reachable* if there exists a q^0-run ρ and $i \in \mathbb{N}$ such that $q = q_i + \delta$ for some $0 \le \delta \le \delta_i$, where $q_i + \delta_i$ is an element of ρ. The set of all the reachable states of $\mathcal{A}$ is denoted by $Reach_{\mathcal{A}}$. Given a run $\rho = q_0 \xrightarrow{\delta_0}_c q_0 + \delta_0 \xrightarrow{a_0}_c q_1 \xrightarrow{\delta_1}_c q_1 + \delta_1 \xrightarrow{a_1}_c \ldots$, we say that a run $\rho' = q'_0 \xrightarrow{\delta'_0}_c q'_0 + \delta'_0 \xrightarrow{a'_0}_c \ldots$ is a *suffix* of ρ if there exists $i \in \mathbb{N}$ and $0 \le \delta \le \delta_i$ such that $q'_0 = q_i + \delta$, $\delta_i = \delta + \delta'_0$, and $a_j = a_{i+j}$ and $\delta_{j+1} = \delta_{i+j+1}$ for each $j \in \mathbb{N}$ (note that a suffix of ρ can start at a state which results from splitting a time step occurring in the run). A run ρ is said to be *progressive* iff $\Sigma_{i\in\mathbb{N}}\delta_i$ is unbounded. Notice that a run is progressive iff all its suffixes are so. A timed automaton is *progressive*[7] iff all its runs starting at the initial state are progressive. For simplicity of the presentation, we restrict our considerations to progressive timed automata only. Note that progressiveness can be checked using for example sufficient conditions of [159], which we discuss in detail in Sect. 2.5. The set of all the progressive (weakly monotonic) q-runs of $\mathcal{A}$ is denoted by $f_{\mathcal{A}}(q)$.

Like for time Petri nets, weakly monotonic runs can be easily used for interpreting untimed temporal logics as well as for checking reachability. In fact, they can be also applied to interpreting timed languages like TCTL or timed μ-calculus. But, it is sometimes more convenient to restrict the runs to contain non-zero time passings only (i.e., $\delta_i > 0$ for all $i \in \mathbb{N}$), which prevents firings of two transitions immediately one after the other. Such runs are called *strongly monotonic.* In what follows, $f^+_{\mathcal{A}}(q)$ is used to denote the set of all the progressive strongly monotonic q-runs of $\mathcal{A}$. By $Reach^+_{\mathcal{A}}$ we denote the set of all the states of $\mathcal{A}$ which are reachable assuming the above definition of q^0-runs.

We refer to the semantics over weakly monotonic runs as to *weakly monotonic,* whereas to the semantics over strongly monotonic runs as to *strongly monotonic.*

Example 2.10. Notice that the run given in Example 2.9 is valid also when time steps are restricted to be non-zero only. However, the sets $Reach_{\mathcal{A}}$ and

[7] Progressiveness is also called a *non-zeno* property.

$Reach^+_{\mathcal{A}}$ are not equal. The state $((1,1),(0,0,0))$, resulting from executing the action a at the initial state, belongs to $Reach_{\mathcal{A}}$, but not to $Reach^+_{\mathcal{A}}$.

□

2.3.1 Alternative (Discrete) Semantics

Similarly to the case of time Petri nets, discrete runs of timed automata can be defined by combining action- and time steps. Again, we do not change the domain $\mathbb{R}_{0+}$ of the clock valuations. Alternatively, one could restrict the time domain to integers [18] getting discrete-time models, so that passing of time would be measured only using integer numbers. It is known that such an approach could be used for modelling accurately synchronous systems, where all the components are driven by one common global clock, but we do not discuss this approach in this book. However, for modelling asynchronous systems it is necessary to use continuous time [7].

In our alternative approach, runs of timed automata are defined such that passing some time and then executing an action (and then, possibly, passing some time again) are combined into a single step. Such an approach, called a *discrete semantics*, is sometimes used for verification of untimed temporal properties or reachability checking [126]. Similarly to the case of time Petri nets, we introduce the following modifications of the successor relation $\rightarrow_c$:

- $\rightarrow_{d_1} \subseteq Q \times A \times Q$, where
  $q \xrightarrow{a}_{d_1} q'$ iff $(\exists q_1 \in Q)(\exists \delta \in \mathbb{R}_{0+})\; q \xrightarrow{\delta}_c q_1 \xrightarrow{a}_c q'$;
- $\rightarrow_{d_2} \subseteq Q \times A \times Q$, where
  $q \xrightarrow{a}_{d_2} q'$ iff $(\exists q_1, q_2 \in Q)(\exists \delta, \delta' \in \mathbb{R}_{0+})\; q \xrightarrow{\delta}_c q_1 \xrightarrow{a}_c q_2 \xrightarrow{\delta'}_c q'$.

Clearly the relation $\rightarrow_{d_2}$ includes $\rightarrow_{d_1}$ (due to the possibility of the δ's to be equal to 0), but to keep compatibility with the TPNs approaches we define both of them. Notice that it is also possible to redefine the relations $\rightarrow_{d_1}$ and $\rightarrow_{d_2}$ by restricting the values of δ, δ' to be non-zero only, but in this case, similarly to the TPNs case, the above inclusion does not need to hold anymore.

For each of the above relations we introduce a notion of a (*discrete*) q_0*-run* of $\mathcal{A}$, which is a maximal sequence of concrete states

$$\rho_r = q_0 \xrightarrow{a_0}_{d_r} q_1 \xrightarrow{a_1}_{d_r} q_2 \xrightarrow{a_2}_{d_r} \cdots,$$

where $r \in \{1,2\}$, and $a_i \in A$ for all $i \in \mathbb{N}$. A state $q \in Q$ is *reachable* (w.r.t. d_r) if there exists a q^0-run ρ_r and $i \in \mathbb{N}$ such that $q = q_i$, where q_i is an element of ρ_r. The set of all the reachable states (w.r.t. d_r) is denoted by $Reach^{d_r}_{\mathcal{A}}$ (notice that the set can vary depending on the successor relation $\rightarrow_{d_r}$). *Concrete (discrete) state space* of $\mathcal{A}$ is now a transition system $C_{d_r}(\mathcal{A}) = (Q, q^0, \rightarrow_{d_r})$, where $r \in \{1,2\}$.

Example 2.11. Consider the product automaton $\mathcal{A}$ shown in Fig. 2.5. For the concrete state space $C_{d_1}(\mathcal{A})$, one of the possible q^0-runs is

$$((0,0),(0,0,0)) \xrightarrow{a}_{d_1} ((1,1),(0.5,0,0.5)) \xrightarrow{b}_{d_1} ((2,2),(1.5,1,1.5)) \xrightarrow{e}_{d_1} ((2,2),(0,1.5,2)) \xrightarrow{e}_{d_1} \ldots.$$

In the case of the concrete state space $C_{d_2}(\mathcal{A})$, we can obtain, for example, the q^0-run

$$((0,0),(0,0,0)) \xrightarrow{a}_{d_2} ((1,1),(1,0.5,1)) \xrightarrow{b}_{d_2} ((2,2),(1.5,1,1.5)) \xrightarrow{e}_{d_2} ((2,2),(0.5,1.5,2)) \xrightarrow{e}_{d_2} \ldots.$$

Its first step corresponds to passing 0.5 unit of time, executing the action a and then passing 0.5 time unit again, whereas the second consists in passing 0.5 unit of time and then executing b with no time delay after that. In the third step, the action e is executed without any previous delay, and then 0.5 unit of time passes.

Notice that the sets $Reach_{\mathcal{A}}^{d_1}$ and $Reach_{\mathcal{A}}^{d_2}$ differ (e.g., the state $((1,1),(1,0.5,1))$, obtained from q^0 by passing 0.5 unit of time, executing a and then passing 0.5 time unit again, belongs to $Reach_{\mathcal{A}}^{d_2}$, but is not an element of $Reach_{\mathcal{A}}^{d_1}$). Moreover, both these sets are different from $Reach_{\mathcal{A}}$, since the state $((0,0),\ (0.5,0.5,0.5)) \in Reach_{\mathcal{A}}$ does belong neither to $Reach_{\mathcal{A}}^{d_1}$ nor to $Reach_{\mathcal{A}}^{d_2}$.

□

Similarly as for time Petri nets, it can be proven that for the weakly monotonic semantics we have $Reach_{\mathcal{A}}^{d_1} \subseteq Reach_{\mathcal{A}}^{d_2} \subseteq Reach_{\mathcal{A}}$.

2.4 Concrete Models for TA

Let PV be a set of propositional variables. In order to reason about systems represented by timed automata, we define a valuation function $V_{\mathcal{A}} : L \longrightarrow 2^{PV}$, which assigns a subset of PV to each location. Then, in order to reason about systems represented by networks of timed automata[8], given a network of timed automata $\mathfrak{A} = \{\mathcal{A}_\mathfrak{i} \mid \mathfrak{i} \in \mathfrak{I}\}$ indexed with $\mathfrak{I} = \{\mathfrak{i}_1, \ldots, \mathfrak{i}_{n_\mathfrak{I}}\}$, and functions $V_{\mathcal{A}_\mathfrak{i}} : L_i \longrightarrow 2^{PV_i}$, where $PV_\mathfrak{i} \cap PV_\mathfrak{j} = \emptyset$ for all $i, j \in \mathfrak{I}$ with $i \neq j$, we define a valuation function $V_\mathfrak{C} : Q \longrightarrow 2^{PV}$, where Q is the set of all the concrete states of the product automaton for $\mathfrak{A}$ and $PV = \bigcup_{i=1}^{n_\mathfrak{I}} PV_i$, such that

$$V_\mathfrak{C}(((l_{\mathfrak{i}_1}, \ldots, l_{\mathfrak{i}_{n_\mathfrak{I}}}), \cdot)) = \bigcup_{\mathfrak{i} \in \mathfrak{I}} V_{\mathcal{A}_\mathfrak{i}}(l_\mathfrak{i})$$

[8] In particular, a network can consist of a single automaton only.

(i.e., $V_{\mathfrak{C}}$ assigns the same propositions to the states with the same locations).

Let $C(\mathcal{A}) \in \{C_c(\mathcal{A}), C_{d_1}(\mathcal{A}), C_{d_2}(\mathcal{A})\}$ be a concrete state space of the timed automaton $\mathcal{A}$. The transition system

$$M_{\mathfrak{c}}(\mathcal{A}) = (C(\mathcal{A}), V_{\mathfrak{C}})$$

is called a *concrete model* for $\mathcal{A}$ (*based on the state space* $C(\mathcal{A})$). The concrete model is called *dense* (*discrete*) if $C(\mathcal{A})$ is so.

2.5 Checking Progressiveness

Similarly to the case of time Petri nets, we have also restricted our considerations to a class of progressive timed automata. But, in order to deal with such automata, we need to check whether a given timed automaton $\mathcal{A}$ falls into this class. Unfortunately, this cannot be easily decided without any prior analysis. Our discussion of this problem below is based on [156, 159, 160].

Consider first a timed automaton which is non-progressive because it contains some "terminal" locations (see for example the automaton like in Fig. 2.3, but without the action labelled with *exit*), which can make some of its runs finite. This, however, can be easily avoided, since if a system $\mathcal{S}$ modelled by an automaton $\mathcal{A}_{\mathcal{S}}$ can terminate its execution in some legal end state, which means that there is a location with no outgoing transitions in $\mathcal{A}_{\mathcal{S}}$, one can add a fictitious loop transition with an appropriate guard and a set of clocks to be reset, which can be executed infinitely many times.

Another possibility for a given automaton $\mathcal{A}$ to be non-progressive is when some of its reachable states are *deadlocks* or *timelocks*. Formally, a concrete state of $\mathcal{A}$ is a *deadlock* if it occurs in a finite q^0-run of $\mathcal{A}$ and no passage of time at this state can enable an action, and is a *timelock* if all infinite runs starting from this state are not progressive. Notice that a deadlock is not necessarily a timelock, neither the reverse:

Example 2.12. Consider the automaton $\mathcal{A}$ over one clock x, shown in Fig. 2.8.

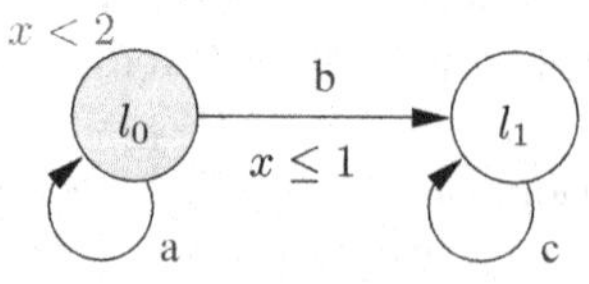

Fig. 2.8. A timed automaton with timelocks

It is easy to see that irrespectively on the definition of runs no state of $\mathcal{A}$ is a deadlock. However, all the concrete states (l_0, v) with $v(x) \in (1, 2)$ are timelocks, since in all the runs starting at them the action a is executed infinitely many times, but the value of the clock x does not exceed two units. On the other hand, if the transition labelled with a was missing, then these states would be deadlocks, but not timelocks, since there would not be infinite runs starting from them at all.

□

A timed automaton is *deadlock-free* (*timelock-free*) if none of its reachable states is a deadlock (timelock, respectively). It should be mentioned that some concrete states are deadlocks only when the strongly monotonic semantics is considered. An example of such a state is $(l_1, (1))$ in the automaton depicted in Fig. 2.9. Notice that different definitions of runs influence also the existence of timelocks: in the automaton in Fig. 2.9 the state $(l_0, (1))$ is a timelock only in the weakly monotonic semantics.

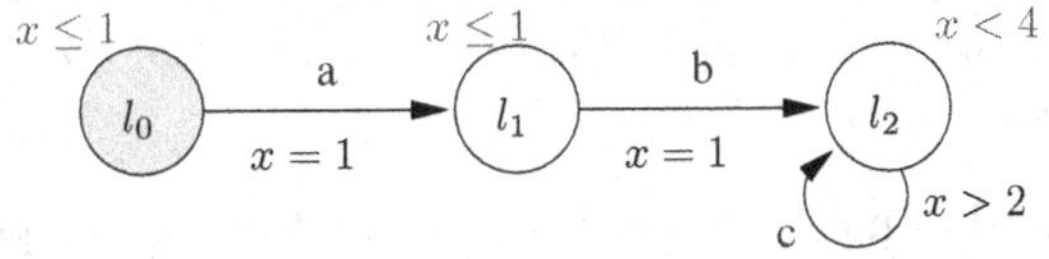

Fig. 2.9. A timed automaton with a deadlock in the strongly monotonic semantics

Unfortunately, the absence of deadlocks and timelocks does not guarantee progressiveness of the automaton. This can be easily derived from the example below, where the state $(l_0, (0))$ is neither deadlock nor timelock, but there are both progressive and non-progressive runs starting at it:

Example 2.13. Consider the automaton which is like that in Fig. 2.8, but with $\mathcal{I}(l_0) = true$. Although all its runs are infinite, some of them are non-progressive. An example of such a run is

$$\rho = (l_0, (0)) \xrightarrow{1}_c (l_0, (1)) \xrightarrow{b}_c (l_1, (1)) \xrightarrow{0.5}_c (l_1, (1.5)) \xrightarrow{c}_c (l_1, (1.5)) \xrightarrow{0.25}_c (l_1, (1.75)) \xrightarrow{c}_c \ldots,$$

where the time delays decrease geometrically as follows: $1, 0.5, 0.25, 0.125, \ldots$. So, their sum is bounded by 2. Notice that the same state is also a beginning of a progressive run, e.g.,

$$(l_0, (0)) \xrightarrow{1}_c (l_0, (1)) \xrightarrow{b}_c (l_1, (1)) \xrightarrow{1}_c (l_1, (2)) \xrightarrow{c}_c (l_1, (2)) \xrightarrow{1}_c (l_1, (3)) \xrightarrow{c}_c \ldots.$$

□

It is also easy to see that the above fact holds irrespectively on the semantics assumed.

In principle there are two methods of checking that a system is progressive. Firstly, one can formulate a sufficient condition on the structure of a timed automaton itself which guarantees this property. Secondly, it is possible to apply verification techniques to establish whether an automaton is progressive.

2.5.1 A Static Technique

One of the possible static (i.e., structural) conditions which can be verified in order to ensure progressiveness of timed automata exploits the notion of strongly progressive structural loops:

Definition 2.14. *A* structural loop *in a timed automaton* $\mathcal{A} = (A, L, l^0, E, \mathcal{X}, \mathcal{I})$ *is a sequence of distinct transitions* $e_1, \ldots e_n$ *such that* $target(e_i) = source(e_{i+1})$ *for all* $i = 1, .., n$ *(where the addition is modulo* n*). A structural loop is* strongly progressive *if there exists a clock* $x \in \mathcal{X}$ *and some* $1 \leq i, j \leq n$ *such that:*

1) x *is a clock that is reset by the transition* e_i*,*
2) x *is bounded from below by at least* 1 *in the guard of the transition* e_j*, i.e.,* $(x < 1 \wedge guard(e_j))$ *evaluates to* $false$*.*

Intuitively, 1) and 2) mean that at least one unit of time elapses in every loop of $\mathcal{A}$. An example of an automaton whose all the structural loops are strongly progressive is shown in Fig. 2.10. Notice that this would not be the case if any of the guards $x \geq 1$ was missing.

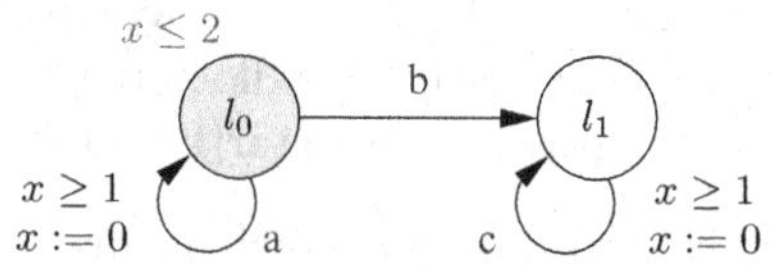

Fig. 2.10. A timed automaton with strongly progressive structural loops

Next, we give a sufficient condition for a timed automaton to be progressive [159]:

Lemma 2.15. $\mathcal{A}$ *is progressive if the following conditions hold:*

1. *every structural loop of* $\mathcal{A}$ *is strongly progressive, and*
2. $\mathcal{A}$ *is deadlock-free.*

Notice that if $\mathcal{A}$ satisfies the above conditions, then each of its q^0-runs is infinite as it is deadlock-free, and is progressive as it is "contained" in some structural loop.

In order to apply the above lemma, we need a method for testing absence of deadlocks in an automaton, for the concrete semantics assumed. A static sufficient condition for checking that a timed automaton $\mathcal{A} = (A, L, l^0, E, \mathcal{X}, \mathcal{I})$

is deadlock-free can be formulated using a local condition on the reachable states of $\mathcal{A}$. To do that, for $l \in L$ we define

$$free(l) = [\![\mathcal{I}(l)]\!] \cap \bigcup_{\{e \in E \mid source(e)=l\}} ([\![guard(e)]\!] \cap [\![\mathcal{I}(l)]\!] \cap [reset(e) := 0][\![\mathcal{I}(target(e))]\!]) \swarrow .$$

Intuitively, the set $free(l) \subseteq \mathbb{R}_{0+}^{n_\mathcal{X}}$ consists of the timed parts of all these concrete states $(l, \cdot)$ of $\mathcal{A}$ from which it is possible to exit l by executing some transition, after passing some (possibly zero) period of time first. It is easy to see that if zero time steps are allowed, then $\mathcal{A}$ is deadlock-free iff for each $(l, v) \in Reach_\mathcal{A}$ we have $v \in free(l)$ [156]. A sufficient static condition for deadlock-freedom in the weakly monotonic dense semantics is given by the following lemma [156]:

Lemma 2.16. *For the weakly monotonic semantics $\mathcal{A}$ is deadlock-free if for each location $l \in L$ and each $e \in E$ with $target(e) = l$ we have*

$$([\![guard(e)]\!][reset(e) := 0]) \nearrow \cap [\![\mathcal{I}(l)]\!] \subseteq free(l),$$

and the initial location $l^0 \in L$ satisfies the condition

$$\{v^0\} \nearrow \cap [\![\mathcal{I}(l^0)]\!] \subseteq free(l^0).$$

The below example illustrates the application of the above lemma:

Example 2.17. Consider the automaton $\mathcal{A}$ shown in Fig. 2.11.

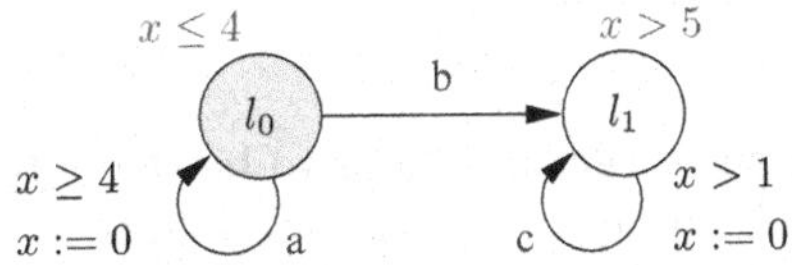

Fig. 2.11. The timed automaton considered in Example 2.17

If the transition labelled with a was missing, then we would have $free(l_0) = [\![x \le 4]\!] \cap ([\![true]\!] \cap [\![x \le 4]\!] \cap [\![x > 5]\!]) \swarrow = \emptyset$. If the initial location had no ingoing transitions, then the first condition of Lemma 2.16 would be satisfied, but the second would not hold due to $\{v^0\} \nearrow \cap [\![\mathcal{I}(l_0)]\!] = [\![x \le 4]\!]$. Therefore, the static test of Lemma 2.16 would return that $\mathcal{A}$ does not satisfy the condition guaranteeing deadlock-freedom in the weakly monotonic semantics. Indeed, it is easy to see that all the states $(l_0, \cdot)$ of the modified automaton are deadlocks, since the invariants of the locations l_0 and l_1 prevent executing the transition labelled with b.

If we consider the automaton as shown in the picture, then $free(l_0) = \emptyset \cup ([\![x \le 4]\!] \cap ([\![x \ge 4]\!] \cap [\![x \le 4]\!] \cap [\{x\} := 0][\![x \le 4]\!]) \swarrow) = [\![x \le 4]\!] \cap ([\![x =

$4\rrbracket \cap \mathbb{R}_{0+}) \swarrow = \llbracket x \leq 4 \rrbracket$. For the transition labelled with a (the only one whose target location is l_0) we compute $\llbracket x \geq 4 \rrbracket[\{x\} := 0] \nearrow \cap \llbracket x \leq 4 \rrbracket = \llbracket x = 0 \rrbracket \nearrow \cap \llbracket x \leq 4 \rrbracket = \llbracket x \leq 4 \rrbracket$. Moreover, we have $\{v^0\} \nearrow \cap \llbracket \mathcal{I}(l_0) \rrbracket = \llbracket x \leq 4 \rrbracket$. The sufficient condition for all the reachable states $(l_0, \cdot)$ not to be deadlocks in the weakly monotonic semantics is then satisfied.

Considering the location l_1, we compute $free(l_1) = \llbracket x > 5 \rrbracket \cap (\llbracket x > 1 \rrbracket \cap \llbracket x > 5 \rrbracket \cap \emptyset) \swarrow = \emptyset$. While testing the conditions of the lemma, for the transition labelled with b we obtain $\mathbb{R}_{0+} \cap \llbracket x > 5 \rrbracket = \llbracket x > 5 \rrbracket \not\subseteq free(l_1)$. Similarly, for the transition labelled with c we get $\llbracket x > 1 \rrbracket[\{x\} := 0] \nearrow \cap \llbracket x > 5 \rrbracket = \mathbb{R}_{0+} \cap \llbracket x > 5 \rrbracket = \llbracket x > 5 \rrbracket \not\subseteq free(l_1)$. Thus, the first condition of Lemma 2.16 does not hold, and therefore one cannot reason that $\mathcal{A}$ is deadlock-free. In fact, all the states $(l_1, \cdot)$ could have been deadlocks, since the invariant of the location prevents executing the transition labelled with c. Notice, however, that all of them are unreachable, and therefore they do not influence the deadlock-freedom of $\mathcal{A}$.

□

In order to ensure deadlock-freedom of $\mathcal{A}$ for the strongly monotonic semantics, one could require that the conditions of Lemma 2.16 hold and the enabling conditions and invariants of $\mathcal{A}$ are built from strict inequalities only. In the below example, we show that the conditions of Lemma 2.16 only are not sufficient.

Example 2.18. Consider again the automaton $\mathcal{A}$ shown in Fig. 2.9. For the locations of $\mathcal{A}$ we compute respectively $free(l_0) = \llbracket x \leq 1 \rrbracket$, $free(l_1) = \llbracket x \leq 1 \rrbracket$ and $free(l_2) = \llbracket x < 4 \rrbracket$. Consequently, it is easy to check that the conditions of Lemma 2.16 hold for $\mathcal{A}$. However, the state $(l_1, (1))$ of $\mathcal{A}$ is a deadlock in the strongly monotonic semantics.

□

The condition for all the structural loops to be strongly progressive is compositional (i.e., if two TA satisfy the condition, then their composition does so as well), whereas the condition given in Lemma 2.16 is not [157, 159]. Therefore, the above structural analysis cannot be applied to component automata in order to reason about properties of their product.

2.5.2 Applying Verification Methods

The static tests described before can be obviously combined with verification techniques applicable to checking progressiveness. Thus, in order to state whether a given automaton $\mathcal{A}$ is deadlock-free (i.e., to test whether the second condition of Lemma 2.15 holds for $\mathcal{A}$) one can built a *bisimulating* abstract model for TA (see Sect. 5.2) and check whether it contains a reachable abstract state with no successors (time self-loops, if exist, are not taken into account). Obviously, the semantics for which deadlocks are searched for has to correspond to the semantics for which a given model is generated. Due to that, the *detailed region graphs* (see Sect. 5.2) allow for tests for both the dense semantics considered, whereas coarser abstract models have been applied so far only to the weakly monotonic semantics.

Another method for checking deadlock-freedom, described in [156], consists in testing reachability of a state $(l_d, \cdot)$, where l_d is a special location corresponding to a deadlock in $\mathcal{A}$. More precisely, given a timed automaton $\mathcal{A} = (A, L, l^0, E, \mathcal{X}, \mathcal{I})$, we construct a new automaton $\mathcal{A}' = (A', L', l^0, E', \mathcal{X}, \mathcal{I}')$, which is like $\mathcal{A}$ besides that

- $A' = A \cup \{a_d\}$, where $a_d \notin A$,
- $L' = L \cup \{l_d\}$ with $l_d \notin L$,
- $\mathcal{I}' : L' \longrightarrow \mathcal{C}_{\mathcal{X}}^{\ominus}$ is given by $\mathcal{I}'(l) = \mathcal{I}(l)$ for $l \in L$, and $\mathcal{I}'(l_d) = true$, and
- $E' = E \cup \{l \xrightarrow{a_d, cc, \emptyset} l_d \mid l \in L \ \wedge [\![cc]\!] \in (\mathbb{R}_{0+}^{n_{\mathcal{X}}} \setminus free(l))\}$.

Intuitively, the construction adds to each location of $\mathcal{A}$ a number of outgoing transitions[9] leading to l_d, which serve as "escape" actions, i.e., can be taken whenever the automaton $\mathcal{A}$ reaches a deadlock state. It is easy to see that from unreachability of $(l_d, \cdot)$ in $\mathcal{A}'$ it follows that $\mathcal{A}$ is deadlock-free. Reachability of such a state is checked using an arbitrary method from these described in the further part of this book. Because of the definition of $free(l)$, the method is applicable to the weakly monotonic semantics.

Example 2.19. Figure 2.13 depicts a timed automaton $\mathcal{A}'$ built for testing deadlocks in the automaton $\mathcal{A}$ shown in Fig. 2.12 (the automaton $\mathcal{A}$, consid-

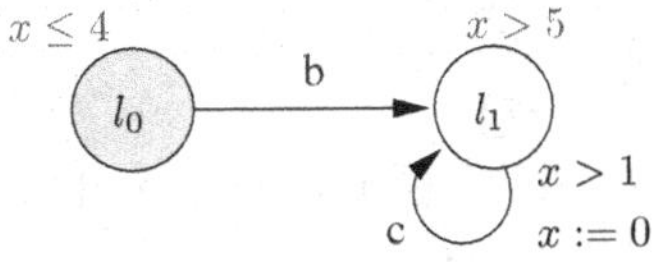

Fig. 2.12. The timed automaton considered in Example 2.19

ered already in Example 2.17, is like that in Fig. 2.11, but without the transition labelled with a). Due to $free(l_0) = \emptyset$, the transition relation of $\mathcal{A}$ is in $\mathcal{A}'$ extended by $e : l_0 \xrightarrow{a_d, true, \emptyset} l_d$. Similarly, since $free(l_1) = [\![x > 5]\!] \cap ([\![x > 1]\!] \cap [\![x > 5]\!] \cap [\{x\} := 0][\![x > 5]\!]) \not= [\![x > 5]\!] \cap ([\![x > 1]\!] \cap [\![x > 5]\!] \cap \emptyset) \not= \emptyset$, the transition relation of $\mathcal{A}'$ contains an element $l_1 \xrightarrow{a_d, true, \emptyset} l_d$. Thus, it is easy to see that all the states $(l_0, \cdot)$ of $\mathcal{A}$ are deadlocks, since the transition labelled with a_d can be executed at all of the states $(l_0, \cdot)$ of $\mathcal{A}'$.

□

[9] The number of the transitions for a location l is equal to the number of the elements of $\mathbb{R}_{0+}^{n_{\mathcal{X}}} \setminus free(l)$.

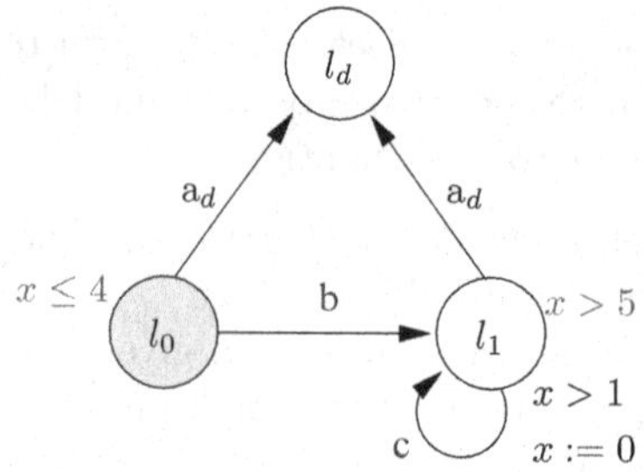

Fig. 2.13. A timed automaton for testing deadlocks in the TA of Fig. 2.12

In the case when not all the structural loops of a given automaton $\mathcal{A}$ are progressive (i.e., if the first condition of Lemma 2.15 is not satisfied) we can check whether the TCTL[10] formula $\varphi = \mathrm{AG}(\mathrm{EF}_{[1,\infty)} true \wedge \mathrm{AF}_{[1,\infty)} true)$ holds in an appropriate model of $\mathcal{A}$ [84,175] (this can be done using a method described in the further part of the book). Intuitively, φ expresses that every state reachable from the initial state can let time pass at least one unit. This ensures that in all the infinite runs time can pass without bound. Notice, however, that this does not guarantee that after entering an arbitrary state an action can be eventually taken, so in order to ensure progressiveness, deadlock-freedom needs to be checked as well. Checking whether the above formula holds can be replaced by testing reachability of an auxiliary location in an augmented automaton $\mathcal{A}''$, similarly to the above-described deadlock detection. The automaton $\mathcal{A}'' = (A'', L'', l^0, E'', \mathcal{X}'', \mathcal{I}'')$ is like $\mathcal{A}$, besides the fact that

- $\mathcal{X}'' = \mathcal{X} \cup \{x_z\}$ with $x_z \notin \mathcal{X}$,
- $L'' = L \cup \{l_p\}$ with $l_p \notin L$,
- $A'' = A \cup \{a_p\}$ with $a_p \notin A$,
- $\mathcal{I}''(l) = \mathcal{I}(l)$ for $l \in L$ and $\mathcal{I}''(l_p) = true$, and
- $E'' = E \cup \{l \xrightarrow{a_p, x_z \geq 1, \emptyset} l_p \mid l \in L\}$.

Intuitively, A'' has an "escape" transition from each location of A, which leads to the auxiliary location l_p. This transition can be taken if the value of the additional clock x_z is at least one. However, instead of the standard reachability analysis, it is checked whether there exists a reachable state q of $\mathcal{A}$ such that in $\mathcal{A}''$ no state $(l_p, \cdot)$ can be reached from a state q' of $\mathcal{A}''$ which is like q, but with the clock x_z set to zero. This is a two-step procedure. A detailed description can be found in [156].

2.5.3 A Solution for Strongly Monotonic Semantics

As it could be seen from the above description, methods for checking progressiveness of TA apply mainly to the weakly monotonic semantics. In order to check progressiveness of an automaton $\mathcal{A} = (A, L, l^0, E, \mathcal{X}, \mathcal{I})$ in the

[10] Timed Computation Tree Logic, see Chap. 4.

strongly monotonic case, one can use, for instance, the method suggested by F. Cassez [51]. This method consists in building an automaton $\mathcal{A}_m$ whose progressiveness in the weakly monotonic semantics implies the same for $\mathcal{A}$ in the strongly monotonic one. $\mathcal{A}_m$ is defined like $\mathcal{A}$ except for the clocks and transitions, which are extended in the following way. The set of clocks of $\mathcal{A}$ is extended by the clock x_s. For each transition e of $\mathcal{A}$, $guard(e)$ is extended by $x_s > 0$, and $reset(e)$ – by x_s.

The weakly monotonic q^0-runs of $\mathcal{A}_m$ correspond to the strongly monotonic q^0-runs of $\mathcal{A}$. Thus, in order to test whether these runs of $\mathcal{A}$ are progressive, one can apply to $\mathcal{A}_m$ a method for testing progressiveness for the weakly monotonic case.

Example 2.20. Figure 2.14 shows the automaton $\mathcal{A}_m$ built for the automaton $\mathcal{A}$ of Fig. 2.11.

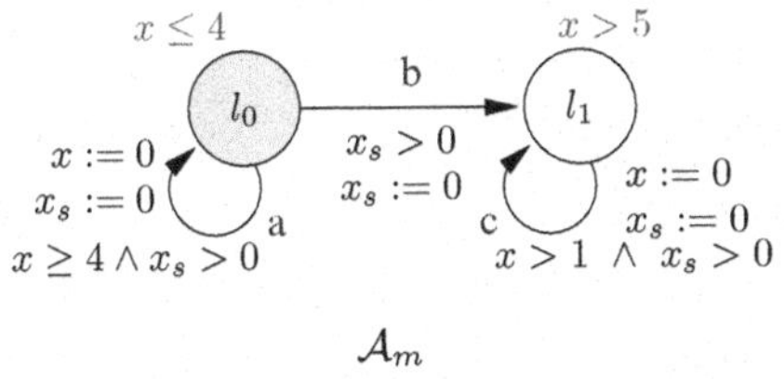

Fig. 2.14. The automaton $\mathcal{A}_m$ used to test progressiveness of $\mathcal{A}$ in Fig. 2.11 in the strongly monotonic semantics

□

Further Reading

Timed automata are discussed in [4, 5, 11]. The notion of progressiveness for timed automata is investigated also in [2, 13, 74].

3

From Time Petri Nets to Timed Automata

There are two approaches to verifying properties of time Petri nets. Either specific algorithms are used, or nets are translated to timed automata in order to exploit verification methods designed for automata. Usually, the concrete state spaces of both the models are required to be *bisimilar* (see [114] or an explanation on p. 17) which ensures preservation of branching temporal properties [49]. Therefore, in the further part of the book we first consider translations from TPNs to TA, and then review the most recent verification methods for both the formalisms.

Several methods for translating time Petri nets to timed automata have been already developed. However, in most cases translations produce automata which extend timed automata. Some of the existing approaches are sketched below.

3.1 Translations to Extended Timed Automata

Sifakis and Yovine [144] presented a translation of a subclass of *time stream Petri nets* [139] to automata whose invariants can be disjunctions of clock constraints. In these nets, each of the arcs (p, t) is assigned a timing interval which specifies when tokens in the place p become available to fire the transition t. An enabled transition t can be fired if

(a) all the places in $\bullet t$ have been marked at least as long as the lower bounds of the corresponding intervals, and
(b) there is at least one place in $\bullet t$ marked no longer than the upper bound of the corresponding interval. (1-safe) time Petri nets can be seen as a subclass of these nets.

In order to translate a net $\mathcal{N}$ to a timed automaton, we define a location for each marking of $\mathcal{N}$, and associate a clock with each of its places. The actions are labelled by the transitions of $\mathcal{N}$. Executing an action labelled by $t \in T$

W. Penczek and A. Półrola: *From Time Petri Nets to Timed Automata*, Studies in Computational Intelligence (SCI) **20**, 51–62 (2006)
www.springerlink.com

resets the clocks of the places in $t\bullet$, whereas its enabling condition corresponds to (a) given above. The invariant of a marking m states that (b) holds for each $t \in en(m)$.

Lime and Roux [103] proposed a method of translating a (general[1]) time Petri net to a timed automaton, using an algorithm of building a *state class graph* G [32] (see Sect. 5.1.1). The nodes of G are *state classes*, i.e., pairs (m, I), where m is a marking, and I is a set of inequalities describing timing constraints under which the transitions of $en(m)$ can be fired. The translation produces an automaton whose locations are the nodes of G, and the transitions, labelled by the names of transitions of the net, correspond to its successor relation (i.e., to the edges of G). Each class (m, I) of G is assigned a set of clocks, each of which corresponds to all the transitions in $en(m)$ which became enabled at the same time. The union of all these sets gives the set of clocks of the automaton. A transition e of the automaton, labelled by a transition t of the net, and with $source(e) = (m, I)$ and $target(e) = (m[t\rangle, I')$, resets all these clocks which in $(m[t\rangle, I')$ are associated with the elements of $newly\_en(m, t)$. It is also possible to assign a value of one clock to another (this goes beyond our standard definition of TA). The invariants and the enabling conditions describe, respectively, when the net can be in a given marking, or when a transition can be fired at a class. Since the number of state classes is finite only if the net is bounded [32], a condition for an on-line checking for unboundedness is provided.

In [52, 77], translations of (general) TPNs to automata equipped with shared variables and urgency modelling mechanisms are provided. The method of [52] generates a product of TA obtained from a network containing the automaton $\mathcal{A}_t$ with one clock for each $t \in T$, and the supervisor automaton $\mathcal{A}_s$. The locations of $\mathcal{A}_t$ correspond to the possible states of t (i.e., enabled, disabled, and being fired). The automaton $\mathcal{A}_s$ with *committed locations* (i.e., locations that have to be left as soon as they are entered) forces the other automata to change synchronously their states when a transition of the net is fired. Shared variables are used to model the number of tokens in the places. In the approach of [77], the transitions are classified according to their number of input, output, and inhibitor[2] places. One automaton with the locations *disabled* and *enabled* is built for each of the classes obtained. Similarly to [52], an additional automaton (with an *urgent transition*, i.e., a transition which has to be taken as soon as it is enabled) ensures a synchronous behaviour of the whole system when a transition is fired, whereas shared variables store the marking of the net.

---

[1] This means that there are no restrictions on the functions F, Eft, and Lft of a TPN.

[2] Informally, an *inhibitor place* is a place in a preset of a transition which, when marked, can disable the transition. We do not consider time Petri nets extended with inhibitor places in the book.

Cortés et al. [60] developed a method of translating slightly extended[3] 1-safe time Petri nets, where the function Lft takes only finite values (called *PRES+ models*) to a set of extended timed automata. When applied to a non-extended net, the translation produces a network of standard TA augmented with variables. The translation of a net is a two-stage procedure. In its first step, an automaton $\mathcal{A}_t$ with one clock is built for each transition t of the net. The locations of $\mathcal{A}_t$ represent the numbers of places in $\bullet t$ which are marked, whereas the transition relation corresponds to firing of the transitions of the net. The clock of $\mathcal{A}_t$ is used to express that t is fired between its earliest and latest firing time. For storing the markings of the net, the variables corresponding to the places are applied. The second stage of the translation consists in reducing the number of clocks to improve on efficiency of verification. To this aim, transitions which cannot be concurrently enabled are grouped together, and the automata corresponding to each group are replaced by the product automaton with one clock renaming these of the components. The product of the system obtained corresponds to the behaviour of the net.

In many cases, the extended automata resulting from the above translations can be further transformed to standard ones (this can be done, e.g., by adding extra clocks or automata to model urgency or shared variables). However, another set of translations from time Petri nets to timed automata results in obtaining standard automata immediately. These translations correspond to the semantics of time Petri nets associating clocks with various components of TPNs. As a result of each of the translations we get one (global) timed automaton rather than a network of timed automata. However, it is important to notice that for efficient verification of time Petri nets we do not need to perform the translation first, but can use directly the transition relation defined by the translation [118], or translate the net in parallel with the verification process, similarly as the product of a network of TA is usually built.

3.2 Translation for "Clocks Assigned to the Transitions"

The first translation [111,163], which besides 1-safe time Petri nets is correct also for TPNs in which there is a bound on the values of all the reachable markings[4], is defined for the semantics assigning clocks to the transitions of the net. Let $\mathcal{N} = (P, T, F, m^0, Eft, Lft)$ be a time Petri net. In order to translate it into a timed automaton $\mathcal{A}_{\mathcal{N}} = (A_{\mathcal{N}}, L_{\mathcal{N}}, l^0, E_{\mathcal{N}}, \mathcal{X}_{\mathcal{N}}, \mathcal{I}_{\mathcal{N}})$, it is straightforward to define locations of $\mathcal{A}_{\mathcal{N}}$ to correspond to the (reachable) markings of the net. However, since a set of all the reachable markings can be found only by analysing the behaviour of the net, the translation sets

[3] The tokens are considered as pairs ($token\_value, token\_type$) and the remaining elements of the definition of the net are adapted to this notion.

[4] Notice that this can result either from a finite maximal capacity of all the places assumed, or from boundedness of the net.

$$L_{\mathcal{N}} = AM_{\mathcal{N}}^k$$

with $k = 1$ if the net is 1-safe, whereas otherwise k is equal to the bound on the number of tokens in the markings. This means that all the markings whose values for all the places do not exceed k are taken as the locations. The initial location is

$$l^0 = m^0,$$

whereas the set of actions is defined as

$$A_{\mathcal{N}} = T,$$

i.e., the actions of the automaton correspond to the transitions of the net. The set of clocks is given by

$$\mathcal{X}_{\mathcal{N}} = \{x_t \mid t \in T\},$$

i.e., a new clock is defined for each of the transitions[5] of $\mathcal{N}$. An invariant of a location $m \in AM_{\mathcal{N}}$ is then given by

$$\mathcal{I}_{\mathcal{N}}(m) = \begin{cases} \bigwedge_{\{t \in T \mid t \in en(m) \wedge Lft(t) < \infty\}} x_t \leq Lft(t) & \\ \qquad\qquad \text{if } \{t \in T \mid t \in en(m) \wedge Lft(t) < \infty\} \neq \emptyset, \\ \mathit{true} \qquad\qquad \text{otherwise.} \end{cases}$$

The set of transitions of $\mathcal{A}_{\mathcal{N}}$ is defined as

$$E_{\mathcal{N}} = \{e_{t,m} \mid t \in en(m) \wedge m \in AM_{\mathcal{N}}^k\}.$$

For a given transition $e_{t,m} \in E_{\mathcal{N}}$, corresponding to firing a transition $t \in en(m)$ at a marking m, the enabling condition is given by

$$guard_{\mathcal{N}}(e_{t,m}) = (x_t \geq Eft(t))$$

(notice that this is sufficient since the invariant for m implies that $x_t \leq Lft(t)$), and the set of clocks to be reset is

$$reset_{\mathcal{N}}(t, m) = \{x_u \in \mathcal{X}_{\mathcal{N}} \mid u \in newly\_en(m, t)\}.$$

The transitions of $\mathcal{A}_{\mathcal{N}}$ are then of the form $e_{t,m} := m \xrightarrow{t,\mathfrak{cc},X} m[t\rangle$, where $\mathfrak{cc} = guard_{\mathcal{N}}(e_{t,m})$ and $X = reset_{\mathcal{N}}(t, m)$.

Next, the propositional variables of PV_P are defined to correspond to the places of the net. This allows to define the valuation function $V_{\mathcal{A}_{\mathcal{N}}} : L_{\mathcal{N}} \longrightarrow 2^{PV_P}$ as

[5] In the case of nets which are not 1-safe, multiple enabledness of a transition are treated separately, i.e., as different transitions. This, obviously, increases the number of clocks.

$$V_{\mathcal{A}_{\mathcal{N}}}(m) = \bigcup_{\{p \in P | m(p) > 0\}} V_{\mathcal{N}}(p).$$

The concrete models $M_c(\mathcal{N}) = (C_c^T(\mathcal{N}), V_{\mathfrak{C}})$ and $M_c(\mathcal{A}_{\mathcal{N}}) = (C_c(\mathcal{A}_{\mathcal{N}}), V'_{\mathfrak{C}})$, where $V'_{\mathfrak{C}}$ extends $V_{\mathcal{A}_{\mathcal{N}}}$ as described in Sect. 2.4, are bisimilar [124]. The above approach was used to define *detailed region graphs* for TPNs [111, 163].

Example 3.1. Consider again the net in Fig. 1.3. It depicts the timed automaton $\mathcal{A}_{\mathcal{N}}$, which is the result of the above translation, restricted to the reachable markings of $\mathcal{N}$. Each of the locations l of $\mathcal{A}_{\mathcal{N}}$, corresponding to a marking m, is annotated by the set of propositions true in l (i.e., by the set of these elements of PV_P which correspond to the places of $\mathcal{N}$ whose values of the function m are non-zero). The initial location is coloured. □

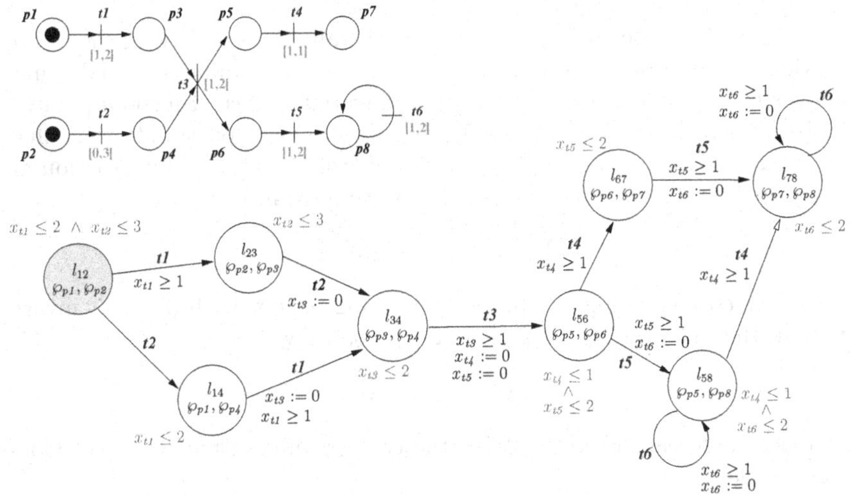

Fig. 3.1. The timed automaton for the net of Fig. 1.3 resulting from the translation for "clocks assigned to the transitions"

3.3 Translation for "Clocks Assigned to the Places"

Let $\mathcal{N} = (P, T, F, m^0, Eft, Lft)$ be a (1-safe) time Petri net. The translation for the semantics assigning clocks to the places of the net $\mathcal{N}$ into a timed automaton $\mathcal{A}_{\mathcal{N}} = (A_{\mathcal{N}}, L_{\mathcal{N}}, l^0, E_{\mathcal{N}}, \mathcal{X}_{\mathcal{N}}, \mathcal{I}_{\mathcal{N}})$, derived from the approach of [144], is more involved. Similarly to the previous approach, the locations of $\mathcal{A}_{\mathcal{N}}$ correspond to the (reachable) markings of the net, but in this case this is not a one-to-one correspondence. Since a set of all the reachable markings can be found only by analysing the behaviour of $\mathcal{N}$, the translation deals with the set of all the markings of $\mathcal{N}$ whose value for any place is not greater than 1 (i.e., with $AM^1_{\mathcal{N}}$) instead. The actions of the automaton correspond to the transitions of the net, i.e.,

$$A_{\mathcal{N}} = T.$$

The set of clocks is

$$\mathcal{X}_{\mathcal{N}} = \{x_p \mid p \in P\},$$

i.e., one clock is defined for each of the places of $\mathcal{N}$. The net can stay in the marking m as long as it satisfies the condition

$$inv_{\mathcal{N}}(m) = \begin{cases} \bigwedge_{\{t \in T \mid t \in en(m) \wedge Lft(t) < \infty\}} \bigvee_{p \in \bullet t} x_p \leq Lft(t) \\ \qquad\qquad\qquad \text{if } \{t \in T \mid t \in en(m) \wedge Lft(t) < \infty\} \neq \emptyset, \\ true \qquad\qquad\qquad\qquad\qquad\qquad \text{otherwise.} \end{cases}$$

Let $[\![inv_{\mathcal{N}}(m)]\!]$ denote the set of all the clock valuations satisfying the above condition. Since $inv_{\mathcal{N}}(m)$ is not necessarily a clock constraint (due to disjunction over $p \in \bullet t$), in order to obtain proper invariants of the timed automaton $\mathcal{A}_{\mathcal{N}}$ we need to define a finite set $\mathfrak{I}_m \subseteq \mathcal{C}^{\ominus}_{\mathcal{X}_{\mathcal{N}}}$ of clock constraints such that $\bigcup_{\mathfrak{cc} \in \mathfrak{I}_m} [\![\mathfrak{cc}]\!] = [\![inv_{\mathcal{N}}(m)]\!]$ and $\bigcap_{\mathfrak{cc} \in \mathfrak{I}_m} [\![\mathfrak{cc}]\!] = \emptyset$ (this is obtained by partitioning $[\![inv_{\mathcal{N}}(m)]\!]$ into disjoint time zones). Then, we introduce the set

$$\mathfrak{L}_m = \{m_{\mathfrak{cc}} \mid \mathfrak{cc} \in \mathfrak{I}_m\}$$

of locations corresponding to the same marking m, but with different invariants, and the function $\mathcal{I}_{\mathcal{N},m} : \mathfrak{L}_m \longrightarrow \mathcal{C}^{\ominus}_{\mathcal{X}_{\mathcal{N}}}$, given by

$$\mathcal{I}_{\mathcal{N},m}(m_{\mathfrak{cc}}) = \mathfrak{cc},$$

which assigns to each element of $\mathfrak{L}_m$ the corresponding element of $\mathfrak{I}_m$. Consequently, define

$$L_{\mathcal{N}} = \bigcup_{m \in AM^1_{\mathcal{N}}} \mathfrak{L}_m$$

and the function $\mathcal{I}_{\mathcal{N}} : L_{\mathcal{N}} \longrightarrow \mathcal{C}^{\ominus}_{\mathcal{X}_{\mathcal{N}}}$ whose value for a given location $l \in L_{\mathcal{N}}$ corresponding to the marking m is given by

$$\mathcal{I}_{\mathcal{N}}(l) = \mathcal{I}_{\mathcal{N},m}(l).$$

As the initial location we assume such a location $l \in \mathfrak{L}_{m^0}$ for which $(0, \ldots, 0) \in [\![\mathcal{I}_{\mathcal{N}}(l)]\!]$. The transition relation is obtained in the following way: given a marking m of $\mathcal{N}$ and a transition $t \in en(m)$, for each $l \in \mathfrak{L}_m$ and each $l' \in \mathfrak{L}_{m[t\rangle}$ we define $e_{l,t,l'} = l \xrightarrow{t,\mathfrak{cc},X} l' \in E_{\mathcal{N}}$, with the enabling condition $\mathfrak{cc} = guard_{\mathcal{N}}(e_{l,t,l'})$ given by

$$guard_{\mathcal{N}}(e_{l,t,l'}) = \bigwedge_{p \in \bullet t} x_p \geq Eft(t)$$

(notice that this is sufficient, since the invariant for l implies that at least one of the places in $\bullet t$ has the value of its clock not greater than $Lft(t)$), and with the set of clocks to be reset $X = reset_{\mathcal{N}}(t, m)$ defined as

$$reset_{\mathcal{N}}(t, m) = \{x_p \in \mathcal{X}_{\mathcal{N}} \mid p \in t\bullet\}.$$

Then, we define

$$E_{\mathcal{N}} = \bigcup_{m \in AM^1_{\mathcal{N}}} \{e_{l,t,l'} \mid l \in \mathfrak{L}_m \ \wedge\ t \in en(m) \ \wedge\ l' \in \mathfrak{L}_{m[t\rangle}\}.$$

Again, the propositional variables of PV_P are defined to correspond to the places of the net. The valuation function $V_{\mathcal{A}_{\mathcal{N}}} : L_{\mathcal{N}} \longrightarrow 2^{PV_P}$ is given by

$$V_{\mathcal{A}_{\mathcal{N}}}(l_m) = \bigcup_{\{p \in P \mid m(p) > 0\}} V_{\mathcal{N}}(p),$$

where l_m is a location of $\mathcal{A}_{\mathcal{N}}$ corresponding to a given marking m. The concrete models $M_c(\mathcal{N}) = (C^P_c(\mathcal{N}), V_{\mathfrak{C}})$ and $M_c(\mathcal{A}_{\mathcal{N}}) = (C_c(\mathcal{A}_{\mathcal{N}}), V'_{\mathfrak{C}})$, where $V'_{\mathfrak{C}}$ extends $V_{\mathcal{A}_{\mathcal{N}}}$ as described in Sect. 2.4, are bisimilar [124].

3.3.1 Supplementary Algorithms

The precise definition of $\mathcal{A}_{\mathcal{N}}$ depends on the algorithm used to construct the set $\mathfrak{I}_m$. Many algorithms for this task can be defined. Below, we present two of them introduced in [124]. The first one computes non-diagonal-free automata, whereas the second one is tuned to output diagonal free automata only.

Obtaining General TA

Consider a marking m of a time Petri net $\mathcal{N}$ and a transition $t \in en(m)$. The set $\mathfrak{I}_m$ is computed in the following way: if for all $t \in en(m)$ we have $Lft(t) = \infty$, then $inv_{\mathcal{N}}(m) = true$, which is already a clock constraint, and therefore $\mathfrak{I}_m = \{inv_{\mathcal{N}}(m)\}$. Otherwise assume a fixed ordering on the set of places P of $\mathcal{N}$. For a transition $t \in \{t' \in T \mid t' \in en(m) \ \wedge\ Lft(t') < \infty\}$, let $p_{j^t_1}, \ldots, p_{j^t_{k_t}}$ be a subsequence of the places of P such that $k_t = |\bullet t|$ and $p_{j^t_1}, \ldots, p_{j^t_{k_t}} \in \bullet t$. Then, we set

$$\mathfrak{I}_m = \{\textstyle\bigwedge_{t \in en(m) \text{ s.t. } Lft(t) < \infty} \mathfrak{cc}^t_i \mid i \in \{1, \ldots, k_t\} \ \wedge$$
$$\mathfrak{cc}^t_i = (x_{p_{j^t_i}} \leq Lft(t) \ \wedge\ \textstyle\bigwedge_{1 \leq r < i} x_{p_{j^t_r}} \geq x_{p_{j^t_i}} \ \wedge\ \textstyle\bigwedge_{i < r \leq k_t} x_{p_{j^t_r}} > x_{p_{j^t_i}})\}.$$

Example 3.2. Consider the net in Fig. 3.2, and assume the lexicographical ordering on the set of its places. For the marking m_{34} with $m_{34}(p_3) = m_{34}(p_4) = 1$ and $m_{34}(p_i) = 0$ for $i = 1, 2, 5, 6, 7, 8$, the set $\mathfrak{I}_{m_{34}}$ consists of

$$\mathfrak{cc}_1 = (0 \leq x_{p_3} \leq 2 \ \wedge\ x_{p_4} > x_{p_3}) \quad \text{and} \quad \mathfrak{cc}_2 = (0 \leq x_{p_4} \leq 2 \ \wedge\ x_{p_3} \geq x_{p_4}).$$

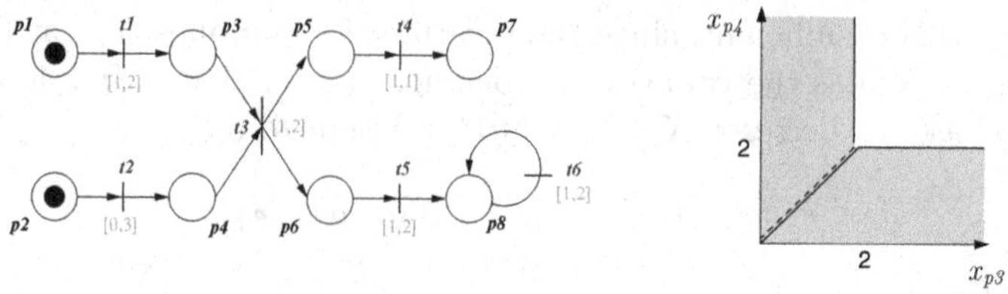

Fig. 3.2. Partitioning $[\![inv_{\mathcal{N}}(m_{34})]\!]$ into disjoint zones – the algorithm for non-diagonal-free TA

The zones partitioning $inv_{\mathcal{N}}(m_{34}) = (x_{p_3} \le 2 \vee x_{p_4} \le 2)$, described by the above constraints, are shown in the right-hand side of Fig. 3.2. The set of clock valuations satisfying $inv_{\mathcal{N}}(m_{34})$ is coloured.

□

Obtaining Diagonal-Free TA

Consider a marking m of a time Petri net $\mathcal{N}$ and a transition $t \in en(m)$. The following method is used to define $\mathfrak{I}_m$ such that $\mathfrak{I}_m \subseteq \mathcal{C}_{\mathcal{X}_{\mathcal{N}}}$ (i.e., $\mathfrak{I}_m$ does not contain clock constraints with differences of clocks). Notice that again if for all $t \in en(m)$ it holds $Lft(t) = \infty$, then $inv_{\mathcal{N}}(m) = true$ (i.e., is a clock constraint), and therefore $\mathfrak{I}_m = \{inv_{\mathcal{N}}(m)\}$. Otherwise assume a fixed ordering on the set of places P of $\mathcal{N}$. For a transition $t \in \{t' \in T \mid t' \in en(m) \wedge Lft(t') < \infty\}$, let $p_{j_1^t}, \ldots, p_{j_{k_t}^t}$ be a subsequence of the places of P such that $k_t = |\bullet t|$ and $p_{j_1^t}, \ldots, p_{j_{k_t}^t} \in \bullet t$. Then, we set

$$\mathfrak{I}_m = \{\textstyle\bigwedge_{t\in en(m) \text{ s.t. } Lft(t)<\infty} \mathfrak{cc}_i^t \mid i \in \{1,\ldots,k_t\} \wedge \mathfrak{cc}_i^t = (\textstyle\bigwedge_{1\le r<i} x_{p_{j_r^t}} > Lft(t) \wedge x_{p_{j_i^t}} \le Lft(t))\}.$$

Example 3.3. Consider the net in Fig. 3.3, and assume the lexicographical ordering on the set of its places. For the marking m_{34} with $m_{34}(p_3) = m_{34}(p_4) = 1$ and $m_{34}(p_i) = 0$ for $i = 1, 2, 5, 6, 7, 8$, the set $\mathfrak{I}_m$ consists of

$$\mathfrak{cc}_1 = (x_{p_3} \le 2) \text{ and } \mathfrak{cc}_2 = (x_{p_3} > 2 \wedge x_{p_4} \le 2).$$

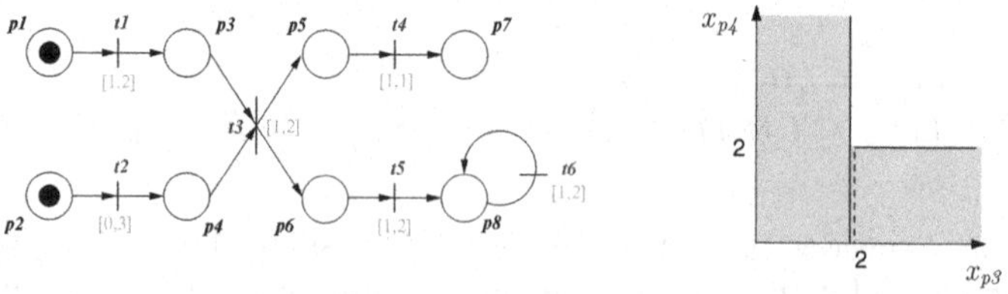

Fig. 3.3. Partitioning $[\![inv_{\mathcal{N}}(m_{34})]\!]$ into disjoint zones – the algorithm for diagonal-free TA

The zones partitioning $inv_{\mathcal{N}}(m_{34}) = (x_{p_3} \leq 2 \ \vee \ x_{p_4} \leq 2)$, described by the above constraints, are shown in the right-hand side of Fig. 3.3. The set of clock valuations satisfying $inv_{\mathcal{N}}(m_{34})$ is coloured.

□

Example 3.4. Consider again the net $\mathcal{N}$ of Fig. 1.3. Fig. 3.4 shows a timed automaton $\mathcal{A}_{\mathcal{N}}$, obtained from $\mathcal{N}$ by the translation for the semantics associating clocks with the places of the net, in which for defining clock constraints used in the enabling conditions and the invariants the above-described algorithm for diagonal-free TA was applied. The set of markings considered was restricted to $RM_{\mathcal{N}}$. Similarly to Example 3.1, the locations are annotated with the sets of propositions true in them (i.e., with the sets of elements of PV_P corresponding to the places of $\mathcal{N}$ for which the values of the given marking are non-zero). The initial location is coloured. Notice that the locations l^1_{34} and l^2_{34} correspond to the same marking of $\mathcal{N}$, but their invariants differ.

□

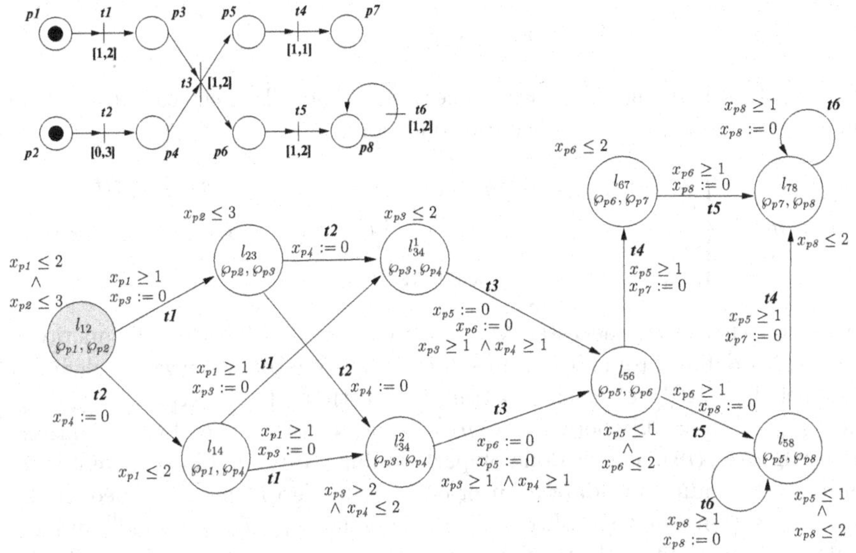

Fig. 3.4. The timed automaton for the net in Fig. 1.3 resulting from the translation for "clocks assigned to the places"

3.4 Translation for Distributed Nets

The next translation is applicable only to distributed time Petri nets, and makes use of the semantics which associates clocks with all their processes.

However, many of its elements are analogous to the translation for the semantics assigning clocks to the places of the net. Due to this, we shall refer to the latter in the description below.

Let $\mathcal{N} = (P, T, F, m^0, Eft, Lft)$ be a distributed time Petri net whose processes are labelled with a finite set of indices $\mathfrak{I} = \{\mathfrak{i}_1, \ldots, \mathfrak{i}_{n_\mathfrak{I}}\}$. For $\mathfrak{i} \in \mathfrak{I}$, let $P_\mathfrak{i}$ denote the set of places of the process $N_\mathfrak{i}$. Similarly to the previous approaches, we translate $\mathcal{N}$ into a timed automaton $\mathcal{A}_\mathcal{N} = (A_\mathcal{N}, L_\mathcal{N}, l^0, E_\mathcal{N}, \mathcal{X}_\mathcal{N}, \mathcal{I}_\mathcal{N})$. The locations of $\mathcal{A}_\mathcal{N}$ correspond in principle to the (reachable) markings of the net, but again this is not a one-to-one mapping. However, for the same reason as before, instead of reachable markings, the set $AM^1_\mathcal{N}$ of all the markings of $\mathcal{N}$ with the upper bound on the number of tokens in the place equal to 1 is considered. The actions of the automaton correspond to the transitions of the net, i.e.,

$$A_\mathcal{N} = T,$$

whereas the set of clocks is given by

$$\mathcal{X}_\mathcal{N} = \{x_\mathfrak{i} \mid \mathfrak{i} \in \mathfrak{I}\},$$

i.e., one clock is defined for each process of the net. The net can stay in the marking m as long as it satisfies the condition

$$inv_\mathcal{N}(m) = \begin{cases} \bigwedge_{\{t \in T \mid t \in en(m) \wedge Lft(t) < \infty\}} \bigvee_{\mathfrak{i} \in \mathfrak{I} \text{ s.t. } \bullet t \cap P_\mathfrak{i} \neq \emptyset} x_\mathfrak{i} \leq Lft(t) & \\ \qquad \text{if } \{t \in T \mid t \in en(m) \wedge Lft(t) < \infty\} \neq \emptyset, & \\ true \qquad \text{otherwise.} & \end{cases}$$

Since $inv_\mathcal{N}(m)$ is not necessarily a clock constraint (due to a disjunction), in order to define proper invariants for the automaton we need to define a finite set $\mathfrak{I}_m \subseteq C^\ominus_{\mathcal{X}_\mathcal{N}}$ of clock constraints such that $\bigcup_{\mathfrak{cc} \in \mathfrak{I}_m} [\![\mathfrak{cc}]\!] = [\![inv_\mathcal{N}(m)]\!]$ and $\bigcap_{\mathfrak{cc} \in \mathfrak{I}_m} [\![\mathfrak{cc}]\!] = \emptyset$, where $[\![inv_\mathcal{N}(m)]\!]$ is the set of all the clock valuations satisfying $inv_\mathcal{N}(m)$. This is done by partitioning $[\![inv_\mathcal{N}(m)]\!]$ into disjoint time zones (to this aim, an adaptation of one of the algorithms described in the previous section can be applied). Then, we define $\mathfrak{L}_m$, $\mathcal{I}_{\mathcal{N},m}$, $L_\mathcal{N}$, l^0 and $\mathcal{I}_\mathcal{N}$ analogously to the translation for clocks assigned to the places. Similarly, the definition of the transition relation $E_\mathcal{N}$ follows the same pattern as that for the previous translation, but now the functions $guard_\mathcal{N}$ and $reset_\mathcal{N}$ are given by

$$guard_\mathcal{N}(e_{l,t,l'}) = \bigwedge_{\mathfrak{i} \in \mathfrak{I} \text{ s.t. } \bullet t \cap P_\mathfrak{i} \neq \emptyset} x_\mathfrak{i} \geq Eft(t)$$

(notice that this is sufficient, since the invariant for l implies that at least one of the processes which contains a place of $\bullet t$ has the valued of its clock not greater than $Lft(t)$), and

$$reset_\mathcal{N}(t, m) = \{x_\mathfrak{i} \in \mathcal{X}_\mathcal{N} \mid \bullet t \cap P_\mathfrak{i} \neq \emptyset\}.$$

Again, the definition of locations allows for introducing the valuation function $V_{\mathcal{A_N}} : L_{\mathcal{N}} \longrightarrow 2^{PV_P}$, with the set PV_P defined as in the previous cases, given by

$$V_{\mathcal{A_N}}(l_m) = \bigcup_{\{p \in P \mid m(p)>0\}} V_{\mathcal{N}}(p),$$

where l_m is a location of $\mathcal{A_N}$ corresponding to the marking m. Similarly to the before-described translations, the concrete models $M_c(\mathcal{N}) = (C_c^N(\mathcal{N}), V_{\mathfrak{C}})$ and $M_c(\mathcal{A_N}) = (C_c(\mathcal{A_N}), V'_{\mathfrak{C}})$, where $V'_{\mathfrak{C}}$ extends $V_{\mathcal{A_N}}$ as described in Sect. 2.4, are bisimilar [124].

Example 3.5. Consider again the (distributed) net $\mathcal{N}$ of Fig. 1.3 consisting of two processes with the sets of places $P_1 = \{p_1, p_3, p_5, p_7\}$ and $P_2 = \{p_2, p_4, p_6, p_8\}$, indexed by $\mathfrak{I} = \{1, 2\}$. Fig. 3.5 shows a timed automaton $\mathcal{A_N}$, obtained from $\mathcal{N}$ by the translation for the semantics associating clocks with the processes of the net, in which for defining clock constraints used in enabling conditions and invariants the above-described algorithm for diagonal-free TA was applied. Similarly to the previous examples, the set of markings considered was restricted to $RM_{\mathcal{N}}$. The locations are annotated with the sets of propositions true in them, and the initial location is coloured. Similarly to the previous example, the locations l_{34}^1 and l_{34}^2 correspond to the same marking of $\mathcal{N}$, but their invariants differ.

□

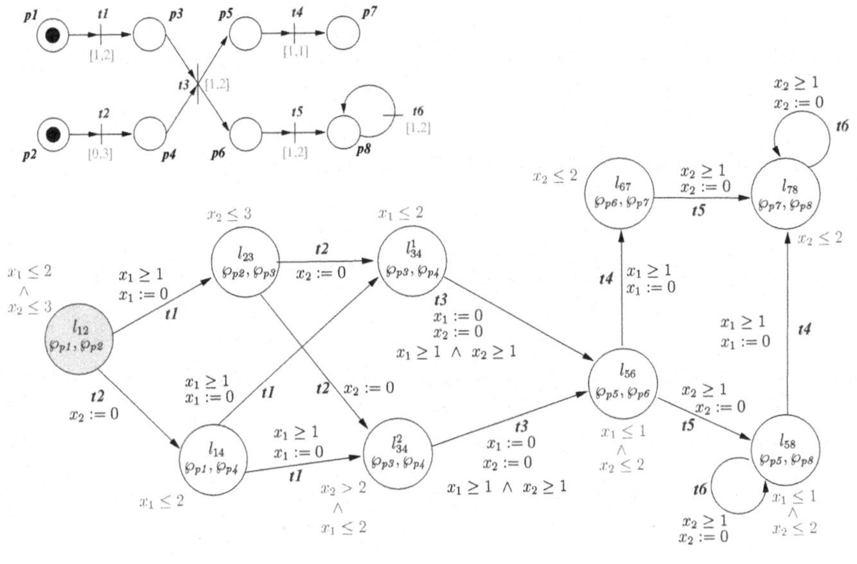

Fig. 3.5. The timed automaton for the net in Fig. 1.3 resulting from the translation for "clocks assigned to the processes"

3.5 Comparing Expressiveness of TPNs and TA

There are several results known about expressiveness of TPNs versus TA. First of all, TPNs form a subclass of TA as for each TPN one can construct a TA such that their state spaces are weak (timed) bisimilar [52,103]. Moreover, there is a TA for which there is no TPN whose state space is weakly (timed) bisimilar [29]. Considering weaker equivalences like timed language acceptance[6], TA and TPNs are equally expressive. In [29], a syntactical subclass of TA is defined which is equally expressive to TPNs w.r.t. (timed) bisimilarity.

Further Reading

We have not dealt with efficiency of the translations in this chapter, but it is obviously interesting to compare the sizes of the resulting timed automata. Partial results concerning this subject are available from [103,124], but a full comparison, in particular between the classes of translations, have not been done so far (to our knowledge at least).

Moreover, in this book we do not consider methods for translating timed automata to time Petri nets. Some descriptions of such techniques can be found in [29,78]. Translations between TA and timed-arcs Petri nets are investigated in [147].

[6] This is an equivalence defined via equality of timed words, i.e., sequences of actions and time delays, generated by timed systems.

4

Main Formalisms for Expressing Temporal Properties

Properties of timed systems are usually expressed using (timed) temporal logics. In this chapter our focus is on the logics that are most commonly used. However, we start with introducing non-temporal logics that are then extended to (timed) temporal logics.

4.1 Non-temporal Logics

In this section we define *propositional logic* (PL) and its quantified version *quantified propositional logic*[1] (QPL), as well as (*quantified*) *separation logic* (Q)SL.

4.1.1 Propositional Logic (PL)

Propositional logic (PL) is briefly recapitulated for two reasons. Firstly, all logics that are discussed are extensions of propositional logic. Secondly, the model checking problem for the discussed temporal logics will be translated to the satisfiability problem of propositional formulas.

Propositional logic formalises reasoning with *not*, *and*, and *or* about the state (world) at the level of *propositions*, i.e., atomic assertions that can be true or false. The definition of the language consists of two parts. The syntax fixes the set of formulas, whereas the semantics provides the meaning for each formula.

Syntax of PL

Formal names for propositions and symbols for *not*, *and*, and *or* are provided. Inductive rules enable to form expressions using these symbols.

The language contains as symbols:

[1] This logic is known also as quantified boolean logic (QBL) or quantified boolean formulas (QBF).

W. Penczek and A. Półrola: *Main Formalisms for Expressing Temporal Properties*, Studies in Computational Intelligence (SCI) **20**, 63–85 (2006)
www.springerlink.com

- a countable set of propositional variables $PV = \{\wp_1, \wp_2, \ldots\}$;
- logical connectives: $\neg$ (logical *not*), $\wedge$ (logical *and*), and $\vee$ (logical *or*);
- auxiliary symbols: (and) (the brackets).

The set of formulas is defined inductively as follows:

$$\varphi := \wp \mid \neg\varphi \mid \varphi \wedge \varphi \mid \varphi \vee \varphi,$$

where $\wp \in PV$.

Where no confusion is likely, parentheses are omitted. Derived connectives are:

$$\varphi \Rightarrow \psi \quad \stackrel{def}{=} \neg\varphi \vee \psi \qquad \text{(logical "if ... then")};$$
$$\varphi \iff \psi \stackrel{def}{=} (\varphi \Rightarrow \psi) \wedge (\psi \Rightarrow \varphi) \quad \text{(logical "if and only if")}.$$

Semantics of PL

The situation in the state, i.e., the semantics of the propositional variables, is modelled as the truth values of the propositional variables. As only one state is considered, this state is not mentioned explicitly in the semantics.

Definition 4.1. *A* model for PL *is a valuation function*

$$V : PV \longrightarrow \{true, false\}$$

from the set of propositional variables to the set containing the logical values true and false.

The reasoning with *not*, *and*, and *or*, i.e., the semantics of $\neg$, $\wedge$, and $\vee$ is given as how the truth of a formula follows from the interpretation of the logical connectives.

A formula φ holds in a model V (denoted $V \models \varphi$) is defined inductively:

$$\begin{array}{ll} V \models \wp & \text{iff } V(\wp) = true, \text{ for } \wp \in PV \\ V \models \neg\varphi & \text{iff not } (V \models \varphi) \text{ (denoted as } V \not\models \varphi), \\ V \models \varphi \wedge \psi & \text{iff } V \models \varphi \text{ and } V \models \psi, \\ V \models \varphi \vee \psi & \text{iff } V \models \varphi \text{ or } V \models \psi. \end{array}$$

A formula φ is *valid* if it holds in all the models (denoted $\models \varphi$). A formula φ is *satisfiable* if it holds in some model.

Example 4.2. Interpreting the formulas in the example with all the possible combinations of truth values immediately yields the following results:

i) $\wp_1 \vee \neg\wp_1$ is valid.
ii) $\wp_1 \wedge \wp_2$ is satisfiable.
iii) $\wp_1 \wedge \neg\wp_1$ is not satisfiable.

□

4.1.2 Quantified Propositional Logic (QPL)

In order to have a more succinct notation for complex operations on propositional formulas, we introduce *quantified propositional logic* (QPL), an extension of propositional logic by means of quantifiers ranging over propositions.

Syntax of QPL

The syntax of QPL is defined as follows:

$$\varphi := \wp \mid \neg\varphi \mid \varphi \wedge \varphi \mid \varphi \vee \varphi \mid \exists\wp.\varphi \mid \forall\wp.\varphi,$$

where $\wp \in PV$.

Semantics of QPL

A *model* V and the semantics of the formulas without quantifiers is defined like for PL. The semantics of the quantified formulas is given below:

$$V \models \exists\wp.\varphi \text{ iff } V[\wp \leftarrow true] \models \varphi \text{ or } V[\wp \leftarrow false] \models \varphi,$$
$$V \models \forall\wp.\varphi \text{ iff } V[\wp \leftarrow true] \models \varphi \text{ and } V[\wp \leftarrow false] \models \varphi,$$

where $\varphi \in$ QPL, $\wp \in PV$, and $V[\wp \leftarrow b]$ denotes the model V' which is like V except for the value of $\wp$ equal to b, for $b \in \{true, false\}$.

4.1.3 (Quantified) Separation Logic ((Q)SL)

Separation logic (SL), known also as *difference logic*, is a quantifier-free fragment of first-order logic. An SL formula is a boolean combination of propositional variables and clock constraints[2], defined in Sect. 2, involving real-valued variables (clocks) from the set $\mathcal{X}$.

Syntax of (Q)SL

Syntax of SL is as follows:

$$\varphi := \wp \mid x_i \sim c \mid x_i - x_j \sim c \mid \neg\varphi \mid \varphi \wedge \varphi \mid \varphi \vee \varphi,$$

where $\wp \in PV$, $x_i, x_j \in \mathcal{X}$, $c \in \mathbb{N}$, and $\sim \in \{\leq, <, =, >, \geq\}$.

In some cases (an example can be found in Sect. 7.4.1) the so-called >-*normalised* SL *formulas* are more convenient than these generated by the grammar presented above. Notice that the formulas of the form $x_i \sim c$ and $x_i - x_j \sim c$ are atomic clock constraints of $\mathcal{C}^{\ominus}_{\mathcal{X}}$, defined in Chap. 2. Thus,

[2] The clock constraints are called also *separation predicates*.

it is possible to augment the set $\mathcal{X}$ by an additional clock $x_0 \notin \mathcal{X}$, and to replace the atomic clock constraints of the form above by the normalised clock constraints of $\mathcal{C}^{\ominus}_{\mathcal{X}^+}$ as shown in Sect. 2.1. In order to obtain >-normalised SL formulas, each inequality $x_i - x_j < c$ ($x_i - x_j \leq c$) with $x_i, x_j \in \mathcal{X}^+$ and $c \in \mathbb{Z}$ is rewritten as $x_j > x_i - c$ ($x_j \geq x_i - c$, respectively). Thus, all the separation predicates in these formulas are of the form $x_i \sim x_j + c$, where $x_i, x_j \in \mathcal{X}^+$, $\sim \in \{>, \geq\}$ and $c \in \mathbb{Z}$.

Quantified Separation Logic (QSL) is an extension of SL and QPL, where quantification over both propositional and real-valued clock variables is allowed.

Semantics of (Q)SL

A model for (Q)SL is defined as an ordered pair

$$(V, v)$$

consisting of two valuation functions, one for the propositions ($V : PV \longrightarrow \{true, false\}$) and one for the real-valued variables ($v : \mathcal{X} \longrightarrow \mathbb{R}$). Note that the values of $\mathcal{X}$ here are from the set of all the real numbers, whereas in some applications, especially when we interpret real-valued variables as clocks, their values can be restricted to non-negative reals. The semantics of the formulas without real-valued quantifiers is like for QPL and the clock constraints. The semantics of the real-valued quantified formulas is given below:

$$\begin{array}{lll} (V, v) \models \exists x.\varphi & \text{iff} & \text{there is } y \in \mathbb{R} \text{ such that } (V, v[x \leftarrow y]) \models \varphi, \\ (V, v) \models \forall x.\varphi & \text{iff} & \text{for all } y \in \mathbb{R}\ (V, v[x \leftarrow y]) \models \varphi, \end{array}$$

where $\varphi \in$ QSL, $x \in \mathcal{X}$, and $v[x \leftarrow y]$ denotes the valuation v' which is like v except for the value of x equal to y.

In the following section we consider untimed formalisms, which are later extended with time constraints.

4.2 Untimed Temporal Logics

All the untimed logics we consider are subsets of CTL* or modal μ-calculus. In fact, CTL* can be considered as a subset of modal μ-calculus, but for simplicity we give semantics for both the logics. We begin with syntax and semantics of CTL*. Next, we define other logics as its restrictions and finally present modal μ-calculus.

4.2.1 Computation Tree Logic* (CTL*)

Syntax of CTL*

Let $PV = \{\wp_1, \wp_2 \ldots\}$ be a set of propositional variables. The language of CTL* is given as the set of all the state formulas φ_s (interpreted at states of a model), defined using path formulas φ_p (interpreted at paths of a model), by the following grammar:

$$\varphi_s := \wp \mid \neg\varphi_s \mid \varphi_s \wedge \varphi_s \mid \varphi_s \vee \varphi_s \mid \mathrm{A}\varphi_p \mid \mathrm{E}\varphi_p$$
$$\varphi_p := \varphi_s \mid \varphi_p \wedge \varphi_p \mid \varphi_p \vee \varphi_p \mid \mathrm{X}\varphi_p \mid \varphi_p \mathrm{U} \varphi_p \mid \varphi_p \mathrm{R} \varphi_p.$$

In the above $\wp \in PV$, A ('for All paths') and E ('there Exists a path') are path quantifiers, whereas X ('neXt'), U ('Until'), and R ('Release') are state operators. Intuitively, the formula $\mathrm{X}\varphi_p$ specifies that φ_p holds in the next state of the path, whereas $\varphi_p \mathrm{U} \psi_p$ expresses that ψ_p eventually occurs and that φ_p holds continuously until then. The operator R is dual to U. So, the formula $\varphi_p \mathrm{R} \psi_p$ says that either ψ_p holds always or it is released when φ_p eventually occurs. Derived boolean and path operators are defined as follows:

$$\begin{aligned}
\varphi_p \Rightarrow \psi_p &\stackrel{def}{=} \neg\varphi_p \vee \psi_p, \\
\varphi_p \iff \psi_p &\stackrel{def}{=} (\varphi_p \Rightarrow \psi_p) \wedge (\psi_p \Rightarrow \varphi_p), \\
\mathrm{G}\varphi_p &\stackrel{def}{=} \mathit{false}\,\mathrm{R}\varphi_p, \text{ and} \\
\mathrm{F}\varphi_p &\stackrel{def}{=} \mathit{true}\,\mathrm{U}\varphi_p, \text{ where} \\
\mathit{true} &\stackrel{def}{=} \wp \vee \neg\wp, \text{ and} \\
\mathit{false} &\stackrel{def}{=} \wp \wedge \neg\wp, \text{ for an arbitrary } \wp \in PV.
\end{aligned}$$

Intuitively, the formula $\mathrm{F}\varphi_p$ specifies that φ_p occurs in some state of the path ('Finally'), whereas $\mathrm{G}\varphi_p$ expresses that φ_p holds in all the states of the path ('Globally').

It is important to mention that the restriction of the language of CTL* such that the negation is applied to propositions only[3] does not change its expressiveness.

Let φ be a state- or path formula of CTL*. By $PV(\varphi)$ we mean a set of all the propositions occurring in φ.

Sublogics of CTL*

In this section we consider restrictions of CTL* resulting in either branching-time logics or linear-time logic. The distinction between branching and linear time consists in how the logic handles branching in the underlying computation tree. In branching-time temporal logic the path quantifiers are used to

[3] This is called a *positive normal form* of CTL*.

quantify over the paths starting from a given state. In linear-time temporal logic, the modalities quantify over states of a single path.

ACTL* (Universal CTL*): the syntax of state formulas is restricted such that negation can be applied to propositions only, and the existential quantifier is not allowed. This is given by the following grammar:

$$\varphi_s := \wp \mid \neg\wp \mid \varphi_s \wedge \varphi_s \mid \varphi_s \vee \varphi_s \mid \mathrm{A}\varphi_p$$
$$\varphi_p := \varphi_s \mid \varphi_p \wedge \varphi_p \mid \varphi_p \vee \varphi_p \mid \mathrm{X}\varphi_p \mid \varphi_p \mathrm{U} \varphi_p \mid \varphi_p \mathrm{R} \varphi_p.$$

ECTL* (Existential CTL*): the syntax of state formulas is restricted such that negation can be applied to propositions only, and the universal quantifier is not allowed. This is given by the following grammar:

$$\varphi_s := \wp \mid \neg\wp \mid \varphi_s \wedge \varphi_s \mid \varphi_s \vee \varphi_s \mid \mathrm{E}\varphi_p$$
$$\varphi_p := \varphi_s \mid \varphi_p \wedge \varphi_p \mid \varphi_p \vee \varphi_p \mid \mathrm{X}\varphi_p \mid \varphi_p \mathrm{U} \varphi_p \mid \varphi_p \mathrm{R} \varphi_p.$$

CTL (Computation Tree Logic): the syntax of path formulas is restricted such that each of state operators must be preceded by a path quantifier, which is given by the following grammar:

$$\varphi_s := \wp \mid \neg\varphi_s \mid \varphi_s \wedge \varphi_s \mid \varphi_s \vee \varphi_s \mid \mathrm{A}\varphi_p \mid \mathrm{E}\varphi_p$$
$$\varphi_p := \mathrm{X}\varphi_s \mid \varphi_s \mathrm{U} \varphi_s \mid \varphi_s \mathrm{R} \varphi_s.$$

Notice that each temporal CTL formula is a boolean combination of $\mathrm{A}(\varphi \mathrm{U} \psi), \mathrm{A}(\varphi \mathrm{R} \psi)$, $\mathrm{AX}\varphi$, and $\mathrm{E}(\varphi \mathrm{U} \psi)$, $\mathrm{E}(\varphi \mathrm{R} \psi)$, $\mathrm{EX}\varphi$ only. Moreover, a restriction of the language of CTL, such that negation is applied to propositions only, does not change its expressiveness.

ACTL (Universal Computation Tree Logic): the temporal formulas of CTL are restricted to positive boolean combinations of $\mathrm{A}(\varphi \mathrm{U} \psi), \mathrm{A}(\varphi \mathrm{R} \psi)$, and $\mathrm{AX}\varphi$ only. Negation can be applied to propositions only.

ECTL (Existential Computation Tree Logic): the temporal formulas of CTL are restricted to positive boolean combinations of $\mathrm{E}(\varphi \mathrm{U} \psi), \mathrm{E}(\varphi \mathrm{R} \psi)$, and $\mathrm{EX}\varphi$ only. Negation can be applied to propositions only.

LTL (Linear-Time Temporal Logic): the formulas of the form $\mathrm{A}\varphi_p$ are allowed only, where φ_p is a path formula which does not contain the path quantifiers A, E. More precisely, the LTL path formulas are defined by the following grammar:

$$\varphi_p := \wp \mid \neg\varphi_p \mid \varphi_p \wedge \varphi_p \mid \varphi_p \vee \varphi_p \mid \mathrm{X}\varphi_p \mid \varphi_p \mathrm{U} \varphi_p \mid \varphi_p \mathrm{R} \varphi_p.$$

$\mathrm{L}_{-\mathrm{X}}$ denotes the logic L without the next-step operator X.

For example, $\mathrm{AFG}(\wp_1 \vee \wp_2)$ is an LTL formula, whereas $\mathrm{AFAG}(\wp_1 \vee \wp_2)$ is an ACTL formula. Each of the above logics can be extended by time constraints (defined later).

For a CTL formula φ we will use a notion of the *length* (or *size*) of φ, denoted $|\varphi|$, which is defined inductively as follows:

- if $\varphi = \wp$ where $\wp \in PV$, then $|\varphi| = 1$,
- if $\varphi = \neg\varphi'$, then $|\varphi| = |\varphi'| + 1$,
- if $\varphi = \varphi' \vee \varphi''$ or $\varphi = \varphi' \wedge \varphi''$, then $|\varphi| = |\varphi'| + |\varphi''| + 1$,
- if $\varphi \in \{\mathrm{EG}\varphi', \mathrm{AG}\varphi', \mathrm{EF}\varphi', \mathrm{AF}\varphi', \mathrm{EX}\varphi', \mathrm{AX}\varphi'\}$, then $|\varphi| = |\varphi'| + 1$,
- if $\varphi \in \{\mathrm{E}(\varphi'\mathrm{U}\varphi''), \mathrm{E}(\varphi'\mathrm{R}\varphi''), \mathrm{A}(\varphi'\mathrm{U}\varphi''), \mathrm{A}(\varphi'\mathrm{R}\varphi'')\}$, then $|\varphi| = |\varphi'| + |\varphi''| + 1$.

Example 4.3. Consider the CTL formula $\varphi = \mathrm{E}(\wp_1\mathrm{U}(\mathrm{EG}\wp_2))$. Then, we have $|\varphi| = 4$.

□

Moreover, for a CTL formula in a positive normal form we will need a notion of a set of the subformulas to be used in automata-theoretic model checking in Sect. 6.1.2. So, let φ be a CTL formula in a positive normal form. The set $SF(\varphi)$ of the subformulas of φ is defined inductively as follows:

- if $\varphi = \wp$ where $\wp \in PV$, then $SF(\varphi) = \{\wp\}$,
- if $\varphi = \neg\wp$, then $SF(\varphi) = \{\wp, \neg\wp\}$,
- if $\varphi = \varphi' \vee \varphi''$ or $\varphi = \varphi' \wedge \varphi''$, then $SF(\varphi) = SF(\varphi') \cup SF(\varphi'') \cup \{\varphi\}$,
- if $\varphi = \mathrm{EX}\varphi'$ or $\varphi = \mathrm{AX}\varphi'$, then $SF(\varphi) = SF(\varphi') \cup \{\varphi\}$,
- if $\varphi = \mathrm{E}(\varphi'\mathrm{U}\varphi'')$ or $\varphi = \mathrm{A}(\varphi'\mathrm{U}\varphi'')$, then $SF(\varphi) = SF(\varphi') \cup SF(\varphi'') \cup \{\varphi\}$,
- if $\varphi = \mathrm{E}(\varphi'\mathrm{R}\varphi'')$ or $\varphi = \mathrm{A}(\varphi'\mathrm{R}\varphi'')$, then $SF(\varphi) = SF(\varphi') \cup SF(\varphi'') \cup \{\varphi\}$.

Example 4.4. Consider the CTL formula $\varphi = \mathrm{E}(\wp_1\mathrm{U}(\mathrm{EX}\wp_2))$. Then, $SF(\varphi) = \{\wp_1, \wp_2, \mathrm{EX}\wp_2, \mathrm{E}(\wp_1\mathrm{U}(\mathrm{EX}\wp_2))\}$. Notice that $\wp_1\mathrm{U}(\mathrm{EX}\wp_2)$ and $\mathrm{X}\wp_2$ are not subformulas of φ, since they do not belong to the language of CTL. Moreover, we have $|\varphi| = 4$.

□

Semantics of CTL*

Semantics of CTL* uses standard Kripke models as defined below.

Definition 4.5. *A* model *is a tuple*

$$M = ((S, s^0, \rightarrow), V),$$

where

- *S is a set of states*
- *$s^0 \in S$ is the initial state,*
- *$\rightarrow \subseteq S \times S$ is a total successor relation*[4], *and*

[4] Totality means that $(\forall s \in S)(\exists s' \in S)\ s \rightarrow s'$. Sometimes a total relation is called *serial*.

- $V : S \longrightarrow 2^{PV}$ *is a valuation function.*

Given a model $M = ((S, s^0, \rightarrow), V)$, by $|S|$ and $|\rightarrow|$ we denote, respectively, the number of states of S and the number of transitions of M. The size of the model M, denoted $|M|$, is given by $|S| + |\rightarrow|$.

For $s_0 \in S$ a *path*

$$\pi = (s_0, s_1, \ldots)$$

is an infinite sequence of states in S starting at s_0, where $s_i \rightarrow s_{i+1}$ for all $i \geq 0$, and

$$\pi_i = (s_i, s_{i+1}, \ldots)$$

is the i-th suffix of π.

Given a model M, a state s, and a path π of M, by $M, s \models \varphi$ ($M, \pi \models \varphi$) we mean that φ holds in the state s (along the path π, respectively) of the model M. However, in what follows the model is sometimes omitted if it is clear from the context. The relation $\models$ is defined inductively below.

$$
\begin{array}{lll}
M, s \models \wp & \text{iff} & \wp \in V(s), \text{ for } \wp \in PV, \\
M, s \models \neg\varphi & \text{iff} & M, s \not\models \varphi, \\
M, x \models \varphi \wedge \psi & \text{iff} & M, x \models \varphi \text{ and } M, x \models \psi, \text{ for } x \in \{s, \pi\}, \\
M, x \models \varphi \vee \psi & \text{iff} & M, x \models \varphi \text{ or } M, x \models \psi, \text{ for } x \in \{s, \pi\}, \\
M, s \models \mathrm{A}\varphi & \text{iff} & M, \pi \models \varphi \text{ for each path } \pi \text{ starting at } s, \\
M, s \models \mathrm{E}\varphi & \text{iff} & M, \pi \models \varphi \text{ for some path } \pi \text{ starting at } s, \\
M, \pi \models \varphi & \text{iff} & M, s_0 \models \varphi, \text{ for a state formula } \varphi, \\
M, \pi \models \mathrm{X}\varphi & \text{iff} & M, \pi_1 \models \varphi, \\
M, \pi \models \varphi \mathrm{U} \psi & \text{iff} & (\exists j \geq 0)\big(M, \pi_j \models \psi \text{ and } (\forall 0 \leq i < j)\ M, \pi_i \models \varphi\big), \\
M, \pi \models \varphi \mathrm{R} \psi & \text{iff} & (\forall j \geq 0)\big(M, \pi_j \models \psi \text{ or } (\exists 0 \leq i < j)\ M, \pi_i \models \varphi\big).
\end{array}
$$

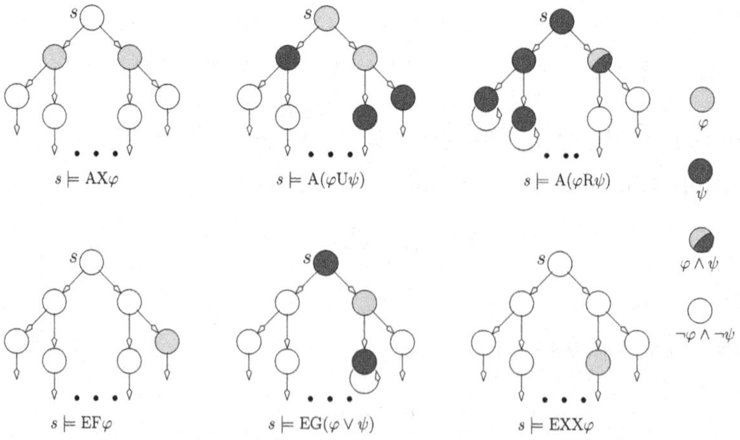

Fig. 4.1. Examples of CTL* formulas which hold in the state s of the model

We adopt the initialised notion of *validity* in a model:

$$M \models \varphi \text{ iff } M, s^0 \models \varphi,$$

where s^0 is the initial state of M. Some examples of CTL* formulas holding in the state s of a given model are presented in Fig. 4.1.

Notice that LTL and CTL are of not comparable expressive power. The LTL formula AFG$\wp$ does not have a counterpart in CTL, whereas the CTL formula AG(EF$\wp$) is not expressible in LTL. Moreover, the CTL* formula AFG$\wp \vee$ AG(EF$\wp$) does not have a counterpart in either LTL or CTL. The relations between sublogics of CTL* are depicted in Fig. 4.2. An arrow from a logic L_1 to L_2 means that L_2 is less expressive than L_1, whereas the logics of the diagram that are not connected with arrows are incomparable.

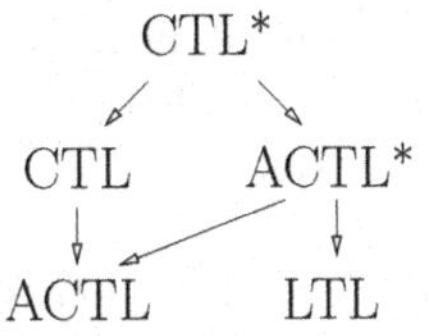

Fig. 4.2. Relations between sublogics of CTL*

When discussing CTL model checking, the following equivalences are used:

- $\mathrm{A}(\varphi \mathrm{U} \psi) \equiv \neg(\mathrm{E}(\neg\psi \mathrm{U}(\neg\varphi \wedge \neg\psi)) \vee \mathrm{EG}(\neg\psi))$,
- $\mathrm{A}(\varphi \mathrm{R} \psi) \equiv \neg\mathrm{E}(\neg\varphi \mathrm{U} \neg\psi)$,
- $\mathrm{E}(\varphi \mathrm{R} \psi) \equiv \neg\mathrm{A}(\neg\varphi \mathrm{U} \neg\psi)$.

4.2.2 Modal μ-Calculus

Propositional modal μ-calculus L_μ was introduced by D. Kozen [95]. In this section we define syntax and semantics of L_μ.

Syntax of Modal μ-Calculus

Let PV be a set of propositional variables and FV be a set of fixed-point variables. The language of modal μ-calculus L_μ is defined by the following grammar:

$$\varphi := \wp \mid \neg\varphi \mid \varphi \wedge \varphi \mid \varphi \vee \varphi \mid \mathrm{AX}\varphi \mid \mathrm{EX}\varphi \mid Z \mid \mu Z.\varphi(Z) \mid \nu Z.\varphi(Z),$$

where $\wp$ ranges over PV, Z – over FV, and $\varphi(Z)$ is a modal μ-calculus formula syntactically monotone in the fixed-point variable Z, i.e., all free occurrences of Z in $\varphi(Z)$ fall under an even number of negations.

Semantics of Modal μ-Calculus

Let

$$M = ((S, s^0, \rightarrow), V)$$

be a model as given in Definition 4.5. Notice that the set 2^S of all subsets of S forms a lattice under the set inclusion ordering. Each element S' of the lattice can also be thought of as a *predicate* on S, where this predicate is viewed as being true for exactly the states in S'. The least element in the lattice is the empty set, which we also refer to as $false$, and the greatest element in the lattice is the set S, which we sometimes write as *true*. A function ζ mapping 2^S to 2^S is called a *predicate transformer*. A set $S' \subseteq S$ is a *fixed point* of a function $\zeta : 2^S \longrightarrow 2^S$ if

$$\zeta(S') = S'.$$

Whenever ζ is *monotonic*, i.e., $S_1 \subseteq S_2$ implies $\zeta(S_1) \subseteq \zeta(S_2)$, where $S_1, S_2 \subseteq S$, it has the least fixed point denoted $\mu Z.\zeta(Z)$ and the greatest fixed point denoted $\nu Z.\zeta(Z)$. When $\zeta(Z)$ is also $\bigcup$-continuous, i.e., $S_1 \subseteq S_2 \subseteq \ldots$ implies $\zeta(\bigcup_{i\geq 0} S_i) = \bigcup_{i\geq 0} \zeta(S_i)$, then

$$\mu Z.\zeta(Z) = \bigcup_{i\geq 0} \zeta^i(false).$$

When $\zeta(Z)$ is also $\bigcap$-continuous, i.e., $S_1 \supseteq S_2 \supseteq \ldots$ implies $\zeta(\bigcap_{i\geq 0} S_i) = \bigcap_{i\geq 0} \zeta(S_i)$, then

$$\nu Z.\zeta(Z) = \bigcap_{i\geq 0} \zeta^i(true)$$

(see [155]).

The semantics of L_μ is given inductively for each formula φ of L_μ, a model M, a valuation $\mathcal{E} : FV \longrightarrow 2^S$ of the fixed-point variables (called an *environment*), and a state $s \in S$:

$$\begin{array}{lll}
M, \mathcal{E}, s \models \wp & \text{iff} & \wp \in V(s), \text{ for } \wp \in PV, \\
M, \mathcal{E}, s \models \neg\varphi & \text{iff} & M, \mathcal{E}, s \not\models \varphi, \\
M, \mathcal{E}, s \models \varphi \wedge \psi & \text{iff} & M, \mathcal{E}, s \models \varphi \text{ and } M, \mathcal{E}, s \models \psi, \\
M, \mathcal{E}, s \models \varphi \vee \psi & \text{iff} & M, \mathcal{E}, s \models \varphi \text{ or } M, \mathcal{E}, s \models \psi, \\
M, \mathcal{E}, s \models \mathrm{AX}\varphi & \text{iff} & (\forall s' \in S)((s \rightarrow s') \Rightarrow (M, \mathcal{E}, s' \models \varphi)), \\
M, \mathcal{E}, s \models \mathrm{EX}\varphi & \text{iff} & (\exists s' \in S)((s \rightarrow s') \wedge M, \mathcal{E}, s' \models \varphi), \\
M, \mathcal{E}, s \models Z & \text{iff} & s \in \mathcal{E}(Z), \text{ for } Z \in FV, \\
M, \mathcal{E}, s \models \mu Z.\varphi(Z) & \text{iff} & s \in \bigcap\{U \subseteq S \mid \\
& & \qquad \{s' \in S \mid M, \mathcal{E}[Z \leftarrow U], s' \models \varphi\} \subseteq U\}, \\
M, \mathcal{E}, s \models \nu Z.\varphi(Z) & \text{iff} & s \in \bigcup\{U \subseteq S \mid \\
& & \qquad U \subseteq \{s' \in S \mid M, \mathcal{E}[Z \leftarrow U], s' \models \varphi\}\},
\end{array}$$

where $\varphi, \psi \in L_\mu$, and $\mathcal{E}[Z \leftarrow U]$ is like $\mathcal{E}$ except that it maps Z to U. Similarly as before we adopt the initialised notion of validity, i.e.,

$$M \models \varphi \text{ iff } M, \mathcal{E}, s^0 \models \varphi$$

for each environment $\mathcal{E}$.

It is known that both CTL and CTL* can be translated into modal μ-calculus [95]. For example, we give characterisations of basic CTL modalities in terms of modal μ-calculus formulas:

- $\mathrm{A}(\varphi \mathrm{U} \psi) \equiv \mu Z.(\psi \vee (\varphi \wedge \mathrm{AX}Z))$,
- $\mathrm{E}(\varphi \mathrm{U} \psi) \equiv \mu Z.(\psi \vee (\varphi \wedge \mathrm{EX}Z))$,
- $\mathrm{AG}\varphi \equiv \nu Z.(\varphi \wedge \mathrm{AX}Z)$,
- $\mathrm{EG}\varphi \equiv \nu Z.(\varphi \wedge \mathrm{EX}Z)$.

The translation of CTL* to modal μ-calculus is more involved and can be found in [70]. It is worth noticing that the translations are important in practice because correctness specifications written in logics such as CTL or CTL* are often much more readable than specifications written directly in modal μ-calculus.

Example 4.6. Consider the formula $\varphi = \mathrm{E}(\wp_1 \mathrm{U} \wp_2)$, and the set of concrete states together with the successor relation, shown in Fig. 4.3(a). Different colours denote valuations of the states as shown in the upper part of the figure.

The formula φ can be expressed as $\varphi' = \mu Z.\wp_2 \vee (\wp_1 \wedge \mathrm{EX}Z)$. It is easy to prove that $\zeta(Z) = \wp_2 \vee (\wp_1 \wedge \mathrm{EX}Z)$ is monotonic: given $S_1 \subseteq S_2 \subseteq S$, we shall show that $\zeta(S_1) \subseteq \zeta(S_2)$. Consider a state $s \in \zeta(S_1)$, from the semantics we have that $s \models \wp_2$, or $s \models \wp_1 \wedge (\exists s' \in S)\ (s \rightarrow s' \wedge s' \in S_1)$. As $S_1 \subseteq S_2$, we have that $s' \in S_2$. Thus, $s \in \zeta(S_2)$ and ζ is monotonic. Since S is finite, ζ is also $\bigcup$-continuous (see [57]), and therefore the least fixed point for $\zeta(Z)$ can be computed as $\bigcup_{i \geq 0} \zeta^i(false)$.

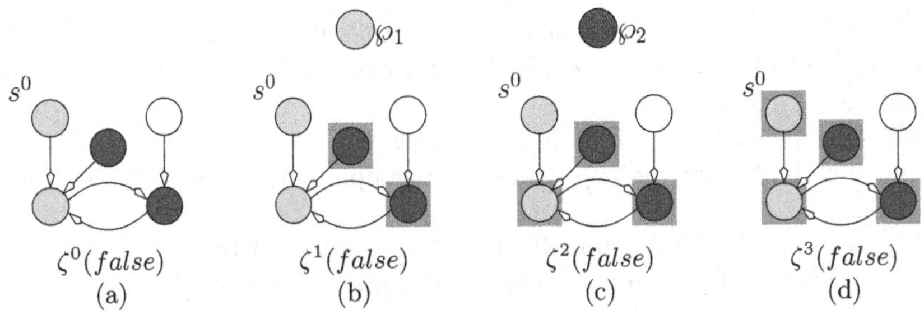

Fig. 4.3. A sequence of approximations for computing $\mathrm{E}(\wp_1 \mathrm{U} \wp_2)$

Parts (a)-(d) of the figure illustrate the process of approximating the set of states of S satisfying the formula φ. The states marked by the grey squares belong to $\zeta^i(false)$ for $i = 0, \ldots, 3$. Notice that in part (a) no states are marked, since $\zeta^0(false) = false$ (the empty set; $\zeta^0(\cdot)$ is an identity). In the next step we compute $\zeta^1(false)$, which returns the set of states satisfying $\wp_2 \vee (\wp_1 \wedge \mathrm{EX} false)$. Thus, in part (b) all the states at which $\wp_2$ holds are marked with squares. The next approximation follows from $\zeta^2(false) = \zeta(\zeta(false)) = \wp_2 \vee (\wp_1 \wedge \mathrm{EX}\zeta(false))$, and therefore in (c) we mark the states which either satisfy $\wp_2$, or satisfy $\wp_1$ and have a successor which belongs to $\zeta(false)$ (i.e., satisfies $\wp_2$). $\zeta^3(false)$ is computed in a similar way. The sets $\zeta^i(false)$ for $i \geq 3$ coincide, so $\bigcup_{i\geq 0} \zeta^i(false) = \zeta^3(false)$.

□

4.2.3 Interpretation of Temporal Logics over Timed Systems Models

For timed systems, untimed temporal logics are typically interpreted over concrete (or abstract) models, where the paths are defined to correspond to either weakly or strongly monotonic runs. In both the cases, we can deal with either a dense or a discrete semantics. When necessary logics are restricted not to use the next-step operator.

The *model checking* problem for CTL* or modal μ-calculus over timed systems is defined as follows: given a CTL* (modal μ-calculus) formula φ and a timed system $\mathcal{T}$ (i.e., a TPN $\mathcal{N}$ or a TA $\mathcal{A}$) together with a valuation function $V_{\mathcal{T}}$, determine whether $M_{\mathfrak{c}}(\mathcal{T}) \models \varphi$ for a selected concrete model $M_{\mathfrak{c}}(\mathcal{T})$.

It is known that most of the untimed properties of systems can be expressed in CTL*, but there are several interesting properties which can be formulated also in CTL:

- $\mathrm{EF}(\wp_1 \wedge \neg\wp_2)$ – it is possible to reach a state where $\wp_1$ holds but $\wp_2$ does not hold.
  One can think of $\wp_1, \wp_2$ as expressing for example *start*, *readiness* of a system.
- $\mathrm{AG}(\wp_1 \Rightarrow \mathrm{AF}\wp_2)$ – always when $\wp_1$ holds, $\wp_2$ will eventually hold.
  One can think of $\wp_1, \wp_2$ as expressing for example *request*, *granted* of a resource of a system.
- $\mathrm{AG}(\mathrm{AF}\wp_1)$ - $\wp_1$ holds infinitely often on each path.
  One can think of $\wp_1$ as expressing for example *enabledness* of a system action.
- $\mathrm{AG}(\mathrm{EF}\wp_1)$ – it is always possible to reach a state satisfying $\wp_1$.
  One can think of $\wp_1$ as expressing for example *restart* of a system.

Besides untimed properties of timed systems, which are directly expressed using the above-defined temporal logics, *reachability* in these systems is usually checked. Given a propositional formula p, the reachability problem for a timed system $\mathcal{T}$ consists in testing whether there is a reachable state satisfying p in a selected concrete model $M_{\mathfrak{c}}(\mathcal{T})$. This problem can be obviously translated to the model checking problem for the CTL formula EFp. However, in spite of that, several efficient solutions, aimed at reachability checking only, exist as well (see Sect. 5.2.3). Recall that the timed systems considered in this book are restricted to these, whose runs starting at the initial states are progressive. However, it is also reasonable to check reachability for systems which do not satisfy that property.

Example 4.7. Consider first Fischer's mutual exclusion example of Fig. 2.4. The most interesting property of this system is mutual exclusion, specified by the formula $\mathrm{AG}(\neg(critical1 \wedge critical2))$. One could also verify that the system does not satisfy the property guaranteeing each process to eventually enter the critical section from its waiting section, which is expressed by $\mathrm{AG}(waiting1 \Rightarrow \mathrm{AF}critical1)$. But, clearly, a weaker property saying that entering the critical section from the waiting section is possible $\mathrm{AG}(waiting1 \Rightarrow \mathrm{EF}critical1)$ holds.

Consider now Train, Gate and Controller example of Fig. 2.6. This system guarantees that if the gate is down, then it will eventually be moved back up $\mathrm{AG}(down \Rightarrow \mathrm{AF}up)$.

□

4.3 Timed Temporal Logics

Timed temporal logics can be interpreted over either discrete[5] or dense models of time [18]. We consider the latter option. Since the model checking problem for TCTL* is undecidable [6], we focus on TCTL [7] and its subsets: TACTL and TECTL, defined analogously to the corresponding fragments of CTL. Next, we discuss timed μ-calculus [84], an extension of TCTL, for which the model checking problem is still decidable.

4.3.1 Timed Computation Tree Logic (TCTL)

Syntax of TCTL

The logic TCTL is an extension of $\mathrm{CTL}_{-\mathrm{X}}$ obtained by subscribing the modalities with time intervals specifying time restrictions on formulas.

Formally, syntax of TCTL is defined inductively by the following grammar:

$$\varphi := \wp \mid \neg\varphi \mid \varphi \wedge \varphi \mid \varphi \vee \varphi \mid \mathrm{A}(\varphi \mathrm{U}_I \varphi) \mid \mathrm{E}(\varphi \mathrm{U}_I \varphi) \mid \mathrm{A}(\varphi \mathrm{R}_I \varphi) \mid \mathrm{E}(\varphi \mathrm{R}_I \varphi),$$

where $\wp \in PV$ and I is an interval in $\mathbb{R}_{0+}$ with integer bounds of the form $[n, n']$, $[n, n')$, $(n, n']$, (n, n'), (n, ∞), and $[n, \infty)$, for $n, n' \in \mathbb{N}$. The derived boolean and temporal operators are defined like for CTL in Sect. 4.2.1.

For example, $\mathrm{A}(false\ \mathrm{R}_{[0,\infty)}(\wp_1 \Rightarrow \mathrm{A}(true\ \mathrm{U}_{[0,5]}\ \wp_2)))$ expresses that for all the runs, always when $\wp_1$ holds, $\wp_2$ holds within 5 units of time.

[5] In a discrete model of time, it is assumed that the flow of time is not continuous.

Semantics of TCTL over Timed Systems

Let $\mathcal{T}$ be a timed system, i.e., either a TPN $\mathcal{N}$ or a TA $\mathcal{A}$, and let

$$M_c(\mathcal{T}) = (C_c(\mathcal{T}), V_{\mathfrak{C}}),$$

where $C_c(\mathcal{T}) = (S, s^0, \rightarrow_c)$ and the relation $\rightarrow_c$ is labelled by the elements of a set $B \cup \mathbb{R}_{0+}$ (where $B = T$ for time Petri nets and $B = A$ for timed automata), be its concrete dense model defined for a semantics in which concrete states are described by valuations of clocks[6]. The semantics of TCTL over timed systems can be defined in two ways, depending on the underlying definition of the run. Below, we provide both the definitions. We start with the *strongly monotonic interval semantics.*

Strongly Monotonic Interval Semantics

Let $\rho = s_0 \xrightarrow{\delta_0}_c s_0 + \delta_0 \xrightarrow{b_0}_c s_1 \xrightarrow{\delta_1}_c s_1 + \delta_1 \xrightarrow{b_1}_c s_2 \xrightarrow{\delta_2}_c \ldots$, where $s_0 \in Reach^+_{\mathcal{T}}$, $\delta_i \in \mathbb{R}_{0+}$ and $b_i \in B$ for $i \in \mathbb{N}$, be an s_0-run of $\mathcal{T}$ such that $\rho \in f^+_{\mathcal{T}}(s_0)$. Recall that for a concrete state $s \in Reach^+_{\mathcal{T}}$, the set $f^+_{\mathcal{T}}(s)$ contains all the progressive strongly monotonic runs of $\mathcal{T}$ starting at s. Notice that this implies that in ρ we have $\delta_i > 0$ for each $i \in \mathbb{N}$. In order to interpret TCTL formulas along a run, we introduce the notion of a *dense path corresponding to* ρ, denoted by π_ρ, which is a mapping from $\mathbb{R}_{0+}$ to a set of states[7], given by

$$\pi_\rho(r) = s_i + \delta$$

for $r = \Sigma_{j=0}^{i-1} \delta_j + \delta$, with $i \geq 0$ and $0 \leq \delta < \delta_i$.

Example 4.8. Consider a run $\rho = s_0 \xrightarrow{1}_c s_0{+}1 \xrightarrow{b_0}_c s_1 \xrightarrow{0.5}_c s_1{+}0.5 \xrightarrow{b_1}_c s_1 \xrightarrow{2}_c \ldots$. We have, e.g., $\pi_\rho(0) = s_0$, $\pi_\rho(0.99) = s_0 + 0.99$, $\pi_\rho(1) = s_1$ etc. Notice that after passing $r = \Sigma_{j=0}^{i}\delta_j$ units of time $\pi_\rho(r)$ corresponds to the state s_{i+1} obtained by executing b_i at the state $s_i + \delta_i$, whereas the state $s_i + \delta_i$ is not represented at the path.

□

Next, for $s_0 \in Reach^+_{\mathcal{T}}$ we define semantics of TCTL formulas in the following way:

$$\begin{array}{lll}
M_c(\mathcal{T}), s_0 \models \wp & \text{iff} & \wp \in V_{\mathfrak{C}}(s_0), \text{ for } \wp \in PV, \\
M_c(\mathcal{T}), s_0 \models \neg\varphi & \text{iff} & M_c(\mathcal{T}), s_0 \not\models \varphi, \\
M_c(\mathcal{T}), s_0 \models \varphi \wedge \psi & \text{iff} & M_c(\mathcal{T}), s_0 \models \varphi \text{ and } M_c(\mathcal{T}), s_0 \models \psi, \\
M_c(\mathcal{T}), s_0 \models \varphi \vee \psi & \text{iff} & M_c(\mathcal{T}), s_0 \models \varphi \text{ or } M_c(\mathcal{T}), s_0 \models \psi,
\end{array}$$

[6] This excludes the firing interval semantics for time Petri nets.
[7] This can be defined thanks to the assumption about $\delta > 0$.

$$
\begin{aligned}
M_c(\mathcal{T}), s_0 \models \mathrm{A}(\varphi \mathrm{U}_I \psi) \ \text{ iff } \ & (\forall\, \rho \in f^+_{\mathcal{T}}(s_0))(\exists r \in I) \\
& \big[M_c(\mathcal{T}), \pi_\rho(r) \models \psi \wedge \\
& (\forall r' < r)\ M_c(\mathcal{T}), \pi_\rho(r') \models \varphi\big], \\
M_c(\mathcal{T}), s_0 \models \mathrm{E}(\varphi \mathrm{U}_I \psi) \ \text{ iff } \ & (\exists\, \rho \in f^+_{\mathcal{T}}(s_0))(\exists r \in I) \\
& \big[M_c(\mathcal{T}), \pi_\rho(r) \models \psi \wedge \\
& (\forall r' < r)\ M_c(\mathcal{T}), \pi_\rho(r') \models \varphi\big], \\
M_c(\mathcal{T}), s_0 \models \mathrm{A}(\varphi \mathrm{R}_I \psi) \ \text{ iff } \ & (\forall\, \rho \in f^+_{\mathcal{T}}(s_0))(\forall r \in I) \\
& \big[M_c(\mathcal{T}), \pi_\rho(r) \models \psi \vee \\
& (\exists r' < r)\ M_c(\mathcal{T}), \pi_\rho(r') \models \varphi\big], \\
M_c(\mathcal{T}), s_0 \models \mathrm{E}(\varphi \mathrm{R}_I \psi) \ \text{ iff } \ & (\exists\, \rho \in f^+_{\mathcal{T}}(s_0))(\forall r \in I) \\
& \big[M_c(\mathcal{T}), \pi_\rho(r) \models \psi \vee \\
& (\exists r' < r)\ M_c(\mathcal{T}), \pi_\rho(r') \models \varphi\big].
\end{aligned}
$$

Again, we adopt the initialised notion of validity in a model:

$$M_c(\mathcal{T}) \models \varphi \ \text{ iff } \ M_c(\mathcal{T}), s^0 \models \varphi,$$

where s^0 is the initial state in $M_c(\mathcal{T})$. Some examples of TCTL formulas holding in a (reachable) state s_0 of the given model are presented in Fig. 4.4.

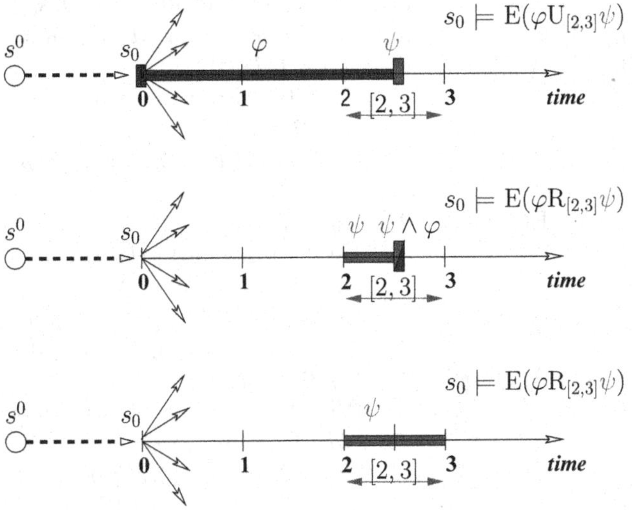

Fig. 4.4. Examples of TCTL formulas which hold in the state s_0 of the model

It is important to mention that there is an alternative semantics for sublogics of $\mathrm{CTL}_{-\mathrm{X}}$, which can be interpreted over dense models by assuming that the semantics of each untimed modality O is like the semantics of the corresponding TCTL modality $\mathrm{O}_{[0,\infty)}$. The *model checking* problem for TCTL over a timed system $\mathcal{T}$ is defined as usual, i.e., given a TCTL formula φ and a timed system together with a valuation function $V_{\mathcal{T}}$, determine whether $M_c(\mathcal{T}) \models \varphi$.

Weakly Monotonic Interval Semantics

An alternative semantics of TCTL over timed systems, called *weakly monotonic interval* one, is defined over weakly monotonic runs. Let $\rho = s_0 \overset{\delta_0}{\to}_c s_0 + \delta_0 \overset{b_0}{\to}_c s_1 \overset{\delta_1}{\to} s_1 + \delta_1 \overset{b_1}{\to}_c s_2 \overset{\delta_2}{\to} \ldots$ be an s_0-run of $\mathcal{T}$, where $s_0 \in S$, $\delta_i \in \mathbb{R}_{0+}$ and $b_i \in B$ for $i \in \mathbb{N}$. Recall that for a concrete state $s \in S$, the set of all the progressive runs of $\mathcal{T}$ starting at s is now denoted by $f_{\mathcal{T}}(s)$. Given a run $\rho \in f_{\mathcal{T}}(s)$, $i \in \mathbb{N}$ and $\delta \in \mathbb{R}_{0+}$, let

$$r(i, \delta) = \Sigma_{j<i}\delta_j + \delta.$$

In this case, we interpret formulas of TCTL along the runs of $\mathcal{T}$ in the following way:

$$
\begin{array}{lll}
M_c(\mathcal{T}), s_0 \models \wp & \text{iff} & \wp \in V_{\mathfrak{C}}(s_0), \text{ for } \wp \in PV, \\
M_c(\mathcal{T}), s_0 \models \neg\varphi & \text{iff} & M_c(\mathcal{T}), s_0 \not\models \varphi, \\
M_c(\mathcal{T}), s_0 \models \varphi \wedge \psi & \text{iff} & M_c(\mathcal{T}), s_0 \models \varphi \text{ and } M_c(\mathcal{T}), s_0 \models \psi, \\
M_c(\mathcal{T}), s_0 \models \varphi \vee \psi & \text{iff} & M_c(\mathcal{T}), s_0 \models \varphi \text{ or } M_c(\mathcal{T}), s_0 \models \psi, \\
M_c(\mathcal{T}), s_0 \models \mathrm{A}(\varphi \mathrm{U}_I \psi) & \text{iff} & (\forall \rho \in f_{\mathcal{T}}(s_0)) \\
& & (\exists i \geq 0)(\exists \delta \leq \delta_i)\ [\, r(i,\delta) \in I \wedge \\
& & (M_c(\mathcal{T}), s_i + \delta \models \psi \ \wedge \\
& & (\forall \delta' < \delta) M_c(\mathcal{T}), s_i + \delta' \models \varphi \ \wedge \\
& & (\forall k < i)(\forall \delta' \leq \delta_k) M_c(\mathcal{T}), s_k + \delta' \models \varphi\,)], \\
M_c(\mathcal{T}), s_0 \models \mathrm{E}(\varphi \mathrm{U}_I \psi) & \text{iff} & (\exists \rho \in f_{\mathcal{T}}(s_0)) \\
& & (\exists i \geq 0)(\exists \delta \leq \delta_i)\ [\, r(i,\delta) \in I \ \wedge \\
& & (M_c(\mathcal{T}), s_i + \delta \models \psi \ \wedge \\
& & (\forall \delta' < \delta) M_c(\mathcal{T}), s_i + \delta' \models \varphi \ \wedge \\
& & (\forall k < i)(\forall \delta' \leq \delta_k) M_c(\mathcal{T}), s_k + \delta' \models \varphi\,)], \\
M_c(\mathcal{T}), s_0 \models \mathrm{A}(\varphi \mathrm{R}_I \psi) & \text{iff} & (\forall \rho \in f_{\mathcal{T}}(s_0)) \\
& & (\forall i \geq 0)(\forall \delta \leq \delta_i)\ [\, r(i,\delta) \in I \Rightarrow \\
& & (M_c(\mathcal{T}), s_i + \delta \models \psi \ \vee \\
& & (\exists k \leq i)(\exists \delta' \leq \delta_k)(r(k,\delta') < r(i,\delta) \ \wedge \\
& & \qquad M_c(\mathcal{T}), s_k + \delta' \models \varphi\,)], \\
M_c(\mathcal{T}), s_0 \models \mathrm{E}(\varphi \mathrm{R}_I \psi) & \text{iff} & (\exists \rho \in f_{\mathcal{T}}(s_0)) \\
& & (\forall i \geq 0)(\forall \delta \leq \delta_i)\ [\, r(i,\delta) \in I \Rightarrow \\
& & (M_c(\mathcal{T}), s_i + \delta \models \psi \ \vee \\
& & (\exists k \leq i)(\exists \delta' \leq \delta_k)(r(k,\delta') < r(i,\delta) \ \wedge \\
& & \qquad M_c(\mathcal{T}), s_k + \delta' \models \varphi\,)].
\end{array}
$$

Analogously as before,

$$M_c(\mathcal{T}) \models \varphi \ \text{ iff } \ M_c(\mathcal{T}), s^0 \models \varphi,$$

where s^0 is the initial state in $M_c(\mathcal{T})$. Moreover, the above semantics enables to introduce an alternative (weakly monotonic) semantics of $\mathrm{CTL}_{-\mathrm{X}}$, defined as in the previous case. Similarly, the model checking problem is stated as before. Some examples of the TCTL formulas holding in the state s_0 of a given model are shown in Fig. 4.5. Notice that the states $s_i + \delta_i$ and s_{i+1} correspond to the same time passed from the beginning of the run.

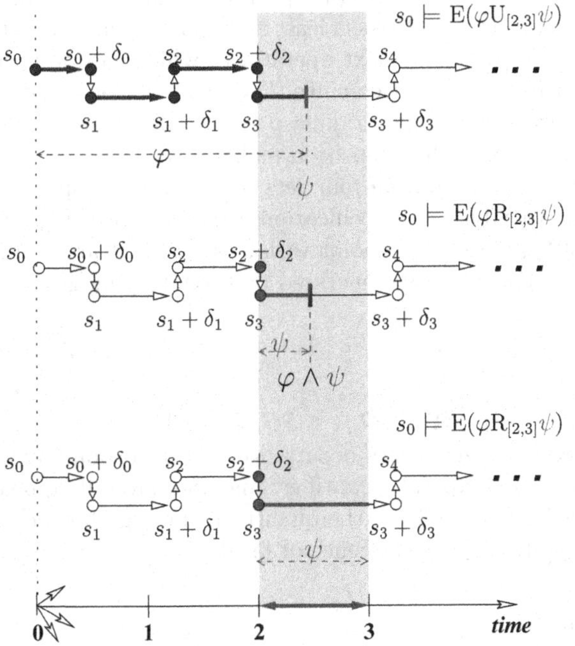

Fig. 4.5. Weakly monotonic interval semantics for TCTL

Example 4.9. Consider first Fischer's mutual exclusion example of Fig. 2.4. The most interesting property for this system is mutual exclusion, specified now by the formula $\mathrm{AG}_{[0,\infty)}(\neg(critical1 \wedge critical2))$. One could also verify that this system does not satify the property guaranteeing each process to eventually enter the critical section from its waiting section, which is expressed by $\mathrm{AG}_{[0,\infty)}(waiting1 \Rightarrow \mathrm{AF}_{[0,\infty)}critical1)$. A weaker property $\mathrm{AG}_{[0,\infty)}(waiting1 \Rightarrow \mathrm{EF}_{[0,c)}critical1)$, for some $c \in \mathbb{N}_+$, holds.

Consider now Train, Gate and Controller example of Fig. 2.6. This system guarantees that if the gate is down, then it will eventually by moved back up within K seconds, for some $K \in \mathbb{N}_+$: $\mathrm{AG}_{[0,\infty)}(down \Rightarrow \mathrm{AF}_{[0,K)}up)$.

□

4.3.2 Timed μ-Calculus ($T\mu$)

Timed μ-calculus ($T\mu$) can be viewed as an extension of modal μ-calculus [95] defined in Sect. 4.2.2.

Syntax of $T\mu$

The formulas of $T\mu$ are built from propositional variables by boolean connectives, a reset quantifier for clocks, a least fixed point quantifier, and a temporal next operator $\triangleright$. The binary next operator $\triangleright$, which can be best viewed as a "single step until", is introduced since there is no notion of "next step" when time is dense. Intuitively, the formula $p \triangleright q$ specifies that p holds until some transition is taken, and this transition makes q true.

The formulas of $T\mu$ contain four sets of variables: propositional variables PV, free clock variables $\mathcal{X}$, specification clock variables SC that are bound by reset quantifiers, and fixed-point variables FV that are bound by fixpoint quantifiers. The formulas are generated by the following grammar:

$$\varphi := \wp \mid \neg\varphi \mid \varphi \wedge \varphi \mid \varphi \vee \varphi \mid x_i \sim c \mid x_i - x_j \sim c \mid \varphi \triangleright \varphi \mid z.\varphi \mid Z \mid \mu Z.\varphi \mid \nu Z.\varphi,$$

where $\wp \in PV$, $x_i, x_j \in \mathcal{X} \cup SC$, $z \in SC$, $Z \in FV$, $\sim \in \{\leq, <, =, >, \geq\}$, and $c \in \mathbb{N}$. It is required that every occurrence of a fixed-point variable $Z \in FV$ in φ is bound by a fixpoint quantifier, and moreover it appears within an even number of negations from the quantifier binding it, which is a standard condition guaranteeing the existence of fixed points.

Semantics of $T\mu$

Let $\mathcal{T}$ be a timed system, i.e., either a TPN $\mathcal{N}$ or a TA $\mathcal{A}$, and let

$$M_c(\mathcal{T}) = (C_c(\mathcal{T}), V_{\mathfrak{C}}),$$

where $C_c(\mathcal{T}) = (S, s^0, \rightarrow_c)$ and the relation $\rightarrow_c$ is labelled by the elements of a set $B \cup \mathbb{R}_{0+}$, be its concrete dense model defined for a semantics in which concrete states are described by valuations of clocks[8].

Assume that the concrete states of $\mathcal{T}$ are represented by pairs $(\cdot, v_{\mathcal{X}})$, where $v_{\mathcal{X}}$ is a valuation of clocks[9]. The formulas of $T\mu$ are interpreted over the states of a given concrete model over the weakly monotonic semantics. To this aim, we need to define the environment to provide values for both specification clocks and fixed-point variables. So, in addition to clocks' valuation $v_{\mathcal{X}}$ and propositions' valuations $V_{\mathfrak{C}}$ we define a two-component function

$$\mathcal{E} = (\mathcal{E}_1, \mathcal{E}_2)$$

---

[8] This, again, excludes the firing interval semantics for time Petri nets.
[9] This can be either a function like in TPNs, or a tuple as in the case of TA.

such that $\mathcal{E}_1 : SC \longrightarrow \mathbb{R}_{0+}$ is a partial function and $\mathcal{E}_2 : FV \longrightarrow 2^S$, where for all specification clocks $z \in SC$ either $\mathcal{E}_1(z) \in \mathbb{R}_{0+}$, or $\mathcal{E}_1(z)$ is undefined. The environment $\mathcal{E}$ is empty when its clock component is empty. By $\mathcal{E} + \delta$, for $\delta \in \mathbb{R}_{0+}$, we denote $\mathcal{E}' = (\mathcal{E}_1', \mathcal{E}_2)$, where $\mathcal{E}_1' = \mathcal{E}_1 + \delta$, i.e., such that the value of each specification clock is increased by δ. We write $\mathcal{E}[\mathrm{e} \leftarrow a]$ for the environment that agrees with the environment $\mathcal{E}$ on all specification and fixed-point variables except for $\mathrm{e} \in SC \cup FV$, which is mapped to a. Given a model $M_c(\mathcal{T})$, a state $s = (\cdot, v_{\mathcal{X}}) \in S$ and an environment $\mathcal{E}$, we define the semantics of $T\mu$ formulas in the following way:

$$
\begin{array}{lll}
M_c(\mathcal{T}), \mathcal{E}, s \models \wp & \text{iff} & \wp \in V_{\mathfrak{C}}(s), \text{ for } \wp \in PV, \\
M_c(\mathcal{T}), \mathcal{E}, s \models \neg\varphi & \text{iff} & M_c(\mathcal{T}), \mathcal{E}, s \not\models \varphi, \\
M_c(\mathcal{T}), \mathcal{E}, s \models \varphi \wedge \psi & \text{iff} & M_c(\mathcal{T}), \mathcal{E}, s \models \varphi \text{ and } M_c(\mathcal{T}), \mathcal{E}, s \models \psi, \\
M_c(\mathcal{T}), \mathcal{E}, s \models \varphi \vee \psi & \text{iff} & M_c(\mathcal{T}), \mathcal{E}, s \models \varphi \text{ or } M_c(\mathcal{T}), \mathcal{E}, s \models \psi, \\
M_c(\mathcal{T}), \mathcal{E}, s \models x_i \sim c & \text{iff} & v_{\mathcal{X}}(x_i) \sim c, \\
M_c(\mathcal{T}), \mathcal{E}, s \models x_i - x_j \sim c & \text{iff} & v_{\mathcal{X}}(x_i) - v_{\mathcal{X}}(x_j) \sim c, \\
M_c(\mathcal{T}), \mathcal{E}, s \models \varphi \rhd \psi & \text{iff} & (\exists s' \in S)(\exists \delta \in \mathbb{R}_{0+})(\exists b \in B) \\
& & ((s \xrightarrow{\delta}_c s{+}\delta \xrightarrow{b}_c s' \wedge M_c(\mathcal{T}), \mathcal{E}{+}\delta, s' \models \psi) \\
& & \wedge \ (\forall 0 \leq \delta' \leq \delta) \\
& & \qquad M_c(\mathcal{T}), \mathcal{E} + \delta', s + \delta' \models \varphi \vee \psi\ ), \\
M_c(\mathcal{T}), \mathcal{E}, s \models z.\varphi & \text{iff} & M_c(\mathcal{T}), \mathcal{E}[z \leftarrow 0], s \models \varphi, \\
M_c(\mathcal{T}), \mathcal{E}, s \models Z & \text{iff} & s \in \mathcal{E}(Z), \\
M_c(\mathcal{T}), \mathcal{E}, s \models \mu Z.\varphi & \text{iff} & s \in \bigcap\{U \subseteq S \mid \{s' \in S \mid \\
& & \qquad M_c(\mathcal{T}), \mathcal{E}[Z \leftarrow U], s' \models \varphi\} \subseteq U\}, \\
M_c(\mathcal{T}), \mathcal{E}, s \models \nu Z.\varphi & \text{iff} & s \in \bigcup\{U \subseteq S \mid U \subseteq \{s' \in S \mid \\
& & \qquad M_c(\mathcal{T}), \mathcal{E}[Z \leftarrow U], s' \models \varphi\}\}.
\end{array}
$$

A state $s \in S$ satisfies the formula φ, which is denoted by $M_c(\mathcal{T}), s \models \varphi$, if $M_c(\mathcal{T}), \mathcal{E}, s \models \varphi$ for all empty $T\mu$-environments[10] $\mathcal{E}$. A formula φ holds in a model $M_c(\mathcal{T})$ ($M_c(\mathcal{T}) \models \varphi$) if

$$M_c(\mathcal{T}), s^0 \models \varphi.$$

Some examples of $T\mu$ formulas are shown in Fig. 4.6.

The logic $T\mu$ is more expressive than TCTL, defined over progressive runs (see Sect. 4.3.1), where time is weakly monotonic [84]. Clearly, the logics defined over different semantics, i.e., weakly and strongly monotonic, are not comparable.

---

[10] This means that the specification clocks are assigned no values. In fact, equivalently, validity could have been defined for all $T\mu$-environments.

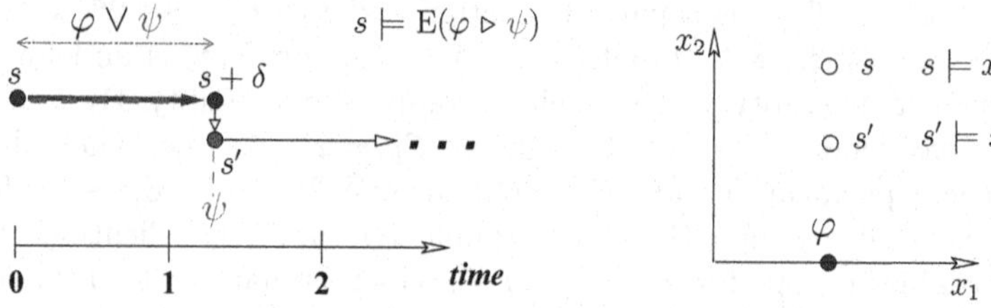

Fig. 4.6. Examples of $T\mu$ formulas

In order to show how TCTL basic operators are definable in $T\mu$ we first give an alternative syntax of TCTL, called $\mathrm{TCTL}_{\mathcal{C}}$[11]. Next, we show a translation from TCTL to $\mathrm{TCTL}_{\mathcal{C}}$, and finally a translation from $\mathrm{TCTL}_{\mathcal{C}}$ to $T\mu$.

4.3.3 Syntax and Semantics of $\mathrm{TCTL}_{\mathcal{C}}$

The formulas of $\mathrm{TCTL}_{\mathcal{C}}$ are built over three sets of variables: clock variables $\mathcal{X}$, specification clock variables SC and propositional variables PV. They are generated by the following grammar:

$$\varphi := \wp \mid \neg\varphi \mid \varphi \wedge \varphi \mid \varphi \vee \varphi \mid x_i \sim c \mid x_i - x_j \sim c \mid$$
$$\mathrm{A}(\varphi \mathrm{U} \varphi) \mid \mathrm{E}(\varphi \mathrm{U} \varphi) \mid \mathrm{A}(\varphi \mathrm{R} \varphi) \mid \mathrm{E}(\varphi \mathrm{R} \varphi) \mid z.\varphi,$$

where $\wp \in PV$, $x_i, x_j \in \mathcal{X} \cup SC$, $z \in SC$, $\sim \in \{\leq, <, =, >, \geq\}$, and $c \in \mathbb{N}$. The derived boolean and temporal operators are defined like for CTL in Sect. 4.2.1.

Let $\mathcal{T}$ be a timed system, i.e., either a TPN $\mathcal{N}$ or a TA $\mathcal{A}$, and let

$$M_c(\mathcal{T}) = (C_c(\mathcal{T}), V_{\mathfrak{C}}),$$

where $C_c(\mathcal{T}) = (S, s^0, \rightarrow_c)$ and the relation $\rightarrow_c$ is labelled by the elements of a set $B \cup \mathbb{R}_{0+}$, be its concrete dense model defined for a semantics in which concrete states are described by valuations of clocks[12]. Let $\rho = s_0 \xrightarrow{\delta_0}_c s_0 + \delta_0 \xrightarrow{b_0}_c s_1 \xrightarrow{\delta_1}_c s_1 + \delta_1 \xrightarrow{b_1}_c s_2 \xrightarrow{\delta_2}_c \ldots$ be an s_0-run of $\mathcal{T}$, where $s_0 \in S$, $\delta_i \in \mathbb{R}$ and $b_i \in B$ for $i \in \mathbb{N}$. Notice that here δ_i can be equal 0, which allows for a run having two consecutive action transitions that are not separated by any time delay. Recall that $f_{\mathcal{T}}(s)$, for a concrete state s, denotes the set of all the progressive runs of $\mathcal{T}$ starting at s, and assume that the concrete states of $\mathcal{T}$ are represented by pairs $(\cdot, v_{\mathcal{X}})$, where $v_{\mathcal{X}}$ is a valuation of clocks[13].

---

[11] In order to distinguish between the logic TCTL introduced previously and that introduced in [84], we denote the latter by $\mathrm{TCTL}_{\mathcal{C}}$, where the index refers to clock constrains.

[12] This, again, excludes the firing interval semantics for time Petri nets.

[13] This can be either a function like in TPNs or a tuple as in the case of TA.

The formulas of $\mathrm{TCTL}_\mathcal{C}$ are interpreted over the states S of a given concrete model. In addition to clocks' valuation $v_\mathcal{X}$ and propositions' valuations $V_\mathfrak{C}$ we use a partial function

$$\mathcal{E} : SC \longrightarrow \mathbb{R}_{0+},$$

where for all specification clocks $z \in SC$, either $\mathcal{E}(z) \in \mathbb{R}$ or $\mathcal{E}(z)$ is undefined. The environment $\mathcal{E}$ is empty when it is undefined for all specification clocks. By $\mathcal{E} + \delta$ for $\delta \in \mathbb{R}_{0+}$, we denote $\mathcal{E}'$ such that the value of each specification clock is increased by δ. We write $\mathcal{E}[z \leftarrow a]$ for the environment that agrees with the environment $\mathcal{E}$ on all specification variables except for z, which is mapped to a. Moreover, given a run $\rho \in f_\mathcal{T}(s)$ for some $s \in S$, $i \in \mathbb{N}$ and $\delta \in \mathbb{R}_{0+}$, let

$$r(i, \delta) = \Sigma_{j<i}\delta_j + \delta.$$

Given a model $M_c(\mathcal{T})$, a state $s_0 = (\cdot, v_\mathcal{X}) \in S$ and an environment $\mathcal{E}$, we define the semantics of $\mathrm{TCTL}_\mathcal{C}$ formulas in the following way:

$$
\begin{array}{lll}
M_c(\mathcal{T}), \mathcal{E}, s_0 \models \wp & \text{iff} & \wp \in V_\mathfrak{C}(s_0), \text{ for } \wp \in PV, \\
M_c(\mathcal{T}), \mathcal{E}, s_0 \models \neg\varphi & \text{iff} & M_c(\mathcal{T}), \mathcal{E}, s_0 \not\models \varphi, \\
M_c(\mathcal{T}), \mathcal{E}, s_0 \models \varphi \wedge \psi & \text{iff} & M_c(\mathcal{T}), \mathcal{E}, s_0 \models \varphi \text{ and } M_c(\mathcal{T}), \mathcal{E}, s_0 \models \psi, \\
M_c(\mathcal{T}), \mathcal{E}, s_0 \models \varphi \vee \psi & \text{iff} & M_c(\mathcal{T}), \mathcal{E}, s_0 \models \varphi \text{ or } M_c(\mathcal{T}), \mathcal{E}, s_0 \models \psi, \\
M_c(\mathcal{T}), \mathcal{E}, s_0 \models x_i \sim c & \text{iff} & v_\mathcal{X}(x_i) \sim c, \\
M_c(\mathcal{T}), \mathcal{E}, s_0 \models x_i - x_j \sim c & \text{iff} & v_\mathcal{X}(x_i) - v_\mathcal{X}(x_j) \sim c, \\
M_c(\mathcal{T}), \mathcal{E}, s_0 \models \mathrm{A}(\varphi \mathrm{U} \psi) & \text{iff} & (\forall \rho \in f_\mathcal{T}(s_0))(\exists i \geq 0)(\exists \delta \leq \delta_i) \\
 & & [\, M_c(\mathcal{T}), \mathcal{E} + r(i,\delta), s_i + \delta \models \psi \;\wedge \\
 & & \quad (\forall k \leq i)(\forall \delta' \leq \delta_k) \\
 & & \quad (\, r(k,\delta') \leq r(i,\delta) \Rightarrow \\
 & & M_c(\mathcal{T}), \mathcal{E} + r(k,\delta'), s_k + \delta' \models \varphi \vee \psi) \,], \\
M_c(\mathcal{T}), \mathcal{E}, s_0 \models \mathrm{E}(\varphi \mathrm{U} \psi) & \text{iff} & (\exists \rho \in f_\mathcal{T}(s_0))(\exists i \geq 0)(\exists \delta \leq \delta_i) \\
 & & [\, M_c(\mathcal{T}), \mathcal{E} + r(i,\delta), s_i + \delta \models \psi \;\wedge \\
 & & \quad (\forall k \leq i)(\forall \delta' \leq \delta_k) \\
 & & \quad (\, r(k,\delta') \leq r(i,\delta) \Rightarrow \\
 & & M_c(\mathcal{T}), \mathcal{E} + r(k,\delta'), s_k + \delta' \models \varphi \vee \psi) \,], \\
M_c(\mathcal{T}), \mathcal{E}, s_0 \models \mathrm{A}(\varphi \mathrm{R} \psi) & \text{iff} & (\forall \rho \in f_\mathcal{T}(s_0))(\forall i \geq 0)(\forall \delta \leq \delta_i) \\
 & & [\, M_c(\mathcal{T}), \mathcal{E} + r(i,\delta), s_i + \delta \models \psi \;\vee \\
 & & \quad (\exists k \leq i)(\exists \delta' \leq \delta_k) \\
 & & \quad (\, r(k,\delta') \leq r(i,\delta) \wedge \\
 & & M_c(\mathcal{T}), \mathcal{E} + r(k,\delta'), s_k + \delta' \models \varphi \wedge \psi)], \\
M_c(\mathcal{T}), \mathcal{E}, s_0 \models \mathrm{E}(\varphi \mathrm{R} \psi) & \text{iff} & (\exists \rho \in f_\mathcal{T}(s_0))(\forall i \geq 0)(\forall \delta \leq \delta_i) \\
 & & [\, M_c(\mathcal{T}), \mathcal{E} + r(i,\delta), s_i + \delta \models \psi \;\vee \\
 & & \quad (\exists k \leq i)(\exists \delta' \leq \delta_k) \\
 & & \quad (\, r(k,\delta') \leq r(i,\delta) \wedge \\
 & & M_c(\mathcal{T}), \mathcal{E} + r(k,\delta'), s_k + \delta' \models \varphi \wedge \psi)]. \\
M_c(\mathcal{T}), \mathcal{E}, s_0 \models z.\varphi & \text{iff} & M_c(\mathcal{T}), \mathcal{E}[z \leftarrow 0], s_0 \models \varphi.
\end{array}
$$

The state s satisfies the formula φ, denoted $s \models \varphi$, if $M_c(\mathcal{T}), \mathcal{E}, s \models \varphi$ for all empty TCTL$_C$ environments $\mathcal{E}$. A formula φ holds in a model $M_c(\mathcal{T})$ ($M_c(\mathcal{T}) \models \varphi$) if

$$M_c(\mathcal{T}), s^0 \models \varphi.$$

Some examples of TCTL$_C$ formulas which hold in the state s_0 of a given model are depicted in Fig. 4.7. Again, notice that unlike for the dense paths in the strongly monotonic interval semantics, two states can correspond to the same time point (i.e., to the same time period passed from the beginning of the run).

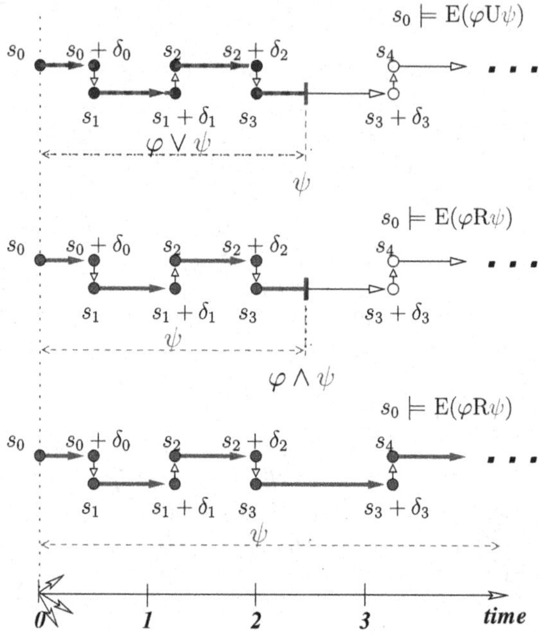

Fig. 4.7. Some examples of TCTL$_C$ formulas

A few comments regarding the above logic TCTL$_C$, a version of TCTL, are in order. First of all, in order to make sure that the formulas of the form $z.\mathrm{A}((z \leq c)\mathrm{U}(z > c))$, where $c \in \mathbb{N}$, hold in each state of every model, we require that in the semantics of $\varphi \mathrm{U} \psi$ the disjunction $\varphi \vee \psi$ – rather than φ only – is true in all the states before ψ holds. Secondly, the formulas are interpreted over weakly monotonic runs rather than over strong monotonic ones. Thirdly, the time-bounded temporal operators used in TCTL are definable in TCTL$_C$. For instance, the time-bounded response property

$$\mathrm{AG}_{[0,\infty]}(\wp_1 \Rightarrow \mathrm{AF}_{[c_1,c_2]}\wp_2)$$

that says every request $\wp_1$ must be answered ($\wp_2$) within the time interval $[c_1, c_2]$, is expressible in $\mathrm{TCTL}_\mathcal{C}$ by the formula

$$\mathrm{AG}z.(\wp_1 \Rightarrow \mathrm{AF}(\wp_2 \wedge c_1 \leq z \wedge z \leq c_2).$$

Similarly, the time-bounded possibility property

$$\mathrm{AG}_{[0,\infty)}(\wp_1 \Rightarrow \mathrm{EF}_{[c_1,c_2]}\wp_2)$$

is expressible in $\mathrm{TCTL}_\mathcal{C}$ by the formula

$$\mathrm{AG}z.(\wp_1 \Rightarrow \mathrm{EF}(\wp_2 \wedge c_1 \leq z \wedge z \leq c_2).$$

Our next step consists in showing that $\mathrm{TCTL}_\mathcal{C}$ is definable in $T\mu$. This result, which is proven in [84], is however much more subtle than it seems to be at the first sight. We showed in Sect. 4.2.2 that in the untimed case, the logic CTL is strictly less expressive than the propositional modal μ-calculus over all our models. It turns out that $T\mu$ is as expressive as $\mathrm{TCTL}_\mathcal{C}$ over the class of real-time systems, which allows us to compute the characteristic sets of $\mathrm{TCTL}_\mathcal{C}$ formulas as fixpoints. In our case this requires to consider only progressive time Petri nets and timed automata, which has already been assumed. Then, we can translate every $\mathrm{TCTL}_\mathcal{C}$ formula φ into a $T\mu$ formula φ' using the following algorithm:

a) Replace each subformula of the form $\mathrm{A}(\varphi \mathrm{U} \psi)$ with the formula

$$\mu Z.(\psi \vee \neg z.(\mathrm{E}((\neg Z)\mathrm{U}(\neg(\varphi \vee Z) \vee z > c))),$$

for any positive constant $c > 0$, and then

b) Replace each subformula of the form $\mathrm{E}(\varphi \mathrm{U} \psi)$ with the formula

$$\mu Z.(\psi \vee (\varphi \rhd Z)).$$

Recall that the remaining subformulas ($\mathrm{A}(\varphi \mathrm{R} \psi)$, $\mathrm{E}(\varphi \mathrm{R} \psi)$) are expressible with the above ones (see p. 71).

Further Reading

For an introduction to modal and temporal logic the reader is referred to [75, 86, 87]. Semantics of TCTL are discussed in [6, 7, 9]. The logics $\mathrm{TCTL}_\mathcal{C}$ as well as $T\mu$ are introduced in [83]. A survey of other real-time logics can be found in [12, 81]. Some comparisons of temporal logics can be found also in [69, 116, 142].

Part II

Model Generation and Verification

5

Abstract Models

In this chapter we define abstract models for timed systems and show how to generate them. Unlike concrete models, abstract ones are (usually) finite and possibly minimal w.r.t. the properties they preserve. Therefore, these models are typically used for verifying properties by means of model checking. In what follows, abstract models for time Petri nets are considered in Sect. 5.1, whereas these for timed automata – in Sect. 5.2.

Let $C(\mathcal{T}) = (S, s^0, \rightarrow_{\mathfrak{c}})$ be a concrete state space of a timed system $\mathcal{T}$, where $\rightarrow_{\mathfrak{c}}$ is a B-labelled transition relation for a given set of labels B,[1] and let $M_{\mathfrak{c}}(\mathcal{T}) = (C(\mathcal{T}), V_{\mathfrak{C}})$ be a concrete model for $\mathcal{T}$ based on the state space $C(\mathcal{T})$ and defined over a set of propositional variables PV. Denote by $Reach(\mathcal{T})$ the set of all the states reachable in $C(\mathcal{T})$. In order to give definitions of some abstract models, and to define some others in a more efficient way, we introduce a notion of *reachability* of a set of concrete states. A set $U \subseteq S$ is then called *reachable* if $U \cap Reach(\mathcal{T}) \neq \emptyset$.[2] Moreover, given a family of sets $\mathcal{U} \subseteq 2^S$, let $Reach(\mathcal{U})$ denote the set of all the reachable elements of $\mathcal{U}$, i.e.,

$$Reach(\mathcal{U}) = \{U \in \mathcal{U} \mid U \cap Reach(\mathcal{T}) \neq \emptyset\}.$$

Then, we introduce the following definition:

Definition 5.1. *An* abstract model *of a timed system $\mathcal{T}$ over a set of propositional variables PV is a structure*

$$M_{\mathfrak{a}}(\mathcal{T}) = ((W, w^0, \rightarrow_{\mathfrak{a}}), V_a),$$

where

[1] $B = A \cup \mathbb{R}_{0+}$ in case of timed automata and $B = T \cup \mathbb{R}_{0+}$ in case of time Petri nets.

[2] Such a definition is essential for introducing pseudo-bisimulating and pseudo-simulating models, whereas in the case of bisimulating, simulating and surjective models it allows to restrict the requirements put on them to a part of the state space only.

W. Penczek and A. Półrola: *Abstract Models*, Studies in Computational Intelligence (SCI) **20**, 89–154 (2006)
www.springerlink.com

- $W \subseteq 2^S$ *is a set of* abstract states, *which are sets of concrete states of* S,
- $\rightarrow_{\mathfrak{a}} \subseteq W \times B \times W$ *is an* abstract transition relation,
- $V_a : W \longrightarrow 2^{PV}$ *is a* valuation function, *and*
- $w^0 \in W$ *is the* initial abstract state,

satisfying the following conditions:

- $s^0 \in w^0$,
- $V_a(w) = V_{\mathfrak{C}}(s)$ *for each* $w \in W$ *and* $s \in w$.

Unless otherwise stated, we consider abstract models such that

$$Reach(\mathcal{T}) \subseteq \bigcup_{w \in W} w,$$

i.e., such that for each reachable concrete state of $\mathcal{T}$ there is an abstract state the concrete one belongs to.

Usually, besides the conditions listed above, we specify also some additional requirements on $\rightarrow_{\mathfrak{a}}$, which depend on the properties to be preserved by $M_{\mathfrak{a}}(\mathcal{T})$. The following two belong to the most widely used:

EE$_1$) for every $w_1, w_2 \in Reach(W)$ and each $b \in B$

$$\mathrm{EE}_1(w_1, w_2): \text{ if } w_1 \xrightarrow{b}_{\mathfrak{a}} w_2 \text{ then } (\exists s_1 \in w_1)(\exists s_2 \in w_2)\ s_1 \xrightarrow{b}_{\mathfrak{c}} s_2;$$

EE$_2$) for every $s_1, s_2 \in S$ and each $b \in B$

$$\mathrm{EE}_2(s_1, s_2): \text{ if } s_1 \xrightarrow{b}_{\mathfrak{c}} s_2 \text{ then } (\forall w_1 \in Reach(W) \text{ s.t. } s_1 \in w_1)\ (\exists w_2 \in W)(s_2 \in w_2 \wedge w_1 \xrightarrow{b}_{\mathfrak{a}} w_2).$$

Both the conditions are typically used together. EE$_1$ guarantees that each two reachable abstract states related by $\xrightarrow{b}_{\mathfrak{a}}$ contain representatives related by $\xrightarrow{b}_{\mathfrak{c}}$. EE$_2$ ensures that each reachable abstract state containing a representative with a concrete b-successor has a $\xrightarrow{b}_{\mathfrak{a}}$-successor, which gives us that all the transitions which can be taken at the elements of an abstract state have their counterparts in the abstract successor relation. It should be noticed, however, that these conditions are formulated such that they apply to the case when the elements of W are not disjoint (e.g., W is an arbitrary *covering*[3] of S, or of some $S' \subseteq S$ satisfying $Reach(\mathcal{T}) \subseteq S'$). In the case when W consists of disjoint elements, EE$_1$ and EE$_2$ are usually reformulated to the one joint condition:

EE) for every $w_1, w_2 \in Reach(W)$ and each $b \in B$

$$\mathrm{EE}(w_1, w_2): \ w_1 \xrightarrow{b}_{\mathfrak{a}} w_2 \text{ iff } (\exists s_1 \in w_1)(\exists s_2 \in w_2)\ s_1 \xrightarrow{b}_{\mathfrak{c}} s_2.$$

[3] By a *covering* of a set D we mean a family of its subsets $\mathcal{D}$ such that $\bigcup_{D' \in \mathcal{D}} D' = D$. Elements of $\mathcal{D}$ do not need to be disjoint.

An example of state spaces satisfying this condition are shown in Fig. 5.1.

Other conditions of our interest are listed below:

EA) for every $w_1, w_2 \in Reach(W)$ and each $b \in B$

$EA(w_1, w_2)$: if $w_1 \xrightarrow{b}_{\mathfrak{a}} w_2$ then $(\forall s_2 \in w_2)(\exists s_1 \in w_1)\ s_1 \xrightarrow{b}_{\mathfrak{c}} s_2$;

AE) for every $w_1, w_2 \in Reach(W)$ and each $b \in B$

$AE(w_1, w_2)$: if $w_1 \xrightarrow{b}_{\mathfrak{a}} w_2$ then $(\forall s_1 \in w_1)(\exists s_2 \in w_2)\ s_1 \xrightarrow{b}_{\mathfrak{c}} s_2$;

U) for every $w_1, w_2 \in Reach(W)$ and each $b \in B$

$U(w_1, w_2)$: if $w_1 \xrightarrow{b}_{\mathfrak{a}} w_2$ then

$$(\forall s_1 \in cor(w_1))(\exists s_2 \in cor(w_2))\ s_1 \xrightarrow{b}_{\mathfrak{c}} s_2,$$

for a function $cor : W \longrightarrow 2^S$ s.t. $cor(w) \subseteq w$, and $s^0 \in cor(w^0)$.

Again, EA, AE and U are formulated such that they apply to the general case, but if the elements of W are disjoint, the conditions are often modified by replacing implications by "if and only if". In what follows, both the versions of EA, AE and U are used.

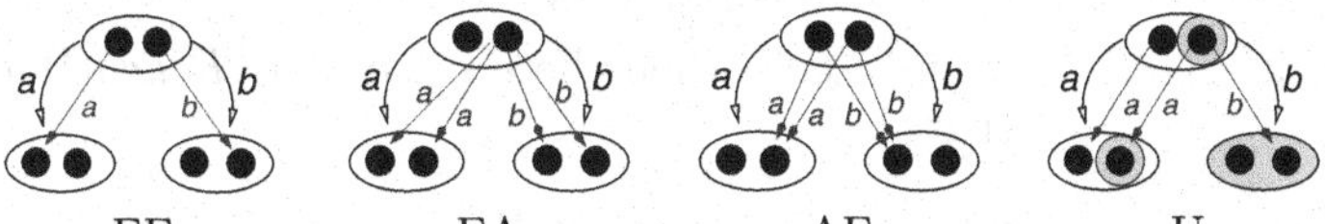

Fig. 5.1. The conditions EE, EA, AE and U in four abstract state spaces. The black dots represent concrete states and the straight lines denote the concrete successor relation, whereas the ellipses and the arcs – the abstract states and the abstract transition relation. The cores of abstract states are coloured

Examples of abstract state spaces satisfying the above conditions are shown in Fig. 5.1. The condition EA restricts the abstract transition relation (on the reachable part of W) to the pairs of states such that for each pair the elements of the successor state have an $\rightarrow_{\mathfrak{c}}^{-1}$-related[4] element in the predecessor state. This condition, together with the requirement that w^0 consists of s^0 and possibly some successors of the initial state, and the requirement of either EE_1 and EE_2 or EE, is put on abstract models to preserve LTL [32, 44, 150][5]. The resulting models are sometimes referred to as *surjective*. The condition AE, specifying the symmetric property w.r.t. the successor and the predecessor of each pair of states in the abstract transition relation, is known as a

[4] For the relation $\mathcal{R}$, by $\mathcal{R}^{-1}$ we denote its inverse.

[5] In some of these papers the proofs are given under the assumption that $w^0 = \{s^0\}$.

bisimulation one. So, it is used (together with EE_1 and EE_2, or with EE) to ensure preservation of CTL* or TCTL [9,49]. The condition U is a weakening of AE, which puts the same restriction as the latter, but only on a subset (called a *core*) of each abstract state. It is known as a *simulation condition.* Similarly, U (again, combined with EE_1 and EE_2, or with EE) is applied to preserve ACTL* or TACTL [117]. The models whose all the reachable states satisfy AE (U) are called *bisimulating* (*simulating*, respectively).

Obviously, both the surjective and (bi)simulating abstract models preserve reachability properties. However, for reachability verification, models with weaker requirements[6] on $\rightarrow_{\mathfrak{a}}$ are also sufficient. Some of them (mainly these relaxing the condition EA) are introduced for specific timed systems [45,63,98], whereas the others (e.g., pseudo-bisimulating models [125]) are defined for the general case. The main idea behind the definition of pseudo-bisimulating models consists in relaxing the condition on the transition relation on bisimulating models, formulated for all the predecessors of each reachable abstract state, such that it applies only to one of them, reachable from w^0 in the minimal number of steps. Let $dpt(w)$ denote the *depth* of $w \in W$, i.e., the minimal $i \in \mathbb{N}$ such that $w = w_i$ for some path $(w_0, w_1, \ldots)$ in $(W, w^0, \rightarrow_{\mathfrak{a}})$ with $w_0 = w^0$. Formally, pseudo-bisimulating models are abstract models satisfying EE_1 and EE_2 (or EE) together with the following condition (see also Fig. 5.2):

pAE) for every $w_1, w_2 \in Reach(W)$ and each $b \in B$
$pAE(w_1, w_2)$: if $w_1 \xrightarrow{b}_{\mathfrak{a}} w_2$ then there exists $w \in Reach(W)$ such that

- $w \xrightarrow{b'}_{\mathfrak{a}} w_2$ for some $b' \in B$,
- the depth of w is minimal in the set

$$\{dpt(w') \mid w' \xrightarrow{b''}_{\mathfrak{a}} w_2 \text{ for some } b'' \in B\},$$

and

- $(\forall s \in w)(\exists s_2 \in w_2)\ s \xrightarrow{b'}_{\mathfrak{c}} s_2$.

Another example of reachability-preserving models are *pseudo-simulating models* [126], combining the definitions of simulating and pseudo-bisimulating ones. Formally, they satisfy either EE_1 and EE_2 or EE, together with the condition below:

pU) for every $w_1, w_2 \in Reach(W)$ and each $b \in B$
$pU(w_1, w_2)$: if $w_1 \xrightarrow{b}_{\mathfrak{a}} w_2$ then there exists $w \in Reach(W)$ such that

- $w \xrightarrow{b'}_{\mathfrak{a}} w_2$ for some $b' \in B$,

[6] Clearly, weaker conditions give rise to smaller abstract models.

- the depth of w is minimal in the set

$$\{dpt(w') \mid w' \xrightarrow{b''}_{\mathfrak{a}} w_2 \text{ for some } b'' \in B\},$$

  and
- $(\forall s \in cor(w))(\exists s_2 \in cor(w_2))\ s \xrightarrow{b'}_{\mathfrak{c}} s_2$, for a function $cor : W \longrightarrow 2^S$ s.t. $cor(w) \subseteq w$ and $s^0 \in cor(w^0)$.

Again, in the case when the elements of W are disjoint, pAE and pU can be reformulated by replacing the implication by the equivalence. Examples of abstract state spaces satisfying the above conditions are shown in Fig. 5.2. Notice that in both the cases the condition $\mathrm{EE}_1(w_1, w_2)$ (or $\mathrm{EE}(w_1, w_2)$) holds only (but neither $\mathrm{AE}(w_1, w_2)$, (nor $\mathrm{U}(w_1, w_2)$, respectively), since there is another predecessor of w_2 of a depth smaller than w_1 for which the condition $\mathrm{AE}(w_1, w_2)$ ($\mathrm{U}(w_1, w_2)$, respectively) holds.

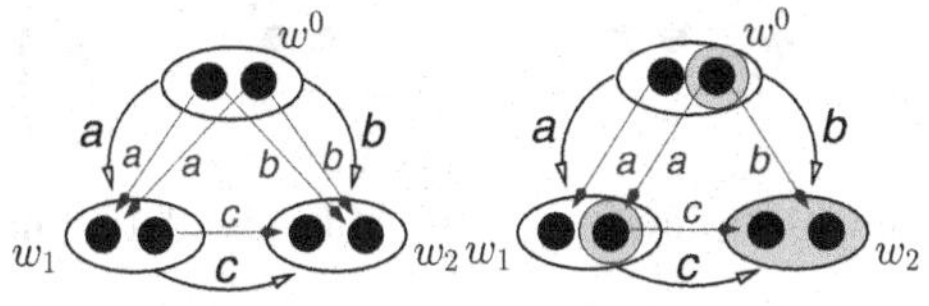

Fig. 5.2. Conditions on transition relation in pseudo-bisimulating and pseudo-simulating models

The next example illustrates the differences between the above-defined abstract models.

Example 5.2. Consider the concrete model shown in Fig. 5.3(a). The dots represent concrete states, and the straight lines denote the concrete successor relation. The set of labels of $\rightarrow_{\mathfrak{c}}$ is given by $B = \{a, b, c, d, e, f, g, h\}$. Assume that for each $b' \in B$ the valuations of all the states reached by executing a b'-labelled transition (i.e., the values of the function $V_{\mathfrak{C}}$ for the b'-successor states) are the same. Moreover, the states obtained by b- and e-labelled transitions share the valuations (this is marked by the light-coloured dots in the picture), whereas for all the other pairs of labels the valuations differ. The figure displays the differences between various kinds of abstract models built for the above-mentioned concrete one. The ellipses and the arcs represent the abstract states and the abstract transition relation. The cores of abstract states are coloured.

□

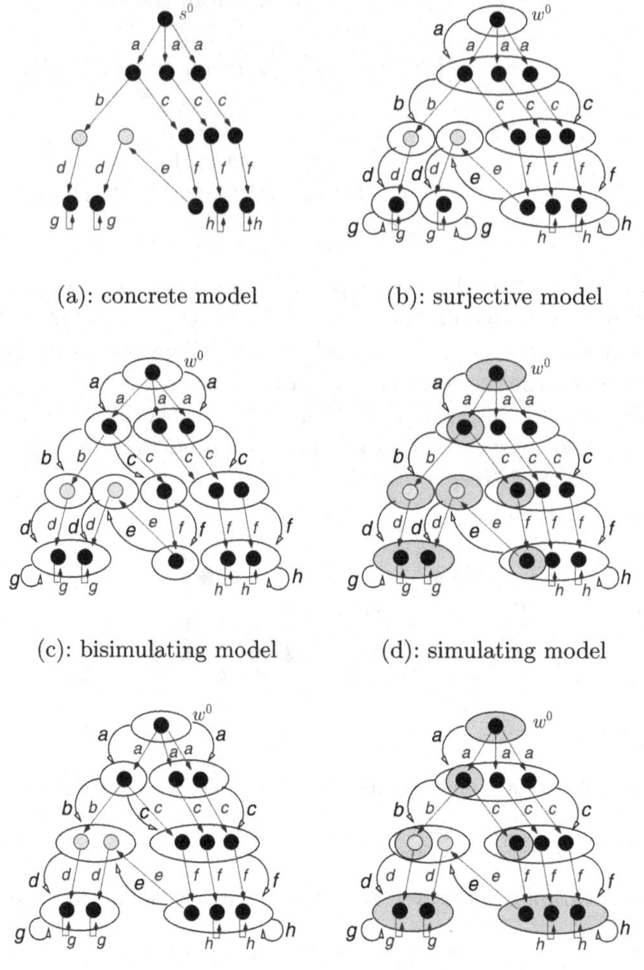

(a): concrete model (b): surjective model

(c): bisimulating model (d): simulating model

(e): pseudo-bisimulating model (f): pseudo-simulating model

Fig. 5.3. Various abstract models built for the same underlying concrete one

It is important to mention that the concrete models underlying the abstract ones can be not only discrete or dense, but also *time-abstracted* dense, i.e., obtained from concrete dense models by a simplification of the successor relation resulting from treating the time transitions in a special way:

- The typical approach consists in abstracting away the exact amount of time passed between any two time-related states and replacing all the labels $\delta \in \mathbb{R}_{0+}$ in the time successor relation by the common label $\tau \notin B$. This gives us a *strong time-abstracted* (*strong ta-*) concrete successor relation and a *strong time-abstracted* (*strong ta-*) concrete model.

Other simplifications introduce also some modifications to the action successor relation:

- The first of them, resulting in a *delay time-abstracted* (*delay ta-*) successor relation and a *delay time-abstracted* (*delay ta-*) model, besides abstracting away the amount of time as described above, combines together passing some time and then an action step, analogously to the relations $\rightarrow_{Rd_1}$ and $\rightarrow_{d_1}$ (see pp. 21, 40).
- Another solution, besides abstracting away the exact amount of time passed, considers as action steps combinations of passing some time, an action step, and then possibly passing some time again, analogously to the relations $\rightarrow_{Rd_2}$ and $\rightarrow_{d_2}$ (see pp. 21, 40). The resulting successor relation and model are called *observational time-abstracted* (*observational ta-*).

Figure 5.4 illustrates the differences between the time-abstracted and discrete successor relations. In each of the pictures (a)-(e) its left-hand part depicts the states related by the concrete dense successor relation $\rightarrow_c$, whereas the right-hand part shows the states which are in the successor relation under consideration. The discrete successor relations $\rightarrow_{d_1}$ and $\rightarrow_{d_2}$ are presented, respectively, in the pictures (d) and (e). The subfigures (a), (b) and (c) illustrate, respectively, the time-abstracted, time-abstracted delay and time-abstracted observational successor relation. Notice that, unlike the discrete ones, all these relations involve time steps.

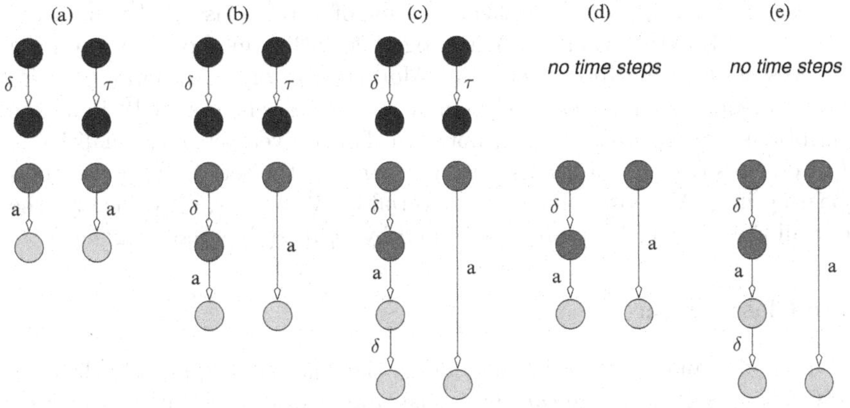

Fig. 5.4. Time-abstracted (a-c) and discrete (d-e) successor relations

In what follows, we adopt the convention that abstract models inherit the names of their underlying concrete ones (e.g., we obtain strong ta-bisimulating models, discrete pseudo-simulating models etc.). It is also important to mention that the kind of the underlying concrete model influences the properties preserved by a given abstract one (e.g., neither delay nor observational ta-bisimulating models preserve $\mathrm{CTL}^*_{-\mathrm{X}}$ [159], but both the models preserve reachability). On the other hand, abstract models preserving the same properties (e.g., reachability) built for different underlying concrete models can be of different sizes. This justifies variety of their kinds.

Next, we present the main approaches to generating various types of abstract models for timed systems.

5.1 Model Generation for Time Petri Nets

We start with discussing *state class* approaches for generating abstract models. Then, we review some remaining approaches to defining these models for time Petri nets.

5.1.1 State Class Approaches

The methods of building abstract models for time Petri nets are usually based on the *state class approach*, which consists in representing an abstract state by a marking and a set of linear inequalities. The algorithms for building such models differ w.r.t. approaches to their concrete semantics (which, however, is usually discrete) and various restrictions on the definition of TPNs. For each of the given approaches, we follow the same pattern of the presentation. Thus, we discuss a semantics applied, a notion of a state class, and a condition on firability of a transition at a class together with a method of computing its successor resulting from this firing. Moreover, in order to obtain a finite abstract model, an equivalence relation on state classes is provided[7]. In all the descriptions, we use the following notation. For a given set of inequalities I, $sol(I)$ denotes the set of its solutions, and $var(I)$ – the set of all the variables appearing in I. Moreover, for a set of variables $\mathcal{V}$, by $Iqs(\mathcal{V})$ we denote the set of all the inequalities built over $\mathcal{V}$, and by $IqSets(\mathcal{V})$ – the set $2^{Iqs(\mathcal{V})}$.

State Class Graph

The method of building this basic model of the state class approach (denoted by SCG, and called also *linear state class graph, LSCG* in [35]) was defined by Berthomieu at al. [32, 33] for unrestricted TPNs (see Definition 1.3) with infinite capacity of the places, and is based on the discrete firing intervals semantics $\rightarrow_{Fd_1}$ (see pp. 15, 21). A *state class* is defined as a pair

[7] Notice, however, that the equivalence relation does not guarantee finiteness of the model; see e.g. Lemma 5.4.

$$C = (m, I),$$

where m is a marking, and I is a set of inequalities built over the set of variables

$$var(I) = \{v_t \mid t \in en(m)\}.$$

All the possible values of a variable v_t in the set $sol(I)$ (called the *firing domain* of C) form the timing interval, relative to the time when C was entered, in which t can be fired (or, in other words, specify the minimal and the maximal period of time to be spent in the class before firing t). Intuitively, a class can be seen as a set of concrete states, the union of the firing intervals of which is equal to the firing domain of the class. However, given a concrete state $\sigma^F = (m', fi)$ and a class $C = (m, I)$, we are only able to say whether σ^F can potentially belong to C. This can happen if $m' = m$ and for each $t \in en(m)$ the firing interval of t (i.e., $fi(t)$) is included in $sol(I)|_{v_t}$. In order to determine whether $\sigma^F \in C$ we have to consider also the sequence of transitions $\eta \in T^*$ whose firing at the initial marking resulted in creating C: we have that $\sigma^F \in C$ if it can be obtained from $(\sigma^F)^0$ by firing η. Notice also that given a pair (m, I) only, we cannot determine which concrete states belong to a class this pair describes (this can be seen from Example 5.3). Thus, the pairs (m, I) define in fact equivalence classes of sets of concrete states [31].

The initial class of the state class graph is given by

$$C^0 = (m^0, \{\text{``}Eft(t) \leq v_t \leq Lft(t)\text{``} \mid t \in en(m^0)\}).$$

A transition $t \in en(m)$ is *firable* at a class $C = (m, I)$ if the set of inequalities

$$I_1 = I \cup \{\text{``}v_t \leq v_{t_*}\text{``} \mid t_* \in en(m) \setminus \{t\}\}$$

is consistent (i.e., $sol(I_1) \neq \emptyset$), which intuitively means that t can fire earlier than any other enabled transition. For the class $C' = (m', I')$, resulting from firing t at C, the marking is $m' = m[t\rangle$, whereas I' is obtained from I in the following four steps:

1. the set I' is initialized to I_1 given above (i.e., to the set obtained from I by adding the firability condition for t),
2. all the variables in I' are substituted by new ones reflecting the fact of firing t at the time given by v_t (i.e., $v'_{t_*} = v_{t_*} - v_t$ for all $t_* \neq t$ with $t_* \in en(m)$), which relates the values of the variables to the time the class C' was entered,
3. the variables corresponding to the transitions disabled by firing of t are eliminated, and
4. the system is extended by the set of inequalities

$$\{\text{``}Eft(t_*) \leq v_{t_*} \leq Lft(t_*)\text{``} \mid t_* \in newly\_en(m, t)\},$$

which adds also the new variables corresponding to the transitions in $newly\_en(m, t)$ (see p. 5 for the definition).

Two classes C, C' are considered as *equivalent* (denoted $C \equiv_S C'$) if their markings and firing domains are equal. Equivalent classes have the same descendant trees in the state class graph, but can correspond to different sets of concrete states (see Example 5.3).

Example 5.3. Consider the net[8] $\mathcal{N}$ shown in Fig. 5.5. Below, we show a part of

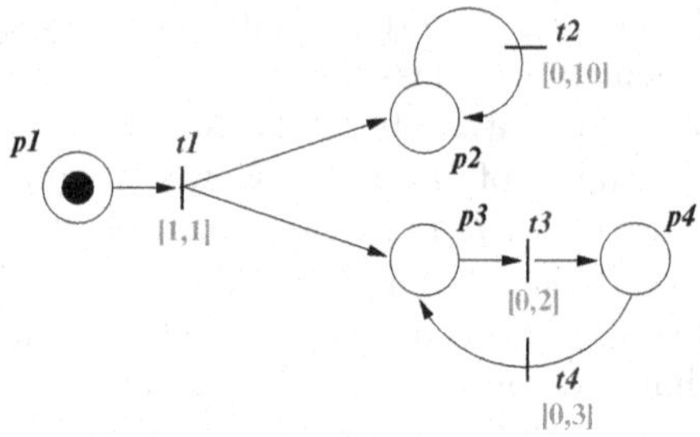

Fig. 5.5. The time Petri net considered in Example 5.3

the state class graph for this net. For simplicity of the presentation, a marking m is displayed as a vector $(m(p_1), m(p_2), m(p_3), m(p_4))$, the set of inequalities of a class (m, I) as the conjunction of its elements, and the function fi of a concrete state (m, fi) – as a vector $(fi(t_1), fi(t_2), fi(t_3), fi(t_4))$ (see p. 15 for the definition of concrete states).

- The initial class of the SCG for $\mathcal{N}$ is given by
  $C_0 = ((1,0,0,0), 1 \le v_{t_1} \le 1)$,
  and corresponds to the concrete state $(\sigma^F)^0 = ((1,0,0,0), ([1,1], [\,], [\,], [\,]))$.
- Firing t_1 at C_0 results in the class
  $C_1 = ((0,1,1,0), 0 \le v_{t_2} \le 10 \ \wedge \ 0 \le v_{t_3} \le 2)$.
  The class corresponds to the concrete state $\sigma_1^F = ((0,1,1), ([\,], [0,10], [0,2], [\,]))$ obtained from $(\sigma^F)^0$ by passing one time unit and then firing t_1.
- Then, firing t_3 at C_1 gives us the class
  $C_2 = ((0,1,0,1), 0 \le v_{t_2} \le 10 \ \wedge \ 0 \le v_{t_4} \le 3)$.
  The class contains all the concrete states obtained by firing t_3 at σ_1^F. Notice that this firing can lead to different states, depending on how much time passes before t_3 fires. Thus, waiting two units of time leads to the state $\sigma_2^F = ((0,1,0,1), ([\,], [0,8], [\,], [0,3]))$, one unit of time – to $\sigma_3^F = ((0,1,0,1), ([\,], [0,9], [\,], [0,2]))$, firing t_3 immediately after its enabling leads to $\sigma_4^F = ((0,1,0,1), ([\,], [0,10], [\,], [0,2]))$ etc. In fact, C_2 corresponds to an infinite set of concrete states given by $\{((0,1,0,1), ([\,], [0,a], [\,], [0,2])) \mid a \in [8,10]\}$.

[8] The net in the figure is not progressive, but this does not influence the method of building the state class graph. Such a net is considered in order to give a simple example.

- Firing t_4 at C_2 results in the class
  $C_3 = ((0,1,1,0), 0 \leq v_{t_2} \leq 10 \ \wedge \ 0 \leq v_{t_3} \leq 2)$,
  which is equivalent to C_1. However, C_3 corresponds to the different set of concrete states than C_1. For example, immediate firing t_4 at σ_2^F gives us the state $\sigma_5^F = ((0,1,1,0)([\,],[0,8],[0,2],[\,]))$, and at σ_4^F – the state $\sigma_6^F = ((0,1,1,0),([\,],[0,10],[\,],[0,2]))$; if one unit of time passes before t_4 fires, then from σ_2^F we obtain $\sigma_7^F = ((0,1,1,0),([\,],[0,7],[0,2],[\,]))$, from σ_4^F - the state $\sigma_8^F = ((0,1,1,0),([\,],[0,9],[0,2],[\,]))$ etc.

The above shows that equivalent classes can contain different sets of concrete states. Moreover, it is even possible that one of them is a singleton, whereas another one corresponds to an infinite number of states.

□

The conditions on finiteness of a state class graph are given by the following lemma [32]:

Lemma 5.4. *The number of the state classes of a TPN is finite (up to the equivalence) if and only if the net is bounded.*

The intuition is that in a bounded net the number of reachable markings is finite, and since the number of different firing domains of a net is finite as well [32], the whole SCG is finite.

Although the boundedness problem for TPNs is undecidable [93], the paper [32] provides some sufficient conditions. One of them, which can be checked on-the-fly while generating the state class graph, is stated below. The notation $m > m'$, where m, m' are markings, means that $(\forall p \in P)(m(p) \geq m'(p)) \wedge (\exists p \in P)(m(p) > m'(p))$.

Lemma 5.5. *A TPN is bounded if no pair of state classes $C = (m, I)$ and $C' = (m', I')$ satisfying the conditions*

- *C' is reachable from C,*
- *$m' > m$,*
- *$sol(I) = sol(I')$,*
- *$(\forall p \in \{p' \in P \mid m'(p') > m(p')\}) \ m(p) > \max_{t \in T}(F(p,t))$*

is reachable from the initial class.

Intuitively, the first three conditions on C, C' express that there is a sequence of transitions the firing of which (starting at the class C) leads to the class C' with the same firing domain, but with a different marking such that C' contains at least the same number of tokens in any place as C does. Additionally, the fourth condition requires that all the places, for which the number of tokens increases in C', contain more tokens in C than can be consumed by firing of any outgoing transition. The same sequence of transitions can thus be fired infinitely many times, giving infinitely many non-equivalent classes (with different markings).

It should be noticed that the method for generating state class graphs is correct also for nets whose maximal capacity of the places is finite. In this case, however, the graph is always finite, which follows from Lemma 5.4 in a straightforward way.

A pseudo-code of the algorithm for generating the state class graph for a time Petri net $\mathcal{N}$ is given in Fig. 5.6. The classes of the graph are stored in the variable `classes`, which is initially equal to the empty set. Calling the recursive procedure `dfs` for the class C^0 results in building the graph in the depth first search order. The method consists in generating a maximal path starting at a given node, then backtracking to the nearest "fork" (i.e.,

---

INPUT ARGUMENTS:
    a time Petri net $\mathcal{N} = (P, T, F, m^0, Eft, Lft)$
GLOBAL VARIABLES:
    $classes$: $2^{RM_{\mathcal{N}} \times IqSets(\{v_t | t \in T\})}$
RETURN VALUES:
    build_SCG(), dfs(): a graph of nodes of $RM_{\mathcal{N}} \times IqSets(\{v_t \mid t \in T\})$

```
1.    procedure build_SCG(𝒩) is
2.    begin
3.        classes := ∅;
4.        dfs(C⁰);
5.    end build_SCG;

6.    procedure dfs(C′) is
7.    begin
8.    for each C ∈ classes do
9.        if the conditions of Lemma 5.5 hold for C and C′ then
10.           terminate;        // the net is likely to be unbounded
11.       end if;
12.   end do;
13.   if (∀C ∈ classes) C′ ≢ₛ C then
14.       classes := classes ∪ {C′};
15.       for each t firable at C′ do
16.           compute the t-successor C″ of C′;
17.           dfs(C″);
18.       end do;
19.   else
20.       handle equivalence of classes;
21.   end if;
22.   end dfs;
```

Fig. 5.6. An algorithm for generating state class graph of a time Petri net

to a node on this path which is of a maximal depth of these which do not have all the successors generated), generating a new maximal path starting at this node etc. The process of building the graph includes also on-the-fly checking of boundedness of the net (i.e., testing whether for a given class C' there is a predecessor C such that C, C' satisfy the conditions of Lemma 5.5) and terminating if the net is likely to be unbounded (and therefore to be of an infinite number of state classes). Moreover, the successors are generated only for these classes which are not equivalent to some of these generated previously. Otherwise, an equivalence-handling procedure is executed. Given a class C' equivalent to an existing class C, the procedure redirects to C' the edges previously incoming C, and then deletes C'.

Example 5.6. Consider the time Petri net shown in Fig. 5.7.

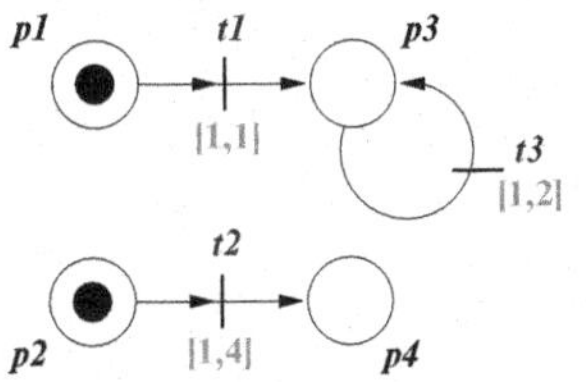

Fig. 5.7. The time Petri net considered in Example 5.6

The process of constructing the state class graph for this net is shown in Fig. 5.8. The graph on the left is an "intermediate" structure, i.e., its classes which are equivalent to others are still present. The right-hand side of the

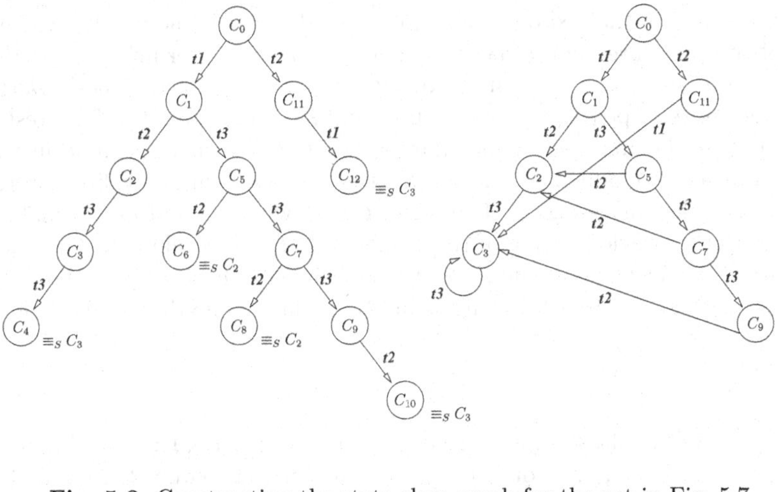

Fig. 5.8. Constructing the state class graph for the net in Fig. 5.7

figure depicts the graph obtained by deleting these classes and redirecting their incoming edges to the appropriate representatives.

Given a class (m, I), for simplicity of the presentation we show the marking m as a vector of the form $(m(p_1), m(p_2), m(p_3), m(p_4))$, whereas the set of inequalities I is given as the conjunction of its elements. The classes of the above state class graph are listed below:

$$\begin{array}{lcl}
C_0 &=& ((1,1,0,0),\ 1 \leq v_{t_1} \leq 1 \ \wedge\ 1 \leq v_{t_2} \leq 4),\\
C_1 &=& ((0,1,1,0),\ 0 \leq v_{t_2} \leq 3 \ \wedge\ 1 \leq v_{t_3} \leq 2),\\
C_2 &=& ((0,0,1,1),\ 0 \leq v_{t_3} \leq 2),\\
C_3 &=& ((0,0,1,1),\ 1 \leq v_{t_3} \leq 2),\\
C_4 &=& ((0,0,1,1),\ 1 \leq v_{t_3} \leq 2),\\
C_5 &=& ((0,1,1,0),\ 0 \leq v_{t_2} \leq 2 \ \wedge\ 1 \leq v_{t_3} \leq 2),\\
C_6 &=& ((0,0,1,1),\ 0 \leq v_{t_3} \leq 2),\\
C_7 &=& ((0,1,1,0),\ 0 \leq v_{t_2} \leq 1 \ \wedge 1 \leq v_{t_3} \leq 2),\\
C_8 &=& ((0,0,1,1),\ 0 \leq v_{t_3} \leq 2),\\
C_9 &=& ((0,1,1,0)),\ 0 \leq v_{t_2} \leq 0 \ \wedge\ 1 \leq v_{t_3} \leq 2,\\
C_{10} &=& ((0,0,1,1),\ 1 \leq v_{t_3} \leq 2),\\
C_{11} &=& ((1,0,0,1),\ 0 \leq v_{t_1} \leq 0),\\
C_{12} &=& ((0,0,1,1),\ 1 \leq v_{t_3} \leq 2).
\end{array}$$

The inequalities describing a class give timing intervals in which the enabled transitions can be fired. For example, the inequalities of C_0 express that the transition t_1 can be fired in one time unit, and t_2 between one and four time units after its enabling (i.e, after the net starts). Firing of t_1 at C_0 gives us the class C_1, at which t_2 can fire either immediately after the class is entered (this follows from passing one unit of time before firing t_1 at C_0), or after some delay not exceeding three time units. The timing interval assigned to t_3 is equal to $[1, 2]$, since t_3 is newly enabled. Firing of t_2 leads to the class C_2, at which t_3 is enabled only, and can be fired either immediately (which is possible if t_2 was fired not earlier than one unit of time after firing t_1) or after a delay of at most two time units since C_2 has been entered (where waiting two time units is possible if t_2 was fired immediately after t_1). The rest of the graph can be analysed in a similar way. Notice also that it can occur that some transitions enabled at the marking of a class cannot by fired at this class. This takes place e.g. in the case of C_9, at which t_2 and t_3 are enabled, but due to the previous firing of the sequence t_1, t_3, t_3, t_3 taking at least (and in this case exactly) four time units, the transition t_2 has to fire without any additional delay, whereas for firing t_3 in C_9 a delay of one time unit would be necessary.

□

Since the state class graph satisfies the conditions EE_1, EE_2 and EA up to the equivalence (i.e., the conditions hold provided the equivalence-handling procedure has not been executed), the model preserves the LTL formulas.

Preservation of the conditions by the state class graph requires an additional explanation. It is easy to see from the above description that the equivalence-handling procedure, called for a class C' and a previously generated C, can influence preservation of the conditions: $\mathrm{EA}(C, C_1)$ is satisfied for each C_1 which is a successor of C, but $\mathrm{EA}(C_2, C)$ can obviously fail to hold for each C_2 which is a predecessor of C and was previously a predecessor of C', since equivalent classes correspond to different sets of concrete states. Moreover, the above operation on equivalent classes can possibly make even the condition EE_1 (or EE, if applies) fail when EA is not satisfied. A similar situation occurs in most of the below-described constructions of graphs exploiting equivalences of classes. However, violation of the conditions is strictly a technical problem, and the models obtained preserve the same properties as these without the equivalence of classes handled[9].

It should be also mentioned that if we are not interested in preservation of all the LTL formulas, the above method of building the state class graph can be improved by checking whether a newly obtained class is included in an existing one, and not generating its successors in this case [31]. The resulting models preserve reachability properties.

Geometric Region Graph

In the terminology of abstract models for time Petri nets, by *atomic* we call a class C such that the condition $\mathrm{AE}(C, C')$ holds for all its successors C'. A model is *atomic* if it satisfies EE_1 and EE_2 (or EE), and all its classes (before combining equivalent ones) are atomic (i.e., the model satisfies AE). However, the information carried by firing domains of the classes of a state class graph is not sufficient to check their atomicity. To this aim, some additional information about the histories of firings needs to be added, which requires a modification of the state class graph. Such a model for 1-safe nets with finite values of the function Lft, introduced by Yoneda and Ryuba in [174], is called a *geometric region graph.* The construction exploits the discrete transition relation $\rightarrow_{Td_1}$ and the semantics in which clocks are associated with the transitions of the net (see pp. 11, 21). Now, the state classes are defined as triples

$$C = (m, I, \eta),$$

where I is a set of inequalities, and m is a marking obtained by firing from m^0 the sequence $\eta \in T^*$ of transitions. The variables in $var(I)$ represent the absolute firing times (i.e., counted since the net started) of the transitions in η. Note that different firings of the same transition are then distinguished; $v_t^j \in var(I)$ corresponds to the j-th firing of $t \in T$. Unlike in the construction

[9] Recall that we check reachability of sets of places only. Reachability of a concrete state obviously does not need to be preserved, since an equivalence of classes does not necessarily mean their equality.

of a state class graph, I can be viewed as describing the history of the states of a given class rather than these states as such. Thus, the inequalities used in the process of building the state class graph are modified in such a way that they specify the timing conditions which were to hold to enter the class they describe. Intuitively, a state class $C = (m, I, \eta)$ represents a set of concrete states which are obtained from $(\sigma^T)^0$ by firing the sequence η of transitions, satisfying the timing constraints represented by I.

The initial state class is given by

$$C^0 = (m^0, \emptyset, \epsilon),$$

where ϵ is the empty sequence of transitions. Firing of $t \in en(m)$ at a class $C = (m, I, \eta)$ is given in terms of the *parents* of the enabled transitions, i.e., the transitions in η which most recently made the transitions in $en(m)$ enabled (or, more precisely, in terms of the variables corresponding to the times at which these transitions were fired). As the parent of the transitions enabled at m^0 one assumes a fictitious transition ν, which denotes the start of the net. The time it fires (often taken to be 0) is represented by the variable v_ν. Let $parent(v_t, C)$ denote the variable corresponding to the parent of t. If t appears $k-1$ times in η, then it is firable at C iff it is enabled and for any transition $t_* \in en(m)$ which appears $l-1$ times in η the set

$$I \cup \{\text{``}parent(v_t^k, C) + Eft(t) \leq parent(v_{t_*}^l, C) + Lft(t_*)\text{``}\}$$

is consistent, which intuitively means that t can be fired earlier than any other enabled transition. If t is fired at C, then the class $C' = (m', I', \eta')$ is reached, where $m' = m[t\rangle$, $\eta = \eta t$, and the set of inequalities I' is equal to

$$\begin{aligned} I \cup \{&\text{``}Eft(t) \leq v_t^k - parent(v_t^k, C) \leq Lft(t)\text{``}\} \\ \cup \{&\text{``}v_t^k \leq parent(v_{t_*}^l, C) + Lft(t_*)\text{``} \mid t_* \in en(m) \\ &\qquad\qquad \text{and } t_* \text{ appears } l-1 \text{ times in } \eta\}. \end{aligned}$$

The set I' describes timing conditions which need to hold for firing t at C (i.e., the history of the class C').

Two classes $C = (m, I, \eta)$, $C' = (m', I', \eta')$ are equivalent (denoted $C \equiv_G C'$) if their markings are equal, the enabled transitions possess the same parents[10], and the sets $sol(I)$ and $sol(I')$ projected on the sets of the parents of the transitions enabled at C and C', respectively, are equal (which intuitively means that the constraints on differences between firing times of the parents of the enabled transitions are the same in both the classes[11]).

[10] This requirement is provided in [174] for technical reasons, but omitted in the implementation used in that paper, since for the further behaviour of the net it is meaningless whether the transitions which enabled a given one were the same in both the classes. The modified equivalence relation is denoted by $\equiv_G'$.

[11] Such an interpretation agrees with the implementation of the method.

The equivalent classes have the same descendant trees in the geometric region graph, but they can correspond to different sets of concrete states (see Example 5.7).

The algorithm for generating the geometric region graph is like that in Fig. 5.6 without the lines 8–12 (since the algorithm is applicable to 1-safe nets only, the number of inequivalent classes is always finite). The geometric region graph satisfies the conditions EE_1, EE_2 and EA before any operation on equivalent classes is applied (see p. 102 for an explanation).

Example 5.7. Consider again the net used in Example 5.6. A geometric region graph for this net is presented in Fig. 5.9. The left-hand part of the picture displays the graph in which equivalent classes are still present, whereas in the graph in the right-hand part the equivalent classes are merged into one.

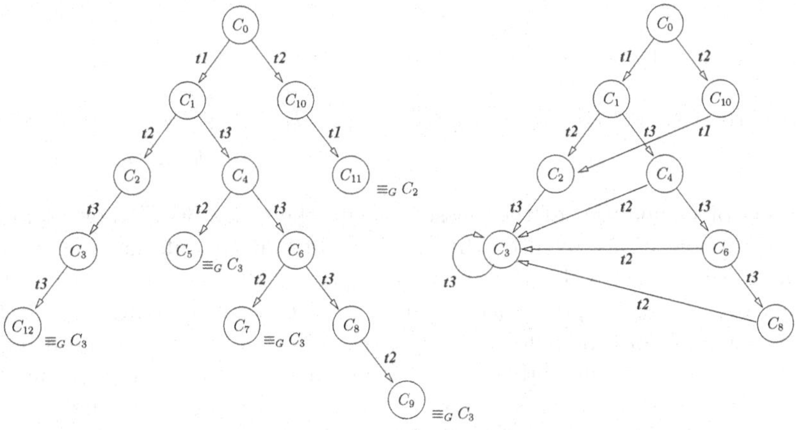

Fig. 5.9. Constructing the geometric region graph for the net in Fig. 5.7

The classes of the graph are listed below. Again, a marking m is given as a vector $(m(p_1), m(p_2), m(p_3), m(p_4))$. The superscripts of the variables corresponding to transitions which cannot be fired more than once (i.e., t_1 and t_2) are omitted for simplicity.

$$
\begin{aligned}
C_0 &= ((1,1,0,0), I_0, \epsilon), && \text{where } I_0 = \emptyset, \\
C_1 &= ((0,1,1,0), I_1, t_1), && \text{where } I_1 = \{\text{``}1 \le v_{t_1} - v_\nu \le 1\text{``}, \\
& && \qquad\qquad \text{``}v_{t_1} \le v_\nu + 4\text{``}\}, \\
C_2 &= ((0,0,1,1), I_2, t_1 t_2), && \text{where } I_2 = I_1 \cup \{\text{``}1 \le v_{t_2} - v_\nu \le 4\text{``}, \\
& && \qquad\qquad \text{``}v_{t_2} \le v_{t_1} + 2\text{``}\},
\end{aligned}
$$

$$
\begin{aligned}
C_3 &= ((0,0,1,1), I_3, t_1t_2t_3), && \text{where } I_3 = I_2 \cup \{\text{“}1 \le v_{t3}^1 - v_{t_2} \le 2\text{“}\},\\
C_{12} &= ((0,0,1,1), I_{12}, t_1t_2t_3t_3), && \text{where } I_{12} = I_3 \cup \{\text{“}1 \le v_{t3}^2 - v_{t3}^1 \le 2\text{“}\},\\
C_4 &= ((0,1,1,0), I_4, t_1t_3), && \text{where } I_4 = I_1 \cup \{\text{“}1 \le v_{t3}^1 - v_{t_1} \le 2\text{“}, \text{“}v_{t3}^1 \le v_\nu + 4\text{“}\},\\
C_5 &= ((0,0,1,1), I_5, t_1t_3t_2), && \text{where } I_5 = I_4 \cup \{\text{“}1 \le v_{t_2} - v_\nu \le 4\text{“}, \text{“}v_{t_2} \le v_{t3}^1 + 2\text{“}\},\\
C_6 &= ((0,1,1,0), I_6, t_1t_3t_3), && \text{where } I_6 = I_4 \cup \{\text{“}1 \le v_{t3}^2 - v_{t3}^1 \le 2\text{“}, \text{“}v_{t3}^2 \le v_\nu + 4\text{“}\},\\
C_7 &= ((0,0,1,1), I_7, t_1t_3t_3t_2), && \text{where } I_7 = I_6 \cup \{\text{“}1 \le v_{t_2} - v_\nu \le 4\text{“}, \text{“}v_{t_2} \le v_{t3}^2 + 2\text{“}\},\\
C_8 &= ((0,1,1,0), I_8, t_1t_3t_3t_3), && \text{where } I_8 = I_6 \cup \{\text{“}1 \le v_{t3}^3 - v_{t3}^2 \le 2\text{“}, \text{“}v_{t3}^3 \le v_\nu + 4\text{“}\},\\
C_9 &= ((0,0,1,1), I_9, t_1t_3t_3t_3t_2), && \text{where } I_9 = I_8 \cup \{\text{“}1 \le v_{t_2} - v_\nu \le 4\text{“}, \text{“}v_{t_2} \le v_{t3}^3 + 2\text{“}\},\\
C_{10} &= ((1,0,0,1), I_{10}, t_2), && \text{where } I_{10} = \{\text{“}1 \le v_{t_2} - v_\nu \le 4\text{“}, \text{“}v_{t_2} \le v_\nu + 1\text{“}\},\\
C_{11} &= ((0,0,1,1), I_{11}, t_2t_1), && \text{where } I_{11} = I_{10} \cup \{\text{“}1 \le v_{t_1} - v_\nu \le 1\text{“}, \text{“}v_{t_1} \le v_\nu + 4\text{“}\}.
\end{aligned}
$$

The sets of inequalities of the classes specify their "histories". For example, the timing conditions given by the set I_1 mean that t_1 could have been fired at the initial marking if one unit of time passed after the net started. The history of the class C_4, obtained by firing t_3 at C_1, says that besides passing exactly one unit of time before firing t_1, the passage of time between firing t_1 and t_3 was between one and two units (this was necessary to fire t_3), but the total time passed between the start of the net and firing t_3 was not greater that four units (which prevented disabling t_2 by passage of time). Notice also that C_8 has no t_3-successor in the graph, since the transition t_3 is not firable (it is easy to check that the set $I_8 \cup \{\text{“}v_{t3}^3 + 1 \le v_\nu + 4\text{“}\}$ is inconsistent).

Some comments on the equivalence of classes are in order. Firstly, the classes C_4, C_6 and C_8 are not equivalent in spite of the same enabled transitions (t_2 and t_3) whose parents are also the same (ν and t_3, respectively). It is easy to compute that in C_4 the difference between the variables corresponding to the parents of the enabled transitions is given by $2 \le v_{t3}^1 - v_\nu \le 3$, in C_6 – by $3 \le v_{t3}^2 - v_\nu \le 4$, and in C_8 – by $4 \le v_{t3}^3 - v_\nu \le 4$. Secondly, if only one transition (here t_3) is enabled at two classes, then these classes are equivalent if the parent of this transition in both of them is the same (e.g., $C_2 \equiv_G C_{11}$ and $C_3 \equiv_G C_{12} \equiv_G C_4 \equiv_G C_7 \equiv_G C_9$). However, applying the less restricted relation $\equiv'_G$ (see Footnote 10) can lead to smaller graphs (cf. Fig. 5.10, which displays such a graph for the net considered above).

Let $((m(p_1), m(p_2), m(p_3), m(p_4)), (clock^T(t_1), clock^T(t_2), clock^T(t_3)))$ denote the state $(m, clock^T)$. It is easy to see from the above example that equiv-

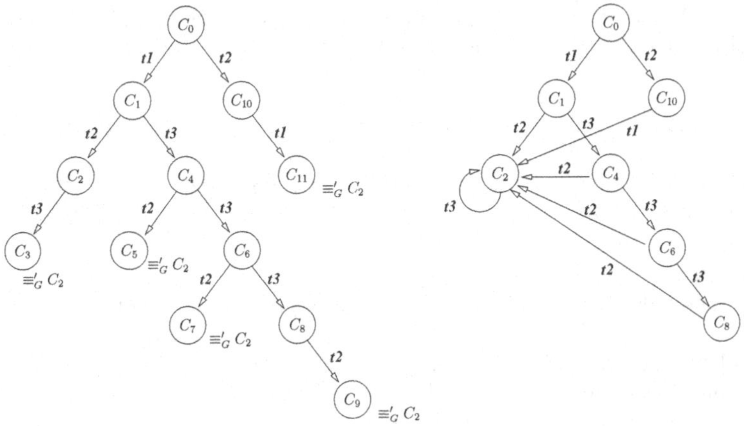

Fig. 5.10. The geometric region graph of the modified equivalence relation, built for the net in Fig. 5.7

alent classes can contain different concrete states. For instance, the class C_2 contains the state $((0,0,1,1),(3,3,2))$ obtained from $(\sigma^T)^0 = ((1,1,0,0),(0,0,0))$ by waiting one unit of time, firing t_1, waiting two units of time and then firing t_2. However, the class C_{11} does not contain such a state, since all the states which belong to this class satisfy $clock^T(t_3) = 0$ (the class is entered by firing t_1, which enables t_3 and therefore sets the "clock" of t_3 to 0).

□

Atomic State Class Graph

In order to obtain a CTL*-preserving structure, a geometric region graph needs to be refined to make all its classes atomic. The model obtained this way is called an *atomic state class graph* (ASCG) [174]. If a class $C = (m, I, \eta)$ is not atomic, then there is some inequality ξ such that satisfaction of ξ is necessary for the concrete states in C to have descendants in a successor C' of C, but the solution sets of both $I \cup \xi$ and $I \cup \neg\xi$ are non-empty. The class C is then split into

$$C_1 = (m, I \cup \xi, \eta) \text{ and } C_2 = (m, I \cup \neg\xi, \eta).$$

The descendants of C_1 and C_2 are computed from copies of the descendants of C, by modifying their sets of inequalities adding ξ and $\neg\xi$, respectively. As a result, the atomic state class graph satisfies (up to the equivalence, see p. 102), besides EE_1 and EE_2, also both the conditions AE and EA.

A pseudo-code of the algorithm for building atomic state class graph is shown in Fig. 5.11. In fact, building the geometric region graph and making

Fig. 5.11. An algorithm for generating atomic state class graph of a (1-safe) time Petri net (continued on the next page)

INPUT ARGUMENTS:
  a (1-safe) time Petri net $\mathcal{N} = (P, T, F, m^0, Eft, Lft)$
GLOBAL VARIABLES:
  $classes$: $2^{RM_{\mathcal{N}} \times IqSets(\{v_t^i | t \in T \wedge i \in \mathbb{N}\} \cup \{v_\nu\})}$
RETURN VALUES:
  build_ASCG(), dfs_a(), partitioning(): a graph of nodes of
      $RM_{\mathcal{N}} \times IqSets(\{v_t^i \mid t \in T \wedge i \in \mathbb{N}\} \cup \{v_\nu\})$
  find_ineqs(): $IqSets(\{v_t^i \mid t \in T \wedge i \in \mathbb{N}\} \cup \{v_\nu\})$
  propagated_ineq(): $Iqs(\{v_t^i \mid t \in T \wedge i \in \mathbb{N}\} \cup \{v_\nu\})$
  parent_class(), the_class_to_be_split():
      $RM_{\mathcal{N}} \times IqSets(\{v_t^i \mid t \in T \wedge i \in \mathbb{N}\} \cup \{v_\nu\})$

```
1.    procedure build_ASCG(N) is
2.    begin
3.        classes := ∅;
4.        dfs_a(C⁰);
5.    end build_ASCG;

6.    procedure dfs_a(C′) is
7.    begin
8.    if (∀C ∈ classes) C′ ≢_G C then
9.        classes := classes ∪ {C′};
10.       for each t firable at C′ do
11.           partitioning(C′,t);        // C′ can be modified or deleted
12.           if C′ deleted or became equivalent to another class then
13.               break;        // backtracking
14.           end if;
15.           if t is firable at C′ then
16.               compute the t-successor C″ of C′;
17.               dfs_a(C″);
18.           end if;
19.       end do;
20.   else
21.       handle equivalence of classes;
22.   end if;
23.   end dfs_a;

24.   procedure partitioning(C′,t) is
25.   begin
26.       iqs := find_ineqs(C′, t);
27.       while (∃ξ ∈ iqs) do
28.           C := the_class_to_be_split(ξ);        // assume C = (m, I, η)
```

```
29.       if sol(I ∪ ξ) ≠ ∅ ∧ sol(I ∪ ¬ξ) ≠ ∅ then
30.         C1 := (m, I ∪ ξ, η); C2 := (m, I ∪ ¬ξ, η);
31.         ξ1 := propagated_ineq(ξ); ξ2 := propagated_ineq(¬ξ);
32.         C'' := parent_class(C);        // assume C'' = (m'', I'', η'')
33.         if sol(I'' ∪ ξ1) ≠ ∅ then      // C1 is a successor of C''
34.           make C1 the successor of C'' in the graph;
35.           if (∃C3 ∈ classes) C3 ≡G C1 then
36.             handle equivalence of classes;
37.           else
38.             replace the edges from C by edges from C1 in the graph;
39.             iqs := iqs ∪ {ξ1};
40.             classes := classes \ {C} ∪ {C1};
41.             C := C1;        // this includes deleting C
42.             use ξ to modify the subtree rooted at C1;        // this involves
43.                   // handling (in)equivalence and possible emptiness of classes
44.           end if;
45.         else
46.           delete C1;
47.         end if;
48.         if sol(I'' ∪ ξ2) ≠ ∅ then      // C2 is a successor of C''
49.           make C2 the successor of C'' in the graph;
50.           iqs := iqs ∪ {ξ2};
51.           classes := classes \ {C}; delete C;
52.           dfs_a(C2);
53.         else
54.           delete C2;
55.         end if;
56.       end if;
57.       iqs := iqs \ {ξ};
58.     end do;
59. end partitioning;
```

its classes atomic is performed in parallel: before firing a transition t at a class C, it is checked whether t can be fired at all the concrete states of C, and the class is split if necessary. Thanks to this, when the successors of C are generated, the class is atomic w.r.t. them. The successors are computed analogously as in the geometric region graph.

The partitioning process is a bit more involved. Firstly, when the procedure `partitioning` is called for a class C' and a transition t, there is possibly more than one inequality which can make the class non-atomic (in fact, each of these inequalities expresses that t can fire earlier than another enabled transition). They are computed in line 26 of the algorithm and stored in the set *iqs* (each element of this set "remembers" also the class to be split w.r.t. it). Moreover, splitting of C' can make some of its predecessors non-atomic. Due to this, when a class C (which can be either C' or one of its predecessors) is partitioned, new inequalities ξ_1, ξ_2 that need to be satisfied for all the concrete states in the "parent" C'' of C to have descendants respectively in C_1 and C_2, can be added to iqs (see lines 31, 39, 50). Partitionings are performed as long as the set iqs is non-empty. Notice, however, that an element of iqs does not necessarily cause partitioning of the corresponding class. This is tested in line 29.

Secondly, when C is partitioned, ξ_1 and ξ_2 are used to check whether C_1 and C_2 are successors of C''. One of the classes (i.e., the one the states of which satisfy ξ) possibly replaces C in the graph (and in the procedure `dfs_a` called for C), whereas for the second the procedure `dfs_a` can be run. Notice, moreover, that since all the modifications, introduced to the graph by the procedure `partitioning` called for a class C' and a transition t, can potentially cause that t is not firable at the class which replaces C', the algorithm tests this firability after backtracking to `dfs_a` (line 15).

Finally, the procedure `partitioning` needs to deal with the equivalence of classes. The class C_1 obtained by splitting C can occur to be equivalent to another class. This is not tested after returning to `dfs_a` (this procedure tests equivalence only once, for the class for which it is called), and therefore needs to be checked and handled in this procedure. Moreover, the subtree rooted at C so far is modified by adding ξ to the sets of inequalities of its classes, and the modified classes need to be checked w.r.t. the equivalence (notice that it is also possible that the previous relations are broken). Empty classes (i.e., that whose sets of inequalities are inconsistent) are removed from the graph.

Example 5.8. Consider again the net $\mathcal{N}$ shown in Fig. 5.7 and the geometric region graph for this net (see Example 5.7). In the process of generating an atomic state class graph, the algorithm splits the class C_6 of the geometric region graph, using the inequality ξ = "$v_{t_3}^2 - v_\nu \leq 3$". The explanation for this is as follows: after firing the sequence of transitions $\eta = t_1t_3t_3$, it is possible to fire both t_2 and t_3. However, t_3 can occur only if the preceding passage of time does not disable t_2, which implies that the previous firing of t_3 (i.e., the second one in η) could take place not later than three time units after the net started. Thus, C_6 is partitioned, which involves adding ξ to I_6, and creating a new class

$$C_{13} = ((0,1,1,0), I_{13}, t_1t_2t_3), \text{ where } I_{13} = I_6 \cup \{\text{"}v_{t_3}^2 - v_\nu > 3\text{"}\},$$

at which the transition t_2 is firable only. The successors of C_6 are modified, and these of C_{13} are generated.

In further steps of the algorithm partitionings are propagated towards the root of the graph. So, the class C_4 is split using the inequality ξ' = "$v_{t_3}^1 - v_\nu \leq 2$". This, in turn, can be explained by the fact that if in the sequence $\eta' = t_1t_3$ the transition t_3 fires not later than two time units after the net starts, then no further passage of time which allows for the next firing of t_3 can disable t_2, and therefore the predecessors of concrete states belonging to the modified class C_6 have to satisfy the condition given by ξ'. Therefore, the sets of inequalities of C_4 and its successors are modified by adding ξ'. Moreover, the algorithm creates a new class

$$C_{15} = ((0,1,1,0), I_{15}, t_1 t_3), \text{ where } I_{15} = I_4 \cup \{\text{“}v^1_{t_3} - v_\nu > 2\text{“}\},$$

and generates its successors. The resulting graph is presented in Fig. 5.12.

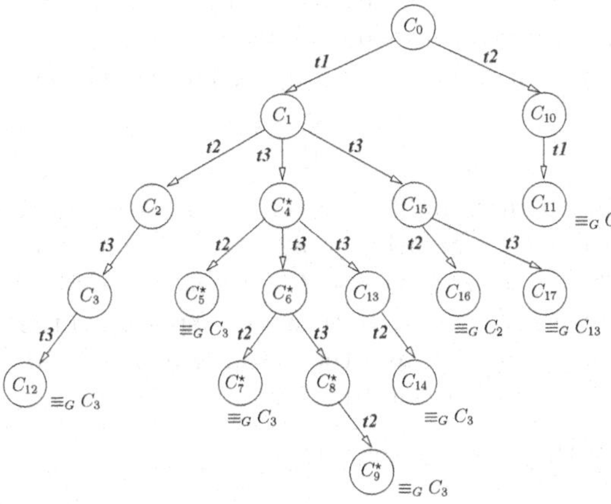

Fig. 5.12. The atomic state class graph for the net in Fig. 5.7

The classes C_i which appear both in the geometric region graph of Fig. 5.9 and in the ASCG, but in the latter their sets of inequalities are modified, are marked with small stars.

□

Pseudo-atomic State Class Graph

The atomic state class graph's construction can be further modified to generate *pseudo-atomic state class graphs*, which (up to the equivalence) satisfy, besides EE_1 and EE_2, the condition U instead of AE [117]. The models are built in a way similar to atomic state class graphs, but in some cases instead of splitting the classes only their cores are refined. A pseudo-code of the algorithm is similar to that in Fig. 5.11, but with a different procedure `partitioning`. The successor of a class in `dfs_a` is computed in the same way as in the geometric region graph, and the core of the newly generated successor is set to equal to the whole class. Moreover, two classes C and C' are equivalent in the pseudo-atomic state class graph if $C \equiv_G C'$ in the atomic state class graph, and if their cores are equal.

Strong State Class Graph

Berthomieu and Vernadat introduced another method of building abstract models, applicable to the general class of TPNs with no restriction on the capacity of the places (transitions with infinite latest firing times require, however, a special treatment) [35]. The models, called *strong state class graphs* (SSCGs), can be then further refined to satisfy the condition AE. The solution can be seen as a combination of the state class graph and the geometric region graph approach. The definition of concrete states, as well as the semantics applied, follow the approach of state class graphs. The model consists of classes of the form

$$C = (m, I),$$

where I is a set of inequalities built over the set of variables corresponding to the transitions in $en(m)$, similarly to [32, 33]. However, the value of the variable v_t corresponding to a transition $t \in en(m)$ gives the time elapsed since t has been last enabled, which, in turn, corresponds to the approach of geometric region graphs. The set I and the marking m enable to compute the set of the concrete states that belong to the class.

The initial class of the graph is given by

$$C^0 = (m^0, \{\text{``}0 \leq v_t \leq 0\text{``} \mid t \in en(m^0)\}).$$

Firability of a transition t at a class, as well as the set of inequalities of the successor, are defined in terms of the times elapsed since the transitions have been enabled, using additional temporary variables denoting possible firing times of t. More precisely, $t \in en(m)$ is firable at a class $C = (m, I)$ if the set of inequalities

$$I_1 = I \cup \{\text{``}\theta \geq 0\text{``}\} \cup \{\text{``}Eft(t) \leq v_t + \theta\text{``}\} \cup \{\text{``}v_{t_*} + \theta \leq Lft(t_*)\text{``} \mid t_* \in en(m)\}$$

is consistent. Intuitively, the variable θ represents possible firing times of the transition t. For the class $C' = (m', I')$ resulting from firing t at C, we have $m' = m[t\rangle$, whereas I' is obtained from I in the following steps:

1. the set I' is initialized to I_1 given above (i.e., to the set obtained from I by adding the firability condition for t). This introduces the new variable θ;
2. for each $t_* \in en(m')$ a new variable v'_{t_*} is introduced, and I' is augmented by the inequalities:[12]
   - $v'_{t_*} = v_t + \theta$ if $t_* \neq t$ and $m(p) - F(p,t) \geq F(p, t_*)$ for each $p \in \bullet t_*$ (this corresponds to passing θ time units since C has been entered);
   - $0 \leq v'_{t_*} \leq 0$ otherwise;
3. the variables θ and v_{t_*}, for $t_* \in en(m)$, are eliminated.

[12] Obviously, an equality is then treated as the conjunction of two inequalities.

Two classes (m, I) and (m', I') are equivalent if they describe the same sets of concrete states. This is easily tested when the values of Lft are finite only: then, the classes are equivalent if $m = m'$ and $sol(I) = sol(I')$ (the opposite case is explained below). Similarly to the previous approaches, all the classes in the strong state class graph satisfy the conditions EE_1, EE_2 and EA. Thus, the model preserves LTL, and so reachability properties.

In the case of a net without infinite latest firing times of the transitions, its SSCG is finite iff the net is bounded [35]. However, if infinite Lfts are allowed, boundedness of the net is not sufficient to ensure finiteness of the SSCG, as the number of possible intervals specifying the time elapsed since a transition t with $Lft(t) = \infty$ became enabled, can be infinite, which, in turn, can result in an infinite number of the classes. However, these classes can correspond to the same sets of concrete states[13]. To cope with this, the transitions with infinite Lfts are treated in a special way: for a given class $C = (m, I)$, let

$$en^{\infty}(m) = \{t \in en(m) \mid Lft(t) = \infty\}.$$

Moreover, for $e \subseteq en^{\infty}(m)$ let I_e denote the set of inequalities obtained from I by applying the following steps:

1. initializing I_e as

$$I_e = I \cup \{\text{“}v_t < Eft(t)\text{“} \mid t \in e\} \cup \{\text{“}v_t \geq Eft(t)\text{“} \mid t \in en^{\infty}(m) \setminus e\};$$

2. eliminating all the variables v_t for $t \in en^{\infty}(m) \setminus e$;
3. adding to I_e the set of inequalities $\{\text{“}v_t \geq Eft(t)\text{“} \mid t \in en^{\infty}(m) \setminus e\}$.

The class C is replaced by

$$Rlx(C) = \{(m, I_e) \mid e \subseteq en^{\infty}(m)\}.$$

Intuitively, applying the first step of the above procedure corresponds to partitioning C into subsets consisting of the concrete states in which the same groups of transitions of infinite latest firing times reached or exceeded their Efts (which means that in these states the same transitions of infinite latest firing times are assigned firing intervals $[0, \infty)$). Then, the steps 2 and 3 relax the lower and upper bound of the intervals constraining all these variables v_t which correspond to a transition t that reached its earliest firing time, i.e., replace these bounds by $Eft(t)$ and ∞, respectively. This is called *relaxation* of the state class C. The operation requires redefining the equivalence of classes. Two classes $C = (m, I)$ and $C' = (m', I')$ are then equivalent iff the unions of the elements obtained by the relaxation of each of them are equal (i.e., if $m = m'$ and $\bigcup_{(m_*, I_*) \in Rlx(C)} sol(I_*) = \bigcup_{(m_*, I_*) \in Rlx(C')} sol(I_*)$), which again

[13] Recall that we are dealing with the firing interval semantics.

means that C and C' correspond to the same set of concrete states[14]. Applying the relaxation while building the SSCG ensures finiteness of the graph if the net is bounded.

Obviously, the method of building SSCGs is applicable also to the nets of a finite capacity of the places. The results on finiteness of the graph for this case come from the above considerations in a straightforward way.

Example 5.9. Consider again the net of Fig. 5.7. The strong state class graph for this net is presented in Fig. 5.13. The classes of the graph are listed below (the description used is analogous to the previous examples).

$$C_0 = ((0,0,1,1), 0 \leq v_{t_1} \leq 0 \ \wedge \ 0 \leq v_{t_2} \leq 0),$$
$$C_1 = ((0,1,1,0), 1 \leq v_{t_2} \leq 1 \ \wedge \ 0 \leq v_{t_3} \leq 0),$$
$$C_2 = ((0,0,1,1), 0 \leq v_{t_3} \leq 2),$$
$$C_3 = ((0,0,1,1), 0 \leq v_{t_3} \leq 0),$$
$$C_4 = ((0,1,1,0), 2 \leq v_{t_2} \leq 3 \ \wedge \ 0 \leq v_{t_3} \leq 0),$$
$$C_5 = ((0,1,1,0), 3 \leq v_{t_2} \leq 4 \ \wedge \ 0 \leq v_{t_3} \leq 0),$$
$$C_6 = ((0,0,1,1), 0 \leq v_{t_3} \leq 1),$$
$$C_7 = ((0,1,1,0), 4 \leq v_{t_2} \leq 4 \ \wedge \ 0 \leq v_{t_3} \leq 0),$$
$$C_8 = ((1,0,0,1), 1 \leq v_{t_1} \leq 1).$$

In this case, the constraints on the variables corresponding to the transitions specify the time which possibly passed since the transitions became enabled.

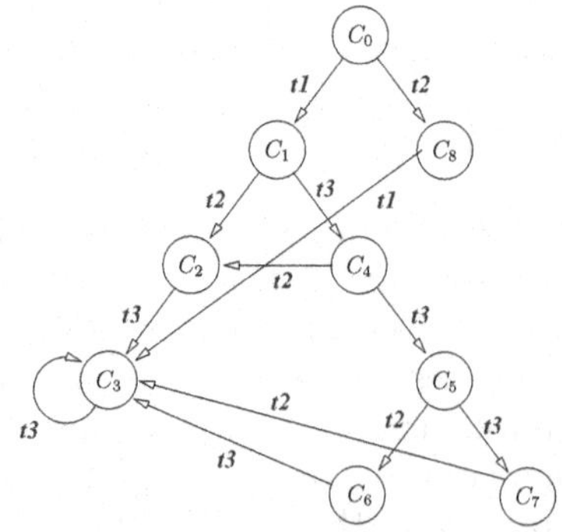

Fig. 5.13. The strong state class graph for the net in Fig. 5.7

[14] It is worth noticing that checking the above conditions is, in practice, not easy to implement. Thus, the SSCG is built in the following way: if a transition t is fired at a class C and the resulting class C' needs to be relaxed, then C gets multiple t-successors, each of which corresponds to one element of $Rlx(C')$. The equivalence of classes is then implemented analogously to the case of finite $Lfts$ (see p. 113).

The class C_0 corresponds to the start of the net (i.e., it contains the initial state $(\sigma^F)^0$ only). The set of inequalities of its successor C_1, obtained by firing t_1, makes the class to contain the state which can be reached from the initial one by passing one time unit and then firing t_1 (which implies that the time elapsed since t_2 became enabled is equal to 1, whereas t_3 became newly enabled). The rest of the classes can be analysed in a similar way.

□

If we are not interested in preserving all the LTL formulas, then the process of building the above graph can be further improved, e.g., by checking whether a new class C is included in an existing one and not generating its successors if so; or by grouping together the classes (with the same marking) whose union is convex [79]. These solutions correspond usually to *abstractions* for timed automata described in Sect. 5.2.3. The resulting models preserve reachability properties.

Strong Atomic State Class Graph

In order to obtain a *strong atomic state class graph*, (i.e., a strong class graph satisfying AE), the above model is refined in a way similar to that of a geometric region graph, i.e., its classes are partitioned until all of them become atomic. However, unlike in an atomic geometric region graph, the classes satisfy EE_1, EE_2 and AE, but not necessarily EA, as an inequality, added to the set I of some class $C = (m, I)$ which is partitioned, is not propagated to the descendants of C.

5.1.2 Other Approaches – an Overview

Besides the state class methods, other approaches to building abstract models for time Petri nets also exist. In many cases they correspond to the solutions known for timed automata and described in the following section. This includes, e.g., the detailed region graph for time Petri nets [111, 163] (see Sect. 5.2.1 for a corresponding definition for TA), or a method for computing the state spaces of time Petri nets based on a construction of the forward-reachability graph for timed automata (see Sect. 5.2.3) [73]. Another branch includes methods defined for time Petri nets which do not have any counterpart for timed automata, like a method for building reachability graph based on the so-called *firing points* of the transitions [43].

5.2 Model Generation for Timed Automata

In this section we define abstract models for timed automata, which can be used for both enumerative and SAT-based symbolic verification. We start with a detailed region graph approach. The main reason for this is that some of SAT-related methods (see Chap. 7) are based on a propositional encoding of detailed regions.

5.2.1 Detailed Region Graphs

Given a timed automaton $\mathcal{A} = (A, L, l^0, E, \mathcal{X}, \mathcal{I})$ and a TCTL or TCTL$_\mathcal{C}$ formula φ. We consider later two interval semantics for TCTL: weakly and strongly monotonic. Let $\mathcal{C}_\mathcal{A} \subseteq \mathcal{C}_\mathcal{X}^\ominus$ be a non-empty set containing all the clock constrains occurring in any enabling condition or in any state invariant of $\mathcal{A}$, and let $c_{max}(\mathcal{A})$ denote the largest constant appearing in them. Moreover, let $c_{max}(\mathcal{A}, \varphi)$ be the largest constant[15] appearing in $\mathcal{C}_\mathcal{A}$ and in any time interval in φ of TCTL and in any clock constraint in φ of TCTL$_\mathcal{C}$. For $\delta \in \mathbb{R}_{0+}$, $frac(\delta)$ denotes the fractional part of δ, and $\lfloor \delta \rfloor$ denotes its integral part.

Definition 5.10 (Equivalence of clock valuations). *For two clock valuations $v, v' \in \mathbb{R}_{0+}^{n_\mathcal{X}}$, $v \simeq_{\mathcal{C}_{\mathcal{A},\varphi}} v'$ iff for all $x, x' \in \mathcal{X}$ the following conditions are met:*

1. $v(x) > c_{max}(\mathcal{A}, \varphi)$ *iff* $v'(x) > c_{max}(\mathcal{A}, \varphi)$,
2. *if* $v(x) \leq c_{max}(\mathcal{A}, \varphi)$ *then*
   a) $\lfloor v(x) \rfloor = \lfloor v'(x) \rfloor$,
   b) $frac(v(x)) = 0$ *iff* $frac(v'(x)) = 0$,
3. *for each clock constraint* $\mathfrak{cc} \in \mathcal{C}_\mathcal{X}^\ominus$ *of the form* $x - x' \sim c$ *with* $c \in \mathbb{N}$, $\sim \in \{<, \leq, =\geq, >\}$, *and* $c \leq c_{max}(\mathcal{A}, \varphi)$ *we have* $(v \models \mathfrak{cc} \iff v' \models \mathfrak{cc})$.

When $\mathcal{A}$ is diagonal-free (i.e., $\mathcal{C}_\mathcal{A} \subseteq \mathcal{C}_\mathcal{X}$), the condition 3. in the above definition is replaced by

3'. if $v(x) \leq c_{max}(\mathcal{A}, \varphi)$ then $frac(v(x)) \leq frac(v(x'))$ iff $frac(v'(x)) \leq frac(v'(x'))$.

Intuitively, the conditions 1., 2., 3'. specify that two clock valuations are equivalent if either

- the values of the same clocks are greater than $c_{max}(\mathcal{A}, \varphi)$, or
- they agree on their integral parts, and the orderings of the fractional parts are the same.

The reason for such a definition is that the integral parts allow to determine whether or not a particular invariant or enabling condition is satisfied, whereas the ordering of the fractional parts is necessary to decide which clock will change its integral part first when the time passes.

[15] Obviously, ∞ is not considered as a constant.

The intuition behind the conditions 1., 2., 3. in Definition 5.10 is similar, besides that the equivalent clock valuations are additionally required to satisfy the same atomic constraints in which the difference of two clocks is compared with a value not exceeding $c_{max}(\mathcal{A}, \varphi)$, even in the case when the value of a single clock is greater than $c_{max}(\mathcal{A}, \varphi)$. This follows from the form of invariants and enabling conditions of the automaton considered. Notice that the condition 3' can be seen as a restriction of the condition 3. of Definition 5.10 to the case of clock valuations in which none of the clocks exceeds $c_{max}(\mathcal{A}, \varphi)$.

Having defined the equivalence of clock valuations, we can formulate the main lemma guaranteeing preservation of TCTL formulas for both the weakly and strongly interval semantics as well as of $\mathrm{TCTL}_\mathcal{C}$ formulas.

Lemma 5.11 (Preserving TCTL **[7, 83]).** *Let $\mathcal{A} = (A, L, l^0, E, \mathcal{X}, \mathcal{I})$ be a timed automaton, $V_\mathcal{A}$ be a valuation function for $\mathcal{A}$, $M_c(\mathcal{A})$ be the concrete dense model of $\mathcal{A}$, and φ be a* TCTL *or* $\mathrm{TCTL}_\mathcal{C}$ *formula. Moreover, let $l \in L$, and $v, v' \in \mathbb{R}_{0+}^{n_\mathcal{X}}$ with $v \simeq_{c_{\mathcal{A},\varphi}} v'$. Then, we have*

$$M_c(\mathcal{A}), (l, v) \models \varphi \textit{ iff } M_c(\mathcal{A}), (l, v') \models \varphi.$$

Next, we define finite abstract models preserving TCTL and $\mathrm{TCTL}_\mathcal{C}$. The equivalence classes of the relation $\simeq_{c_{\mathcal{A},\varphi}}$ are called *detailed zones.* The set of all the detailed zones is denoted by $DZ(n_\mathcal{X})$. As it is easy to see, detailed zones are time zones, so $DZ(n_\mathcal{X}) \subseteq Z(n_\mathcal{X})$. Moreover, it is possible to find an upper bound on the number of detailed zones: e.g., for the diagonal-free case it is not greater than $n_\mathcal{X}! \times 2^{n_\mathcal{X}} \times (2c_{max}(\mathcal{A}, \varphi) + 2)^{n_\mathcal{X}}$.

The set of detailed zones for $c_{max}(\mathcal{A}, \varphi) = 1$ and (respectively) a non-diagonal-free and a diagonal-free TA of two clocks x_1, x_2 is presented in Fig. 5.14.

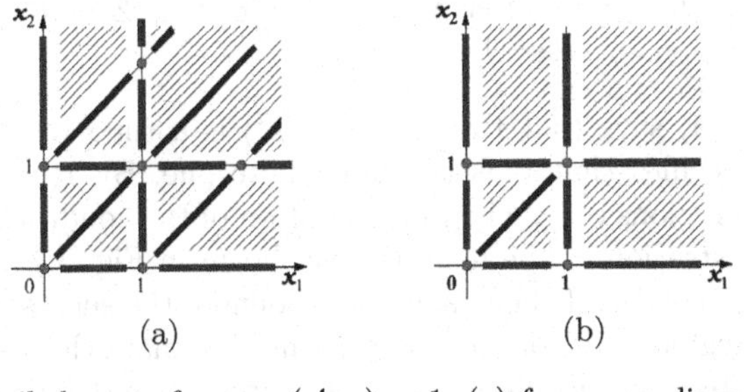

Fig. 5.14. Detailed zones for $c_{max}(\mathcal{A}, \varphi) = 1$: (a) for a non-diagonal free TA (b) for a diagonal-free TA

A detailed zone $Z \in DZ(n_\mathcal{X})$ is *final* if for all $v \in Z$ and $x \in \mathcal{X}$ we have $v(x) > c_{max}(\mathcal{A}, \varphi)$ and for each $c > c_{max}(\mathcal{A}, \varphi)$ there is $v \in Z$ such that $v(x) > c$ for all $x \in \mathcal{X}$. A detailed zone $Z \in DZ(n_\mathcal{X})$ is *open* if there is

$x \in \mathcal{X}$ such that all $v \in Z$ satisfy the condition $v(x) > c_{max}(\mathcal{A}, \varphi)$ and for each $c > c_{max}(\mathcal{A}, \varphi)$ there is $v \in Z$ and $x \in \mathcal{X}$ such that $v(x) > c$. The *initial detailed zone* $Z^0 \in DZ(n_{\mathcal{X}})$ is defined as $Z^0 = \{v^0\}$.

A *detailed region* is a pair (l, Z), where $l \in L$ and $Z \in DZ(n_{\mathcal{X}})$. The action- and time successor relation in the set of the detailed regions can be defined via representatives, which gives us a finite abstract model, called the (*detailed*) *region graph* (DRG). The model preserves TCTL or TCTL$_C$, provided that its transition relation reflects the definition of the weakly or strongly monotonic semantics used to interpret the formulas of the logic. Below we introduce the two alternative approaches.

The definition of the detailed region graph corresponding to the weakly monotonic semantics looks as follows:

Definition 5.12. *The* (detailed) region graph *of a timed automaton $\mathcal{A}$ is a finite transition system*

$$DRG(\mathcal{A}) = (W, w^0, \rightarrow_{\triangleleft}),$$

where

- $W = L \times DZ(n_{\mathcal{X}})$,
- $w^0 = (l^0, Z^0)$ *and*
- $\rightarrow_{\triangleleft} \subseteq W \times (A \cup \{\tau\}) \times W$ *is defined as follows:*
  1. $(l, Z) \xrightarrow{\tau}_{\triangleleft} (l, Z')$ *iff there is* $v \in Z$ *and* $v' \in Z'$ *such that*
     a) $(l, v) \xrightarrow{\delta}_c (l, v')$ *for some* $\delta \in \mathbb{R}_{0+}$, *and*
     b) *if* $(l, v) \xrightarrow{\delta'}_c (l, v'') \xrightarrow{\delta''}_c (l, v')$ *with* $\delta', \delta'' \in \mathbb{R}_{0+}$ *and* $(l, v'') \in w$ *with* $w \in W$, *then* $v \simeq_{C_{\mathcal{A},\varphi}} v''$ *or* $v' \simeq_{C_{\mathcal{A},\varphi}} v''$, *and*
     c) *if* $v \simeq_{C_{\mathcal{A},\varphi}} v'$, *then* $v \simeq_{C_{\mathcal{A},\varphi}} v' + \delta''$ *for each* $\delta'' \in \mathbb{R}_{0+}$
     (time successor),
  2. $(l, Z) \xrightarrow{a}_{\triangleleft} (l', Z')$ *iff there is* $v \in Z$ *and* $v' \in Z'$ *such that* $(l, v) \xrightarrow{a}_c (l', v')$ (action successor).

The condition 1(a) says that the region (l, Z') is a time successor of (l, Z), whereas 1(b) specifies that it is the immediate one. By 1(c) we know that $Z = Z'$ for final regions only. The interpretation of the condition 2 is straightforward. An illustration of the definition is shown in Fig. 5.15, where all the arrows (both the solid and the dashed ones) denote the successor relation between detailed regions. The dashed lines are used to mark the successors of the regions which are treated differently in the alternative definition introduced below.

In order to define a detailed region graph reflecting the strongly monotonic runs we need to distinguish between boundary and non-boundary regions. A detailed region (l, Z) is called *boundary* if for each positive $\delta \in \mathbb{R}$ and each $v \in Z$ we have $\neg(v \simeq_{C_{\mathcal{A},\varphi}} v + \delta)$, which means that the region contains no

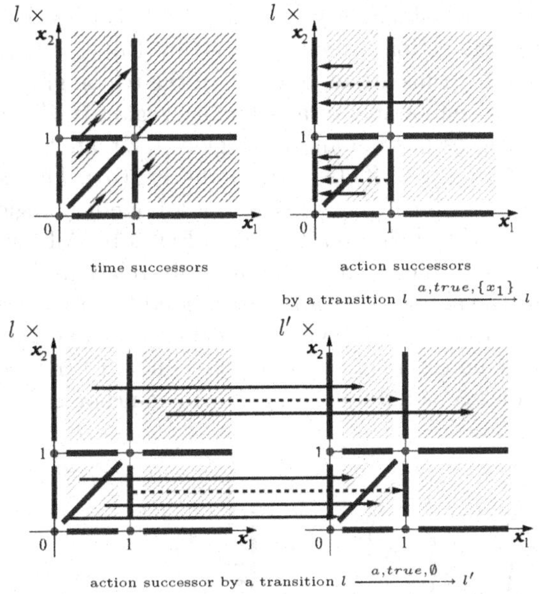

Fig. 5.15. Time- and action successor relation for detailed regions. The invariants of l and l' are assumed to be equal to *true*

time-successors of any of its states. An example of such regions are the points as well as the horizontal and vertical lines shown in Fig. 5.14. In this case the definition of the detailed region graph $DRG_b(\mathcal{A})$ looks as follows:

Definition 5.13. *The* boundary-distinguishing *(*bd-*) (*detailed*)* region graph *of a timed automaton $\mathcal{A}$ is a finite transition system*

$$DRG_b(\mathcal{A}) = (W, w^0, \rightarrow_{\triangleleft b}),$$

where

- $W = L \times DZ(n_{\mathcal{X}})$,
- $w^0 = (l^0, Z^0)$ *and*
- $\rightarrow_{\triangleleft b} \subseteq W \times (A \cup \{\tau\}) \times W$ *is defined by*
  1. $(l, Z) \xrightarrow{\tau}_{\triangleleft b} (l, Z')$ *iff* $(l, Z) \xrightarrow{\tau}_{\triangleleft} (l, Z')$ *in* $DRG(\mathcal{A})$ *(*time successor*)*,
  2. $(l, Z) \xrightarrow{a}_{\triangleleft b} (l', Z')$ *iff the following conditions hold:*
     a) (l, Z) *is not boundary and*
     b) *there is* $v \in Z$ *and* $v' \in Z'$ *such that* $((l, v) \xrightarrow{a}_{c} (l', v')$ *or there is* Z'' *and* $v'' \in Z''$ *such that* $(l, Z) \xrightarrow{\tau}_{\triangleleft} (l, Z'')$ *and* $(l, v'') \xrightarrow{a}_{c} (l', v'))$, *for* $a \in A$

     *(*action successor*)*. —

The condition 1 means that the time successor relation is defined analogously to Definition 5.12. The condition 2(a) expresses the fact that action successors can be executed from non-boundary regions only. This is to ensure that there are no two consecutive action successors in a run. But, to guarantee that all the strongly monotonic runs are represented in the region graph, the a-action successors of the concrete states in a boundary region (l, Z'') that could be reached by a time successor from another region (l, Z), are represented as the a-action successor (l, Z') of (l, Z) (see the condition 2(b)). This is correct since no boundary region can be a time successor of a boundary region.

An illustration of this definition is given in Fig. 5.16. All the arrows (both the solid and the dashed ones) denote the successor relations. The dashed lines in the pictures for the action successors correspond to the relation between a region whose time successor is a boundary region (l, Z) and the region which contains the action successors of concrete states in (l, Z) (notice that each of these lines has its dashed counterpart in Fig. 5.15, "anchored" in the boundary region).

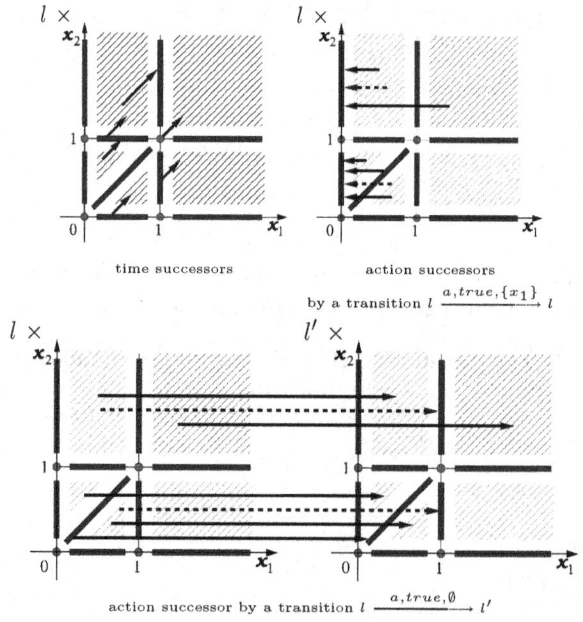

Fig. 5.16. Time- and action successor relation for detailed regions when boundary regions are distinguished. The invariants of l and l' are assumed to be equal to *true*

Examples of the detailed region graphs for both the semantics are given below.

Example 5.14. Fig. 5.17 depicts a timed automaton of three locations (l_0, l_1, l_2) and three clocks x_1, x_2, x_3 (this is the product automaton of Fig. 2.5). In Fig. 5.18, the reachable part of its detailed region graph, built for a formula

φ such that $c_{max}(\mathcal{A}, \varphi) = 1$ and the weakly monotonic semantics, is shown. Fig. 5.19 presents the reachable part of the detailed region graph built for the same value of $c_{max}(\mathcal{A}, \varphi)$ and the strongly monotonic semantics. The initial location of the automaton as well as the initial regions of the graphs are coloured. In the picture of the region graph DRG_b the boundary regions are marked with bold frames.

□

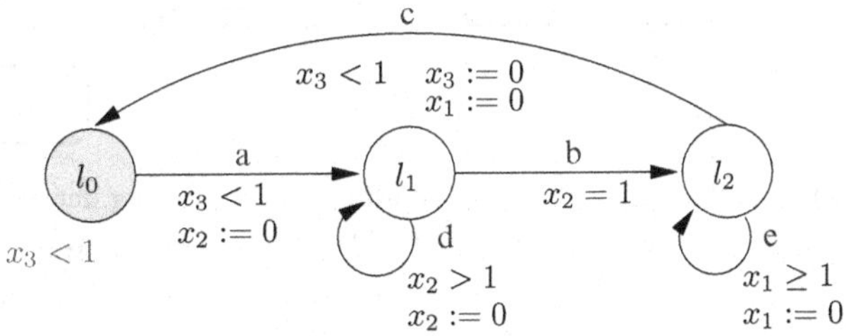

Fig. 5.17. The timed automaton of Example 5.14

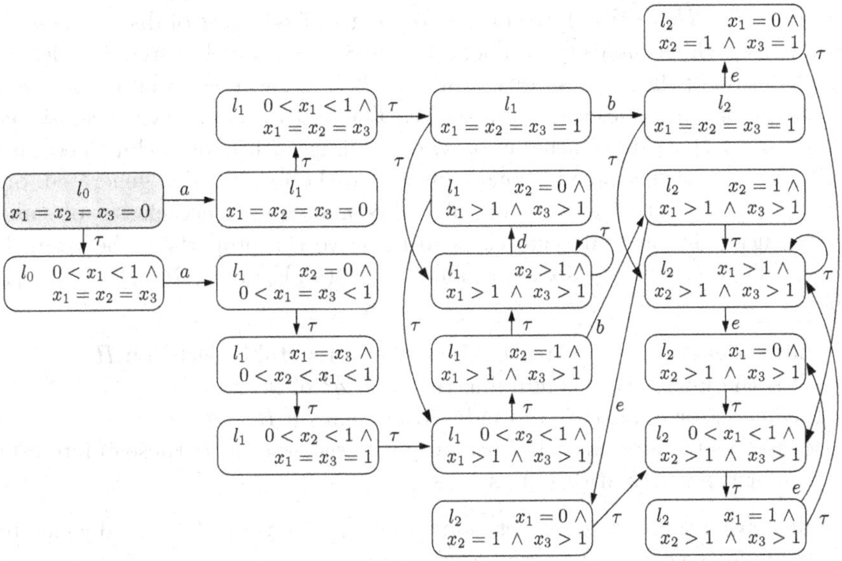

Fig. 5.18. The reachable part of the detailed region graph for the automaton of Fig. 5.17 and a formula φ such that $c_{max}(\mathcal{A}, \varphi) = 1$

In Sect. 7.1 we show how to encode detailed regions and the transition relation in a symbolic way to accomplish bounded model checking.

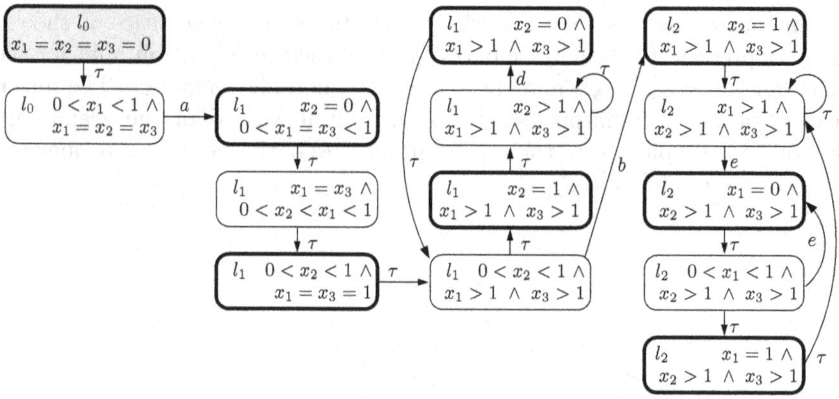

Fig. 5.19. The reachable part of the boundary-distinguishing detailed region graph for the automaton of Fig. 5.17 and a formula φ such that $c_{max}(\mathcal{A}, \varphi) = 1$

5.2.2 Partition Refinement

Partition refinement (minimization) is an algorithmic method for constructing abstract models for timed automata in the *on-the-fly* manner, i.e., without building concrete models first. Typically, the algorithm starts from an initial *partition* $\Pi = \Pi_0$ of the state space Q of $\mathcal{A}$ (i.e., from a set of disjoint *classes* the union of which equals Q, cf. Footnote 3 in Sect. 2), which respects (at least) the valuation of the propositions of interest[16]. The successor relation between the classes of Π satisfies the condition EE. The partition Π is then successively refined until Π (or its reachable part, depending on the algorithm) becomes *stable*, i.e., satisfies the conditions required on the model to be generated, or, more precisely, until the elements of Π become equivalence classes of some equivalence relation that guarantees to preserve the property to be verified. As a result, we obtain an abstract model $M_{\mathfrak{a}} = (G, V_{\mathfrak{a}})$ with $G = (W, w^0, \rightarrow_{\mathfrak{a}})$, where

- the elements of W are (reachable) classes of a stable partition Π,
- w^0 is the initial class containing the state q^0 of $\mathcal{A}$,
- the successor relation $\rightarrow_{\mathfrak{a}}$ is induced by that on Π, and
- $V_{\mathfrak{a}}$ assigns to each class of Π the propositions (or at least these of interest) true at the states of this class.

Usually, either a concrete discrete model for $\mathcal{A}$, or a time-abstracted concrete dense model (see p. 94) is exploited as an underlying concrete model for defining the valuation function and the successor relation in Π. Moreover, weakly monotonic semantics is assumed.

[16] This means that for any proposition $\wp \in PV$ of $\mathcal{A}$ (or at least for these $\wp \in PV$ which are used to express the properties to be checked) and for any class $Y \in \Pi$ the proposition holds either in all the concrete states $q \in Y$, or in none of them.

The classes of partitions are usually represented by *regions*. If the valuations of the clocks in a region do not exceed $c_{max}(\mathcal{A})$,[17] then the region is a union of detailed regions. Formally, given a timed automaton $\mathcal{A} = (A, L, l^0, E, \mathcal{X}, \mathcal{I})$ of $n_{\mathcal{X}}$ clocks, a region $R \in L \times Z(n_{\mathcal{X}})$, denoted by (l, Z) for $l \in L$ and $Z \in Z(n_{\mathcal{X}})$, is a set of concrete states

$$R = \{(l, v) \in L \times \mathbb{R}_{0+}^{n_{\mathcal{X}}} \mid v \in Z\}.$$

The set of all the regions is denoted by $R(n_{\mathcal{X}}, L)$. A concrete state $q' = (l', v')$ of $\mathcal{A}$ is an element of (l, Z) (denoted $q' \in (l, Z)$) iff $l' = l$ and $v \in Z$. For two regions $R = (l, Z)$ and $R' = (l, Z')$ we define their difference as

$$R \setminus R' = \{(l, Z'') \mid Z'' \in Z \setminus Z'\}.$$

Computing $\setminus$ is of an exponential complexity in the number of clocks [159], which means that it can result in the exponential number of regions. This operation potentially returns a set of regions which consists of more than one element. This can cause an important inefficiency of partition refinement algorithms.

Below, we present the main partitioning algorithms and characterize abstract models generated by them. To this aim, we need some additional notions. Given a timed automaton $\mathcal{A}$ and its (discrete or a time-abstracted dense) concrete model $M_{\mathfrak{c}}(\mathcal{A}) = ((Q, q^0, \to_{\mathfrak{c}}), V_{\mathfrak{C}})$ whose successor relation is labelled by the elements of a set B. Let

$$b(q), b^{-1}(q) \subseteq Q,$$

where $q \in Q$ and $b \in B$, denote, respectively, the set of concrete states which are b-successors of q, and the set of concrete states for which q is a b-successor. For a partition Π of Q and $Y, Y' \in \Pi$ let

$$post_b(Y, Y') = \{q' \in Y' \mid (\exists q \in Y)\ q' \in b(q)\},$$

and

$$pre_b(Y, Y') = \{q \in Y \mid (\exists q' \in Y')\ q \in b^{-1}(q')\}.$$

Moreover, we establish a successor relation $\to_{\Pi} \subseteq \Pi \times B \times \Pi$ for the classes of the partition Π, given by

$$Y \xrightarrow{b}_{\Pi} Y' \text{ iff there is } q \in Y \text{ such that } b(q) \cap Y' \neq \emptyset$$

(which corresponds to the condition $\mathrm{EE}(Y, Y')$). Next, for a class $Y \in \Pi$ we define the sets of all its successors and predecessors in Π, i.e.,

$$Post_{\Pi}(Y) = \{Y' \in \Pi \mid (\exists b \in B)\ Y \xrightarrow{b}_{\Pi} Y'\}$$

and

$$Pre_{\Pi}(Y) = \{Y' \in \Pi \mid (\exists b \in B)\ Y' \xrightarrow{b}_{\Pi} Y\}.$$

[17] The formula to be tested is not taken into account at this stage.

Bisimulating Models

Three main minimization algorithms were introduced by Paige and Tarjan [112], Bouajjani at al [40], and Lee and Yannakakis [101]. They are aimed at generating *bisimulating models*, i.e., models which satisfy EE and AE.

The Algorithm of Bouajjani et al.

Since the method of [112] stabilizes all the classes of the partition, whereas these of [40, 101] – the reachable part only, mainly the last two are applied to timed systems in the literature. The algorithm of [40] serves building all the types of bisimulating models mentioned above (i.e., discrete and time-abstracted dense ones). Its generic pseudo-code is presented in Fig. 5.20. The algorithm maintains two variables which store subsets of classes of a current partition Π: reachable, which keeps the classes considered as reachable in a given step, and stable, which collects the elements Y of reachable satisfy-

---

INPUT ARGUMENTS:
  a timed automaton $\mathcal{A} = (A, L, l^0, E, \mathcal{X}, \mathcal{I})$ of $n_{\mathcal{X}}$ clocks
  an initial partition Π_0
GLOBAL VARIABLES:
  Π, reachable, stable: $2^{R(n_{\mathcal{X}}, L)}$
RETURN VALUES:
  minimization_bisim(), Split(): $2^{R(n_{\mathcal{X}}, L)}$

```
1.    procedure minimization_bisim(𝒜, Π₀) is
2.    begin
3.        Π := Π₀; reachable := {[q⁰]}; stable := ∅;
4.        while (∃Y ∈ reachable \ stable) do
5.          C_Y := Split(Y, Π);
6.          if C_Y = {Y} then
7.            stable := stable ∪ {Y};
8.            reachable := reachable ∪ Post_Π(Y);
9.          else
10.           P_Y := {Y' ∈ Π | Y' has been split};
11.           reachable := reachable \ P_Y ∪ {Y' ∈ P_Y | q⁰ ∈ Y'};
12.           stable := stable \ {Y' ∈ Π | (∃Y'' ∈ P_Y) Y' ∈ Pre_Π(Y'')};
13.           Π := (Π \ P_Y) ∪ C_Y;
14.         end if;
15.       end do;
16.   end minimization_bisim;
```

Fig. 5.20. A generic minimization algorithm

ing the condition $\mathrm{AE}(Y, Y')$ for all their $\rightarrow_\Pi$-successors Y' in the current partition. It starts from the class $[q^0] \in \Pi_0$ containing the initial state of $\mathcal{A}$, and then successively searches and refines reachable classes. A class $Y_1 \in \Pi$ that is unstable w.r.t. its successor Y_2 (in the case of bisimulating models unstability of Y_1 w.r.t. Y_2 means that for some $b \in B$ we have $Y_1 \xrightarrow{b}_\Pi Y_2$, but the condition $\mathrm{AE}(Y_1, Y_2)$ does not hold, i.e.,

$$(\exists b \in B)\ pre_b(Y_1, Y_2) \notin \{Y_1, \emptyset\})$$

is partitioned into Y_1' containing all the concrete states which have b-successors in Y_2 (i.e., $Y_1' = pre_b(Y_1, Y_2)$), and $Y_1 \setminus Y_1'$ (see Fig. 5.21). The function `Split(`Y, Π`)`, used in the algorithm, returns either the classes obtained by splitting $Y \in \Pi$ due to its unstability w.r.t. a selected successor, or the class Y if it satisfies $\mathrm{AE}(Y, Y')$ for all the successor classes Y'. Notice that partitioning of a class can result in unstability of its predecessors, and that some new classes obtained by splitting of a given (reachable) one can be unreachable. This is handled in the lines 11–12 of the algorithm.

(a): unstability of Y_1 w.r.t. Y_2 (b): partitioning of the class Y_1

Fig. 5.21. Partitioning of the classes to preserve AE

Example 5.15. The pictures in Fig. 5.22 illustrate the partitioning process while building a bisimulating model, for a part of a partition consisting of the following four classes: Y_1, Y_2, Y_3, Y_4.

□

Implementations of the above algorithm require to define how to compute the image and the inverse image of a class w.r.t. $b \in B$ (or, in other words, pre_b and $post_b$ for $b \in B$), as well as computing Pre_Π and $Post_\Pi$. These constructions depend on the type of a bisimulating model to be generated (i.e., whether it is a discrete [66] or a time-abstracted one [8, 9, 66, 157, 159]). Typically, all the implementations deal with a partition of the part of the state space of the automaton which could possibly be reachable, i.e., with the set

$$\{(l, v) \in Q \mid v \in [\![\mathcal{I}(l)]\!]\}.$$

Moreover, to improve on efficiency, in the case of time-abstracted models they consider as time successors of a given class $(l, Z) \in \Pi$ only these time successors $(l, Z') \in \Pi$ of (l, Z) which satisfy $Z \Uparrow Z' \neq \emptyset$ (so-called *immediate time successors*).

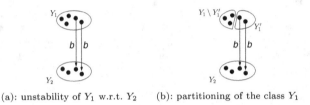

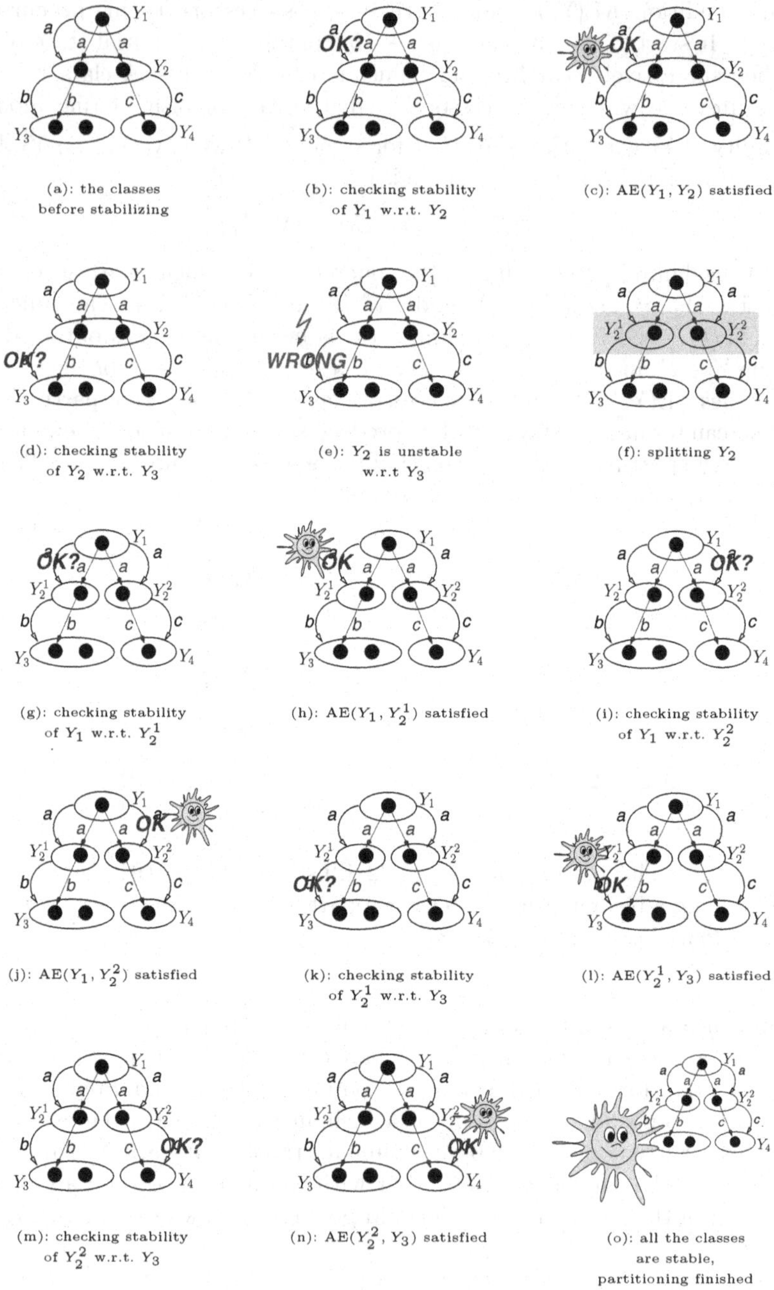

(a): the classes before stabilizing

(b): checking stability of Y_1 w.r.t. Y_2

(c): AE(Y_1, Y_2) satisfied

(d): checking stability of Y_2 w.r.t. Y_3

(e): Y_2 is unstable w.r.t Y_3

(f): splitting Y_2

(g): checking stability of Y_1 w.r.t. Y_2^1

(h): AE(Y_1, Y_2^1) satisfied

(i): checking stability of Y_1 w.r.t. Y_2^2

(j): AE(Y_1, Y_2^2) satisfied

(k): checking stability of Y_2^1 w.r.t. Y_3

(l): AE(Y_2^1, Y_3) satisfied

(m): checking stability of Y_2^2 w.r.t. Y_3

(n): AE(Y_2^2, Y_3) satisfied

(o): all the classes are stable, partitioning finished

Fig. 5.22. A partitioning process for bisimulating models

A Convexity-Preserving Technique

As it has been already stated, computing differences of classes while generating a model, together with the requirement of convexity of zones[18], can lead to inefficiency of the partitioning algorithm, since the result of the above operation is exponential in the number of clocks. A solution to this problem, which is applicable to building strong ta-bisimulating models for timed automata, was presented by Tripakis and Yovine in [158, 159]. Given a timed automaton $\mathcal{A} = (A, L, l^0, E, \mathcal{X}, \mathcal{I})$ of $n_{\mathcal{X}}$ clocks, the main idea consists in starting from an initial partition Π_0 which *respects* the invariants and the enabling conditions of $\mathcal{A}$, i.e., is chosen such that for each $Y \in \Pi_0$, each $e \in E$ and each $l \in L$ we have

$$Y \cap [\![\mathcal{I}(l)]\!], Y \cap [\![guard(e)]\!] \in \{Y, \emptyset\}$$

(notice that for obtaining such an initial partition, computing differences may be necessary), and then refining it such that each class is stabilized simultaneously w.r.t. all its time successors, or w.r.t. all the action successors for a given action. Since for each $Y \in \Pi$, where Π is a refinement of Π_0, if an action $a \in A$ can be performed at a state $q \in Y$, then a can be performed at all the states of the class, the set

$$\{pre_a(Y, Y') \mid Y \xrightarrow{a}_{\Pi} Y'\}$$

constitutes a partition of Y. Thus, the class is stabilized w.r.t. all its a-successors by computing the inverse images and intersections, but with no need of computing differences. A similar result can be obtained in the case of time successors, but to this aim Π_0 has to contain all the concrete states of $\mathcal{A}$, even these which violate the invariants (i.e., the optimisation mentioned on p. 125 cannot be applied). For a partition Π the classes $(l, Z) \in \Pi$ which satisfy $Z \cap [\![\mathcal{I}(l)]\!] = \emptyset$ are called *pseudoclasses*, and Π^{ps} denotes the set of all the pseudoclasses of Π. Thus, given a class $(l, Z) \in \Pi$, when stabilizing the class w.r.t. the time successor relation we take into account all these $(l, Z') \in \Pi$ which satisfy $Z \Uparrow Z' \neq \emptyset$, even if, in fact, they are not time successors of (l, Z) (i.e., there are no $\delta \in \mathbb{R}_{0+}$, $q \in (l, Z)$, $q' \in (l, Z')$ such that $q \xrightarrow{\delta}_c q'$, due to $Z' \cap [\![\mathcal{I}(l)]\!] = \emptyset$). The reason for this construction will become clear in Example 5.16.

A pseudo-code of the algorithm is shown in Fig. 5.23. Similarly to that of Fig. 5.20, the algorithm deals with two sets of classes `stable` and `reachable`, starts from an initial class containing the initial state, and then successively searches and refines reachable classes (notice that pseudoclasses are not added to `reachable`). However, in this case testing stability of a class is a two-stage procedure. Stability of a class Y w.r.t. its time-successors is always checked first, and then, if no unstability occurs, the algorithm checks whether the class satisfies AE(Y, Y') for all its action successors Y'. The above tests and

[18] Recall that the classes of a partition are usually represented by regions.

INPUT ARGUMENTS:
  a timed automaton $\mathcal{A} = (A, L, l^0, E, \mathcal{X}, \mathcal{I})$ of $n_{\mathcal{X}}$ clocks
  an initial partition Π_0
GLOBAL VARIABLES:
  Π, reachable, stable: $2^{R(n_{\mathcal{X}}, L)}$
RETURN VALUES:
  minimization_bisconvex(), $TimeSplit()$, $ActionSplit()$: $2^{R(n_{\mathcal{X}}, L)}$

---

1.   **procedure** minimization_bisconvex($\mathcal{A}, \Pi_0$) **is**
2.   **begin**
3. $\Pi := \Pi_0$; $reachable := \{[q^0]\}$; $stable := \emptyset$;
4.   **while** $(\exists Y \in reachable \setminus stable)$ **do**
5. $C_Y := TimeSplit(Y, \Pi)$;
6.   **if** $C_Y \in \{Y, \emptyset\}$ **then**
7.   **for each** $a \in A$ **do**
8. $C_Y := ActionSplit(Y, a, \Pi)$;
9.   **if** $C_Y \notin \{Y, \emptyset\}$ **then** break; **end if**;
10.   **end do**;
11.   **end if**;
12.   **if** $C_Y = \{Y\}$ **then**
13. $stable := stable \cup \{Y\}$;
14. $reachable := reachable \cup (Post_{\Pi}(Y) \setminus \Pi^{ps})$;
15.   **else**
16. $P_Y := \{Y' \in \Pi \mid Y' \text{ has been split}\}$;
17. $reachable := reachable \setminus P_Y \cup \{Y' \in C_Y \mid q^0 \in Y'\}$;
18. $stable := stable \setminus \{Y' \in \Pi \mid (\exists Y'' \in P_Y)\ Y' \in Pre_{\Pi}(Y'')\}$;
19. $\Pi := (\Pi \setminus P_Y) \cup C_Y$;
20.   **end if**;
21.   **end do**;
22.   **end** minimization_bisconvex;

Fig. 5.23. A minimization algorithm for building strong ta-bisimulating models for TA without computing differences

the partitionings (if necessary) are performed, respectively, by the functions TimeSplit and ActionSplit.

- TimeSplit(Y, Π) is a function which for $Y = (l, Z) \in \Pi$, where $l \in L$ and $Z \in Z(n_{\mathcal{X}})$, returns either Y if it satisfies AE w.r.t. all its immediate time successors $Y' \in \Pi \setminus \Pi^{ps}$, or

$$\{pre_{\tau}(Y, Y') \mid Y' = (l, Z') \wedge Z \Uparrow Z' \neq \emptyset\}$$

otherwise. Similarly,

- `ActionSplit`(Y, a, Π) is a function which for $Y = (l, Z) \in \Pi$ and $a \in A$ returns either Y if it satisfies AE w.r.t. all its a-action successors $Y' \in \Pi \setminus \Pi^{ps}$, or
$$\{pre_a(Y, Y') \mid Y' \in \Pi \wedge Y \xrightarrow{a}_{\Pi} Y'\}$$
otherwise.

Example 5.16. Figure 5.24 displays the partitionings caused by the functions `TimeSplit` and `ActionSplit`. The part (a) shows a set of time zones corresponding to a location l with $\mathcal{I}(l) = (x_1 \leq 5)$. The class (l, Z_{ps}) belongs to the set Π^{ps}. It is easy to see that (l, Z_1) is unstable w.r.t. its time successors (the condition AE$((l, Z_1), (l, Z_j))$ for $j \in \{2, 3\}$ does not hold), and therefore it needs to be refined. Applying the function `TimeSplit` results in the set of classes consisting of (l, Z_1^1), (l, Z_1^2), (l, Z_1^3) and shown in the part (b) of the figure. The classes obtained form a partition of (l, Z). Notice that for computing this partition, taking pseudoclasses into account was necessary.

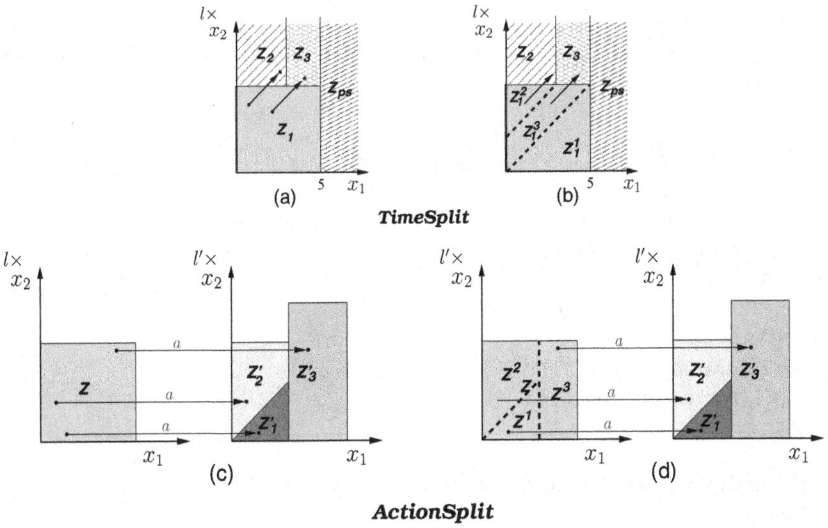

Fig. 5.24. The functions `TimeSplit` and `ActionSplit`

The bottom part of the figure shows the way the function `ActionSplit` works. The class (l, Z) has the three a-successors (l', Z_1'), (l', Z_2') and (l', Z_3') (see (c)), and is unstable w.r.t. each of them. Applying `ActionSplit` gives us the set of classes shown in the part (d) of the figure, consisting of (l, Z^1), (l, Z^2) and (l, Z^3) and constituting a partition of (l, Z).

□

The Lee–Yannakakis Algorithm

Another general partitioning algorithm was introduced in [101]. Its adaptation to the case of timed automata and strong ta-bisimulating models was presented in [172] (the general algorithm, however, seems also to be applicable to generating other kinds of bisimulating models for TA). Given a timed automaton $\mathcal{A} = (A, L, l^0, E, \mathcal{X}, \mathcal{I})$ of $n_{\mathcal{X}}$ clocks, and its concrete model $M_{\mathfrak{c}}(\mathcal{A}) = ((Q, q^0, \rightarrow_{\mathfrak{c}}), V_{\mathfrak{C}})$ whose successor relation is labelled by the elements of a set B, the algorithm starts from an initial partition Π_0 of Q which respects the invariants[19] and the enabling conditions of $\mathcal{A}$, and stabilizes only the reachable classes. To ensure this, each reachable class Y is *marked* with a *representative* $r_Y \in Q$, which is guaranteed to be a reachable state of $\mathcal{A}$. Unlike the approaches described before, the algorithm gives priority to searching than to splitting: its first stage consists in searching the initial partition and marking the reachable classes. Initially, only the initial class $[q^0]$ is marked with $r_{[q^0]} = q^0$.

Fig. 5.25. A generic minimization algorithm by Lee and Yannakakis (continued on the next page)

INPUT ARGUMENTS:
  a timed automaton $\mathcal{A} = (A, L, l^0, E, \mathcal{X}, \mathcal{I})$ of $n_{\mathcal{X}}$ clocks
  an initial partition Π_0
GLOBAL VARIABLES:
  Π: $2^{R(n_{\mathcal{X}}, L)}$, *stack*, *queue*: $(R(n_{\mathcal{X}}, L))^*$
RETURN VALUES:
  minimization_LY(): $2^{R(n_{\mathcal{X}}, L)}$
  pop(), *remove*(): $R(n_{\mathcal{X}}, L)$

```
procedure minimization_LY(A, Π0) is
begin
   Π := Π0; stack := ∅; queue := ∅;
   r[q0] := q0; stack := [q0];

   SEARCH:
   while stack ≠ ∅ do
     Y := pop(stack);
     for each b ∈ B do
       for each Y1 ∈ Π s.t. b(rY) ∩ Y1 ≠ ∅ do
```

---

19 Our approach extends the automata of [172], defined with no invariant function.

10.   **if** $pre_b(Y, Y_1) \neq Y$ **then** $queue := queue + Y$; **end if**;

11.   **if** Y_1 is not marked **then**

12.   select $q \in Y_1 \cap b(r_Y)$; $r_{Y_1} := q$;

13. $stack := stack + Y_1$;

14.   **end if**;

15.   add edge $Y \xrightarrow{b}_{\Pi L} Y_1$;

16.   **end do**;

17.   **end do**;

18.   **end do**;

19.   <u>*SPLIT:*</u>

20.   **while** $queue \neq \emptyset$ **do**

21. $Y := remove(queue)$;

        // computing the set of states which have the successors in the same classes r_Y has ...

22. $Y' = Y$;

23.   **for each** $b \in B$ **do**

24.   **for each** $Y_1 \in \Pi$ s.t. $Y \xrightarrow{b}_{\Pi L} Y_1$ **do**

25. $Y' := pre_b(Y', Y_1)$;

26.   **end do**;

27.   **end do**;   *// ... computed.*

28. $r_{Y'} := r_Y$;

29. $P_Y := Y \setminus Y'$;

30. $\Pi := \Pi \cup P_Y \cup \{Y'\}$;

31.   **for each** $b \in B$ **do**

32.   **for each** $Y_1 \in \Pi$ s.t. $Y_1 \xrightarrow{b}_{\Pi L} Y$ **do**

33.   **if** $b(r_{Y_1}) \cap Y' \neq \emptyset$ **then**

34.   add edge $Y_1 \xrightarrow{b}_{\Pi L} Y'$;

35.   **if** $pre_b(Y_1, Y') \neq Y_1$ **then** $queue := queue + Y_1$; **end if**;

36.   **end if**;

37.   delete edge $Y_1 \xrightarrow{b}_{\Pi L} Y$;

38.   **for each** $Y_2 \in P_Y$ **do**

39.   **if** $b(r_{Y_1}) \cap Y_2 \neq \emptyset$ **then**

40.   **if** Y_2 is not marked **then**

41.   select $q \in b(r_{Y_1}) \cap Y_2$; $r_{Y_2} := q$;

42. $stack := stack + Y_2$;

43.   **end if**;

44.   add edge $Y_1 \xrightarrow{b}_{\Pi L} Y_2$;

45.   **if** $pre_b(Y_1, Y_2) \neq Y_1$ **then** $queue := queue + Y_1$; **end if**;

46.   **end if**;

47.   **end do**;

48.   **end do**;

49.   **end do**;

50. $\Pi := \Pi \setminus \{Y\}$;

51.   **if** $stack \neq \emptyset$ **then goto** *SEARCH*; **end if**;

52.   **end do**;

53. **end** minimization_LY;

For a partition Π of Q, the searching procedure looks as follows: given a representative r_Y of a marked class $Y = (l, Z) \in \Pi$, where $l \in L$ and $Z \in Z(n_{\mathcal{X}})$, the algorithm computes the image $b(r_Y)$ for each $b \in B$, and if some $Y' \in \Pi$ with $Y' \cap b(r_Y) \neq \emptyset$ is not marked, then it becomes marked with an element of $b(r_Y) \cap Y'$. In the case when $B = A \cup \{\tau\}$ (i.e., if a ta-bisimulating model is to be generated) and the time successor relation is considered, the algorithm marks only these classes $(l, Z') \in \Pi$ for which $Z \Uparrow Z'$ is non-empty. The procedure establishes also "step by step"a successor relation $\rightarrow_{\Pi \mathrm{L}} \subseteq \Pi \times B \times \Pi$ between classes,, defined by

$$Y \xrightarrow{b}_{\Pi \mathrm{L}} Y' \text{ iff } ((\exists q' \in Y')\ r_Y \xrightarrow{b}_{\mathfrak{c}} q' \ \wedge\ Y' \text{ is marked})$$

(this means that only marked classes can be related). Again, in the case of $B = A \cup \{\tau\}$ only immediate time successors of a class are taken into acount. This relation is used in the algorithm besides $\rightarrow_{\Pi}$ introduced before.

When no other existing class can be marked, the algorithm starts the stage aimed at partition refinements. If for a marked class $Y \in \Pi$ and some $Y' \in \Pi$ such that $Y \xrightarrow{b}_{\Pi \mathrm{L}} Y'$ for some $b \in B$ the condition $\mathrm{AE}(Y, Y')$ does not hold, then Y is split into a class Y_1, which contains the representative r_Y and all the states which for each $b' \in B$ have their b'-successors in the same classes as r_Y does, and $Y \setminus Y_1$ (see Fig. 5.26). Notice that $Y \setminus Y_1$ can possibly be a set of classes. The class Y_1 is computed as

$$Y_1 = \bigcap \{pre_{b'}(Y, Y') \mid (\exists b' \in B)\ Y \xrightarrow{b'}_{\Pi \mathrm{L}} Y'\}$$

and marked by r_Y, whereas each of the elements of $Y \setminus Y_1$ is marked only if it contains a state which can be its representative, i.e., a successor of a representative of another class of Π. This, obviously, results in rebuilding $\rightarrow_{\Pi \mathrm{L}}$. Partitioning of a class can cause unstability of its predecessors, whereas creating new marked classes requires executing the searching procedure. The above two steps are repeated until all the classes become stable.

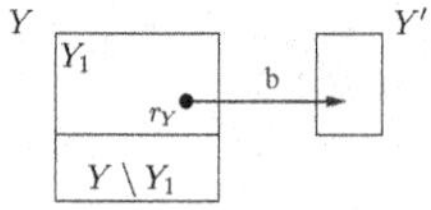

Fig. 5.26. Partitioning of a class in the Lee-Yannakakis algorithm

A pseudo-code of the algorithm, based on [101], is presented in Fig. 5.25. The algorithm maintains two additional structures: `stack` collecting classes of a current partition Π for which the successors are to be searched for, and `queue` which keeps unstable classes of Π. The functions `pop` and `remove` return, respectively, the element (class) removed from top of the stack or from the beginning of the queue. The operation $+$ denotes adding an element to the above structures (i.e., putting it on the top of the stack or at the end of the queue) if it is not present there. The relation $\to_{\Pi L}$ and its updated versions obtained when the model is generated are represented by edges between classes, which are removed and created when the algorithm operates. A representative of a class $Y \in \Pi$ is denoted by r_Y. $[q^0] \in \Pi_0$ denotes the class of the initial partition which contains the initial state of $\mathcal{A}$.

The paper [172] proposes also a solution to the problem of computing differences of regions. This solution is based on a *forest structure*, which keeps the history of partitionings. Therefore, instead of computing a difference of classes and representing the result explicitly, one can only remember that such a difference exists in the partition.

The first two of the above algorithms were modified to build other kinds of abstract models which preserve more restricted classes of properties. The solutions differ in the information stored together with the partition of Q, in the stability conditions and in the methods of refining unstable classes. Below, we sketch the main of them.

Simulating Models

The paper [66] provides a modification of the minimization algorithm of [40], aimed at building *simulating models*, i.e., the models satisfying the conditions EE and U. The new algorithm requires to store, together with the partition Π, also a function $cor : \Pi \to 2^Q$ assigning cores to the classes. The pseudo-code differs from that in Fig. 5.20 in two lines:

- line 10 is replaced by

10. $\quad P_Y := \{Y' \in \Pi \mid Y' \text{ has been split or } cor(Y') \text{ changed } \};$

- line 12 is replaced by

12. $\quad stable := stable \setminus \{Y' \in \Pi \mid (\exists Y'' \in P_Y)\ Y' \in Pre_\Pi(Y'')\} \setminus P_Y;$

the latter following from the fact that a stable class can be split if another class is unstable w.r.t. it (see below). Moreover, the stability condition and the way classes are partitioned differ. Unstability of a class $Y_1 \in \Pi$ w.r.t. its successor Y_2 means that the condition $\mathrm{U}(Y_1, Y_2)$ does not hold, i.e.,

$$(\exists b \in B)\ (pre_b(Y_1, Y_2) \neq \emptyset \wedge pre_b(cor(Y_1), cor(Y_2)) \neq cor(Y_1)).$$

Stabilizing Y_1 w.r.t. Y_2 can result in splitting one or both the classes, or modifying their cores (see Fig. 5.27).

More formally, handling the possible cases can be described by the function $Sp_s(Y_1, Y_2, b, \Pi)$, defined for the classes $Y_1, Y_2 \in \Pi$ satisfying $pre_b(Y_1, Y_2) \neq \emptyset$ and $pre_b(cor(Y_1), cor(Y_2)) \neq cor(Y_1)$, and given by

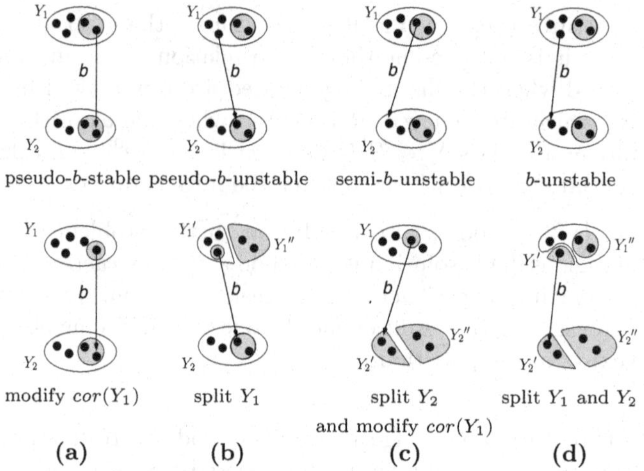

Fig. 5.27. Partitioning of the classes to preserve U

(a) if Y_1 is *pseudo-b-stable* w.r.t. Y_2, i.e., $pre_b(cor(Y_1), cor(Y_2)) \neq \emptyset$, then

$$Sp_s(Y_1, Y_2, b, \Pi) = \{Y_1 \text{ with the core equal to } pre_b(cor(Y_1), cor(Y_2))\},$$

(b) if Y_1 is *pseudo-b-unstable* w.r.t. Y_2, i.e., $pre_b(cor(Y_1), cor(Y_2)) = \emptyset \ \wedge \ pre_b(Y_1, cor(Y_2)) \neq \emptyset$, then

$$\begin{aligned} Sp_s(Y_1, Y_2, b, \Pi) = \{Y_1 \setminus cor(Y_1) &\text{ with the core equal to } pre_b(Y_1, cor(Y_2)), \\ cor(Y_1) &\text{ with the core equal to } cor(Y_1)\}, \end{aligned}$$

(c) if Y_1 is *semi-b-unstable* w.r.t. Y_2, i.e., $pre_b(cor(Y_1), cor(Y_2)) = pre_b(Y_1, cor(Y_2)) = \emptyset \wedge pre_b(cor(Y_1), Y_2) \neq \emptyset$, then

$$\begin{aligned} Sp_s(Y_1, Y_2, b, \Pi) = \{Y_1 &\text{ with the core equal to } pre_b(cor(Y_1), Y_2), \\ cor(Y_2) &\text{ with the core equal to } cor(Y_2), \\ Y_2 \setminus cor(Y_2) &\text{ with the core equal to } Y_2 \setminus cor(Y_2)\}, \end{aligned}$$

(d) if Y_1 is *b-unstable* w.r.t. Y_2, i.e., $pre_b(cor(Y_1), cor(Y_2)) = pre_b(Y_1, cor(Y_2)) = pre_b(cor(Y_1), Y_2) = \emptyset$, then

$$\begin{aligned} Sp_s(Y_1, Y_2, b, \Pi) = \{pre_b(Y_1, Y_2) &\text{ with the core equal to } pre_b(Y_1, Y_2), \\ Y_1 \setminus pre_b(Y_1, Y_2) &\text{ with the core equal to } cor(Y_1), \\ cor(Y_2) &\text{ with the core equal to } cor(Y_2), \\ Y_2 \setminus cor(Y_2) &\text{ with the core equal to } Y_2 \setminus cor(Y_2)\}. \end{aligned}$$

Therefore, the function `Split`(Y, Π) used in the algorithm returns either the class $Y \in \Pi$ if Y satisfies the condition $\mathrm{U}(Y, Y')$ for all its successor classes Y', or the value of the function $Sp_s(Y, Y', b, \Pi)$, where Y' is a selected successor of Y w.r.t. which the class is unstable. Again, partitioning of a class can result in unstability of its predecessors.

Similarly to that in Fig. 5.20, the algorithm can be implemented to generate various kinds of discrete and time-abstracted models [66]. However, the problem of computing differences while generating them has not been solved so far.

Pseudo-bisimulating Models

The paper [125] introduces a method for building *pseudo-bisimulating models* (i.e., models satisfying EE and pAE), preserving reachability properties. This, again, is a modification of the algorithm by Bouajjani et al. [40]. In this case, together with a partition Π, a function $dpt : \Pi \to \mathbb{N} \cup \{\infty\}$ is stored. The function assigns to each class in the set `reachable` its potential depth in the model (or, roughly speaking, its depth in the part of Π which has been visited so far), whereas the classes in $\Pi \setminus$ `reachable` are assigned the value ∞. Unstability of a class Y_1 w.r.t. its successor Y_2 means that Y_2 has no predecessor Y' of the depth not greater than Y_1 such that $\mathrm{AE}(Y', Y_2)$ holds, i.e.,

$$(\exists b \in B)\ pre_b(Y_1, Y_2) \neq \emptyset \ \wedge \ (\forall b' \in B)(\forall Y' \in \Pi \text{ s.t. } Y' \xrightarrow{b'}_{\Pi} Y_2 \ \wedge$$

$$dpt(Y') = \min\{dpt(Y'') \mid Y'' \in Pre_{\Pi}(Y_2)\})\ pre_{b'}(Y', Y_2) \neq Y'.$$

Stabilizing Y_1 w.r.t. Y_2 is performed by selecting a predecessor Y' of Y_2 of a smallest depth, and $b' \in B$ such that $Y' \xrightarrow{b'}_{\Pi} Y_2$, and then partitioning Y' into Y'_* containing all the concrete states which have successors in Y_2 (i.e., $Y'_* = pre_{b'}(Y', Y_2)$), and $Y' \setminus Y'_1$. However, in practice the algorithm can operate in a mode similar to breadth first search (BFS), which allows to partition exactly the class Y_1, without testing depths of other predecessors of Y_2. Such a behaviour is achieved by sorting the set `reachable` w.r.t. the values of the function *dpt*, such that while selecting a class from the set `reachable` $\setminus$ `stable` a class of a smallest depth is always chosen first. This, additionally, enables an *on-the-fly reachability analysis*: when a state satisfying the property of interest is reached, generating the model can be stopped (recall that a property is expressed in terms of propositional variables and logical connectives, and therefore its satisfiability can be tested locally in a given state).

A pseudo-code of the algorithm for generating a model and performing on-the-fly reachability analysis is presented in Fig. 5.28. The differences between this code and that in Fig. 5.20 are underlined. Notice that the algorithm

---

INPUT ARGUMENTS:
   a timed automaton $\mathcal{A} = (A, L, l^0, E, \mathcal{X}, \mathcal{I})$ of $n_{\mathcal{X}}$ clocks
   an initial partition Π_0
   a propositional formula p
GLOBAL VARIABLES:
   Π, reachable, stable: $2^{R(n_{\mathcal{X}},L)}$
RETURN VALUES:
   reachability_minim_pb(): $\{REACHABLE,\ UNREACHABLE\}$
   $Split() : 2^{R(n_{\mathcal{X}},L)}$

```
1.    function reachability_minim_pb(𝒜, Π₀, p) is
2.    begin
3.        Π := Π₀; reachable := {[q⁰]}; dpt([q⁰]) := 0; stable := ∅;
4.        while (∃Y ∈ reachable \ stable) do
5.            C_Y := Split(Y, Π);
6.            if C_Y = {Y} then
7.                stable := stable ∪ {Y};
8.                for Y' ∈ Post_Π(Y) do
9.                    dpt(Y') := min{dpt(Y) + 1, dpt(Y')}; end do;
10.               reachable := reachable ∪ Post_Π(Y);
11.               if (∃Y' ∈ Post_Π(Y) s.t. Y' ⊨ p) then
12.                   return REACHABLE; end if;
13.           else
14.               P_Y := {Y' ∈ Π | Y' has been split};
15.               for Y₁ ∈ {Y₂ ∈ C_Y | q⁰ ∉ Y₂} do dpt(Y₁) := ∞; end do;
16.               reachable := reachable \ P_Y ∪ {Y' ∈ P_Y | q⁰ ∈ Y'};
17.               stable := stable \ {Y' ∈ Π | (∃Y'' ∈ P_Y) Y' ∈ Pre_Π(Y'')};
18.               Π := (Π \ P_Y) ∪ C_Y;
19.           end if;
20.       end do;
21.       return UNREACHABLE;
22.   end reachability_minim_pb;
```

Fig. 5.28. A minimization algorithm for testing reachability on pseudo-bisimulating models

stops if a state satisfying the property p is reached (obviously, the pseudo-bisimulating model is not generated in this case). Otherwise, it returns the answer $UNREACHABLE$, and the set **classes** contains a stable partition of Q, whose classes constitute a pseudo-bisimulating model for $\mathcal{A}$.

Similarly to the algorithm for bisimulating models, also this one can be implemented for discrete or time-abstracted dense cases.

A Convexity-Preserving Technique

The algorithm for building pseudo-bisimulating models suffers from the same drawback as other minimization algorithms, i.e., computing differences of classes leads to a meaningful inefficiency. However, for strong time-abstracted pseudo-bisimulating models the solution of [159] can be applied. To this aim, the algorithm of Fig. 5.23 is modified in a way similar to that in Fig. 5.20: the function *dpt* is stored together with the partition, the set `reachable` is sorted w.r.t. the values of the function *dpt* which makes the algorithm operate in a BFS-like mode, and the underlined parts of the pseudo-code in Fig. 5.28 are added to the corresponding places of the code in Fig. 5.23. Additionally, the functions `TimeSplit` and `ActionSplit` are modified. Recall that unstability of a class Y_1 w.r.t. its successor Y_2 means that Y_2 has no predecessor Y' which satisfies $\mathrm{AE}(Y', Y_2)$ and $dpt(Y') \leq dpt(Y_1)$ (see p. 135). Given a class $Y = (l, Z) \in \Pi$ with $l \in L$ and $Z \in Z(n_{\mathcal{X}})$,

- the function `TimeSplit`(Y, Π) returns either Y if no unstability of Y w.r.t. its immediate time successor $Y' \in \Pi \setminus \Pi^{ps}$ occurs, or

$$\{pre_{\tau}(Y, Y') \mid Y' = (l, Z) \ \wedge Z \Uparrow Z' \neq \emptyset\}$$

  otherwise. Similarly,
- for $a \in A$ the function `ActionSplit`(Y, a, Π) returns either Y if there is no unstability of Y w.r.t. its a-successors, or

$$\{pre_a(Y, Y') \mid Y' \in \Pi \ \wedge \ Y \xrightarrow{a}_{\Pi} Y'\}$$

  otherwise.

Note that since the algorithm operates in the BFS-like mode the functions are defined correctly, since if unstability of Y w.r.t. its successor Y' occurs, Y is a predecessor of Y' of the smallest depth.

Pseudo-simulating Models

As it has been already stated, the definitions of simulating and pseudo-bisimulating models were combined, resulting in the notion of reachability-preserving pseudo-simulating models. Similarly, the above-described minimization algorithms for these two kinds of models can be put together, giving a method for building pseudo-simulating ones and on-the-fly reachability analysis [126]. In this book this is left to the reader as an easy exercise.

Pseudo-simulating models generated by the minimization algorithm can be discrete or time-abstracted dense, in both the cases for the weakly monotonic semantics. A solution of the problem of explosion which occurs while computing differences of classes has not been found for these models so far.

Other Minimization Techniques

Besides the methods based on the minimization algorithms presented above, some other solutions exist. One of the approaches exploits splitting histories instead of a region-based representation, and operates on a product of the specification of a system and a property [146]. Another technique [94] builds reachability-preserving abstract models, exploiting (timed and untimed) histories of concrete states. The solution is based on a different definition of a concrete state of an automaton, which is represented by a location and a sequence of pairs $(transition, time)$ forming a (timed) history of the state. An abstract model is generated by collecting into classes the concrete states with the same untimed histories (i.e., obtained by executing the same transitions), and with the same future (i.e., the ones which are in a *transition bisimulation* relation, see [94]).

5.2.3 Forward-reachability Graphs

Verification based on reachability analysis is usually performed on an abstract model known as *simulation graph* or *forward-reachability graph.* The nodes of this graph can be defined as (not necessarily convex) sets of detailed regions (see Sect. 5.2.1) [41] or as regions (see Sect. 5.2.2) [63, 98, 173]. Usually, the latter approach is used, which follows from a convenient representation of zones by Difference Bound Matrices (DBMs)[20] [68].

Given a timed automaton $\mathcal{A} = (A, L, l^0, E, \mathcal{X}, \mathcal{I})$ of $n_{\mathcal{X}}$ clocks, let $a \in A$, $l \in L$, $e : l \xrightarrow{a,cc,X} l' \in \mathrm{E}$, and $Z \in Z(n_{\mathcal{X}})$, and let

$$Succ_a((l, Z)) = (l', ((Z \cap [\![cc]\!])[X := 0]) \nearrow \cap [\![\mathcal{I}(l')]\!])$$

The simulation graph can be defined[21] as the smallest transition system $G = (W, w^0, \rightarrow_{\mathsf{a}})$ such that

- $w^0 = (l^0, Z_0)$ with $Z_0 = v^0 \nearrow \cap [\![\mathcal{I}(l^0)]\!]$;
- for any $w = (l, Z) \in W$, and any $a \in A$ such that $e : l \xrightarrow{a,cc,X} l' \in \mathrm{E}$, if $w' = Succ_a(w) \neq \emptyset$, then $w' \in W$ and $w \xrightarrow{a}_{\mathsf{a}} w'$.

The pair (G, V_a), where V_a is a valuation function defined as on p. 90, is a surjective[22] abstract model for the timed automaton $\mathcal{A}$, for the weakly monotonic semantics. The model preserves the LTL formulas (and so reachability properties).

---

[20] We discuss DBMs in details in Sect. 5.3

[21] This definition follows [63]. There are also other approaches, with different notions of Z_0 and the successor relation. For example, in [98], time- and action successors of a node are distinguished, whereas [45] reports $Z_0 = (l^0, v^0)$ and $Succ_a((l, Z)) = (l', ([\![cc]\!] \cap (Z \nearrow))[X := 0])$.

[22] Recall that this means that the model satisfies the conditions EE_1, EE_2 and EA, and its initial state is required to contain q^0 and successors of q^0 only; see p. 91.

INPUT ARGUMENTS:
    a timed automaton $\mathcal{A} = (A, L, l^0, E, \mathcal{X}, \mathcal{I})$ of $n_{\mathcal{X}}$ clocks
    a propositional formula p
GLOBAL VARIABLES:
    $visited, waiting$: $2^{R(n_{\mathcal{X}}, L)}$
RETURN VALUES:
    reachability_forward(): $\{REACHABLE, \ UNREACHABLE\}$

```
1.   function reachability_forward(𝒜, p) is
2.   begin
3.       visited := ∅; waiting := {w⁰};
4.       while waiting ≠ ∅ do
5.           get w from waiting;
6.           if w ⊨ p then return REACHABLE; end if;
7.           if (∀w′ ∈ visited) w ≠ w′ then
8.               visited := visited ∪ {w};
9.               for each a ∈ A s.t. (w′ = Succₐ(w) ∧ w′ ≠ ∅) do
10.                  waiting := waiting ∪ {w′};
11.              end do;
12.          end if;
13.      end do;
14.      return UNREACHABLE;
15.  end reachability_forward;
```

Fig. 5.29. A general reachability algorithm

Simulation graphs are usually generated using a forward-reachability algorithm which, starting from w^0, successively computes all the successors $Succ_a(w)$ for all $w \in W$ generated in earlier steps. A pseudo-code of the algorithm is presented in Fig. 5.29[23]. The algorithm maintains two sets of abstract states: waiting (initially equal to $\{w^0\}$), which contains all these abstract states obtained in the earlier steps which neither have been tested w.r.t. satisfaction of a propositional property p nor have their successors generated, and visited (initially empty), which underwent both the above procedures. Notice that the algorithm can be terminated if a state satisfying p is reached before the whole graph is built. If no state satisfying p can be reached, the algorithm tries to build the whole model for the automaton (i.e., it works as long as the set waiting is non-empty). However, such a model can be infinite (see the example below). Therefore, practical implementations are usually augmented with a termination condition which bounds the number of abstract states to be generated.

---

[23] This algorithm (slightly modified) is also used in backward reachability analysis [173].

Example 5.17. Figure 5.30 shows a timed automaton [63], for which the forward reachability graph is infinite. The graph is presented in Fig. 5.31. The initial state is coloured.

□

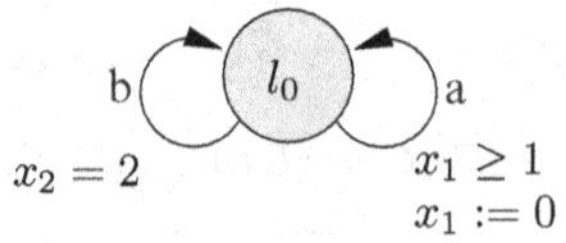

Fig. 5.30. The timed automaton used in Example 5.17

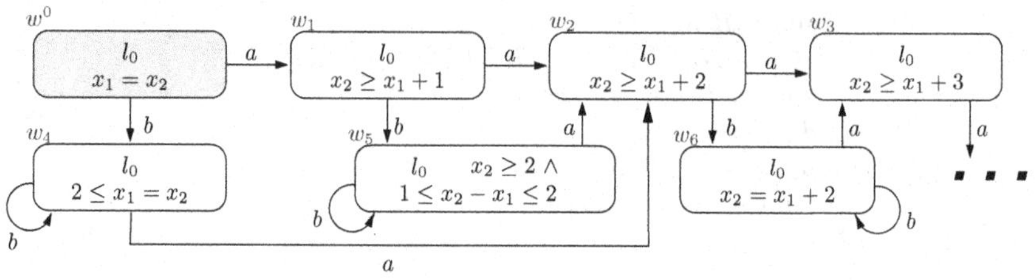

Fig. 5.31. The forward-reachability graph for the automaton of Fig. 5.30

Abstractions for Forward-reachability Graphs

In order to ensure a more efficient reachability verification (but at the cost of loosing preservation of all the LTL formulas), various *abstractions*, enabling to reduce the size of the above model, were defined [25, 26, 63, 98]. Some of them are sketched below.

Inclusion Abstraction

The first solution, which allows to reduce the size of the forward-reachability graph still enabling reachability checking, is called an *inclusion abstraction* [63, 98, 173]. It consists in replacing $w \neq w'$ by

$$w \not\subseteq w'$$

in the line 7. of the algorithm in Fig. 5.29. The idea behind it is that if w is included in an already generated abstract state w', then its successors are a subset of these of w', and therefore generating them adds nothing to the reachability information. An example of a graph obtained this way is shown in the example below:

Example 5.18. Figure 5.32 displays a forward reachability graph for the automaton of Fig. 5.30 after applying the inclusion abstraction. Comparing it with that of Fig. 5.31, the abstract state (region) $w_4 = (l_0, [\![2 \le x_1 = x_2]\!])$ is included in the initial one, and therefore is not present here. Similarly, the state $w_6 = (l_0, [\![x_2 = x_1 + 2]\!])$ is included in $w_5 = (l_0, [\![x_2 \ge 2 \wedge 1 \le x_2 - x_1 \le 2]\!])$. Moreover, all the states of the form $(l_0, [\![x_2 \ge x_1 + c]\!])$ with $c \ge 2$ are included in the state $w_1 = (l_0, [\![x_2 \ge x_1 + 1]\!])$. Notice, however, that the graph is not of the minimal size, since a further reduction is possible.

□

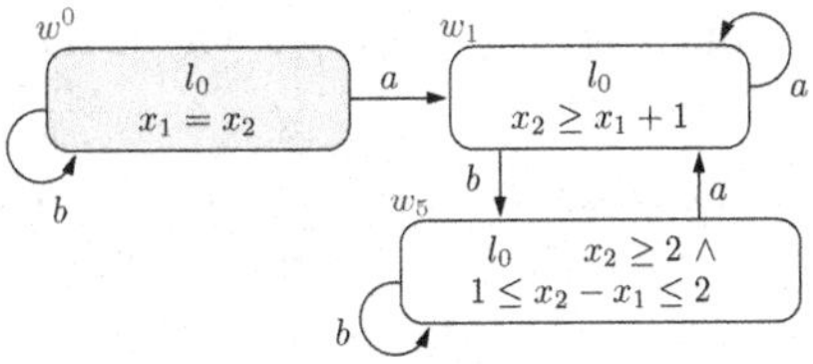

Fig. 5.32. A forward-reachability graph with the inclusion abstraction built for the automaton of Fig. 5.30

It is worth noticing that applying this abstraction does not necessarily make a forward-reachability graph finite (see Example 5.19) and can result in different models for the same automaton, depending on the order the classes are processed.

Example 5.19. Consider the automaton shown in Fig. 5.33 over the three clocks x_1, x_2, x_3. The initial state of its forward-reachability graph is given

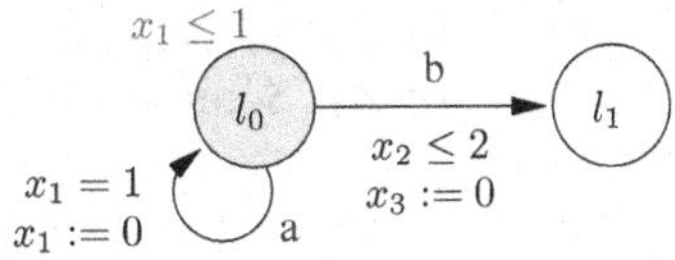

Fig. 5.33. A timed automaton whose forward-reachability graph with the inclusion abstraction is infinite

by $w^0 = (l_0, [\![x_1 = x_2 = x_3 \wedge x_1 \le 1]\!])$. The a-successor w_1 of w^0 is equal to $Succ_a(w^0) = (l_0, [\![x_1 = x_2 = x_3 = 1]\!][\{x_1\} := 0] \nearrow \cap [\![x_1 \le 1]\!]) = (l_0, [\![x_1 \le 1 \wedge 1 \le x_2 = x_3 \le 2]\!])$. Next, $Succ_a(w_1) = (l_0, [\![x_1 = 1 \wedge x_2 = x_3 = 2]\!][\{x_1\} := 0] \nearrow \cap [\![x_1 \le 1]\!]) = (l_0, [\![x_1 \le 1 \wedge 2 \le x_2 = x_3 \le 3]\!])$. It is easy to see that the i-th execution of the action labelled with a gives us the abstract state $(l_0, [\![x_1 \le 1 \wedge i \le x_2 = x_3 \le i + 1]\!])$. The number of the abstract states is thus infinite, and since they are not included one in another, applying the inclusion abstraction does not change this feature of the forward-reachability graph.

□

Extrapolation Abstraction

Another technique, aimed at ensuring finiteness of the model, is a so-called *extrapolation abstraction* (known also as *maximization*, *normalisation*, or *k-approximation*, where k stands for the maximal constant appearing in the constraints of $\mathcal{A}$, i.e., $c_{max}(\mathcal{A})$). Definitions of this abstraction are different for diagonal-free and non-diagonal-free timed automata.

Given a zone $Z = [\![\mathfrak{cc}]\!]$ defined by a clock constraint $\mathfrak{cc} \in \mathcal{C}_{\mathcal{X}}^{\ominus}$ of such a form that none of the atomic constraints in $\mathfrak{cc}$ can be strengthened[24] without reducing $[\![\mathfrak{cc}]\!]$. In the case of diagonal-free automata the extrapolation of Z, denoted $extr(Z)$, is a zone $extr(Z) = [\![\mathfrak{cc}_1]\!]$, where $\mathfrak{cc}_1$ is a clock constraint obtained from $\mathfrak{cc}$ by

- removing all the atomic constraints of the form $x \sim c$ and $x - y \sim c$, where $\sim \in \{\leq, <\}$ and $c > c_{max}(\mathcal{A})$ (i.e., upper bounds greater than $c_{max}(\mathcal{A})$ are eliminated),
- replacing all the atomic constraints of the form $x \sim c$ and $x - y \sim c$, where $\sim \in \{\geq, >\}$ and $c > c_{max}(\mathcal{A})$, by $x > c_{max}(\mathcal{A})$ and $x - y > c_{max}(\mathcal{A})$, respectively (i.e., the lower bounds greater than $c_{max}(\mathcal{A})$ are replaced by $c_{max}(\mathcal{A})$),
- leaving the rest of atomic constraints in $\mathfrak{cc}$ untouched.

The intuition behind this approach, and simultaneously an explanation for the correctness of this abstraction, is twofold. Firstly, for any constraint of $\mathcal{A}$ involving a clock x, the exact value of $v(x)$ is insignificant if greater than the maximal value this clock is compared with. Secondly, if the intersection of a region R and a detailed region r is non-empty, then extending R to contain all the states of r does not change the reachability information. Thus, the abstraction modifies the zones whose intersections with open detailed zones (defined for a diagonal-free automaton and $c_{max}(\mathcal{A}, \varphi) = c_{max}(\mathcal{A})$) are non-empty. It is easy to see that $Z \subseteq extr(Z)$. Some examples of extrapolations of zones are presented in Fig. 5.34.

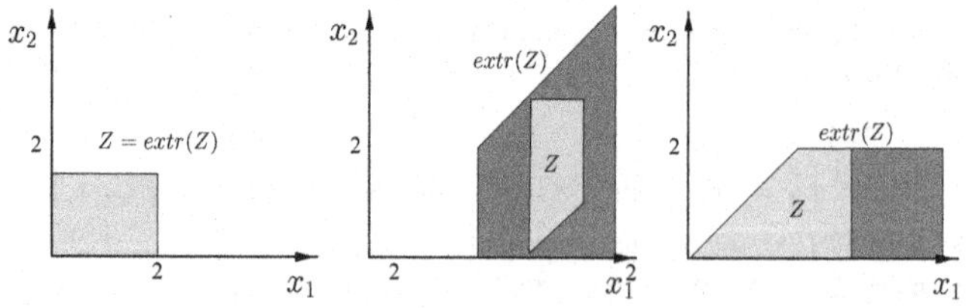

Fig. 5.34. Extrapolations of zones for $c_{max}(\mathcal{A}) = 2$ (diagonal-free TA)

[24] By the strengthening of an atomic constraint $x_i - x_j \sim c$ or $x_i \sim c$, where $\sim \in \{\leq, <\}$, we mean replacing $\leq$ by $<$, or c by c' with $c' < c$.

It is also important to notice that in practice, instead of one common constant $c_{max}(\mathcal{A})$, a maximal constant for each clock can be used.

When the extrapolation abstraction is applied, in the generated simulation graph:

- Z_0 is replaced by $extr(Z_0)$, and
- $Succ_a((l, Z))$ is redefined to be of the form

$$Succ_a((l, Z)) = (l', extr(((Z \cap [\![cc]\!])[X := 0]) \nearrow \cap [\![\mathcal{I}(l')]\!])).$$

This makes the number of nodes of the abstract states in the graph finite, since the number of extrapolated zones is so.

Example 5.20. Figure 5.35 shows a forward-reachability graph for the automaton $\mathcal{A}$ in Fig. 5.30 obtained by applying the extrapolation abstraction. Since $c_{max}(\mathcal{A}) = 2$, the zone $[\![x_2 \geq x_1 + 3]\!]$, obtained while generating the graph, is replaced with $[\![x_2 > x_1 + 2]\!]$.

□

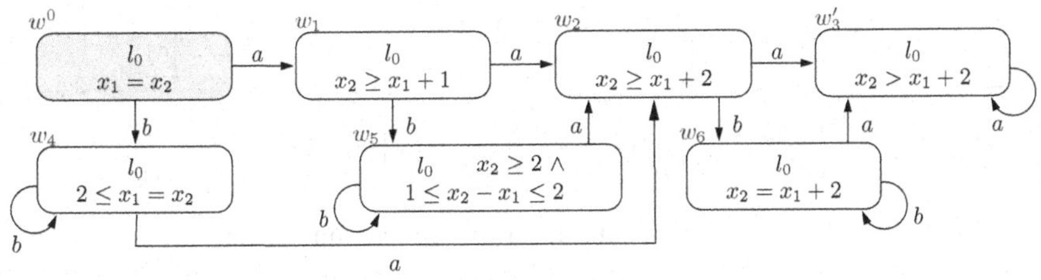

Fig. 5.35. A forward-reachability graph with the extrapolation abstraction built for the automaton of Fig. 5.30

In [45] it was proven that the above method does not work for automata whose constraints contain comparisons of clocks (i.e., incorrect results for reachability analysis can be obtained). This is shown in the example below.

Example 5.21. Figure 5.36 shows a timed automaton whose forward-reachability graph is presented in Fig. 5.37. The location l_6 of the automaton is not reachable, which can be easily derived from the clock constraint describing the final state of the model: if we assume that the condition $x_4 < x_3 + 2$ (i.e., the second conjunct of the guard of the transition labelled with a_6) holds, then from the above and from the constraint describing the state w_5 we have

$$x_2 - x_1 = (x_2 - x_4) + (x_4 - x_3) + (x_3 - x_1) < -5 + 2 + 5 = 2,$$

and therefore the first conjunct of the guard of the action labelled with a_6 is not satisfied. However, if the extrapolation abstraction described above is applied, then the atomic constraints $x_2 - x_4 \leq -5$ and $x_3 - x_1 \leq 5$ occurring

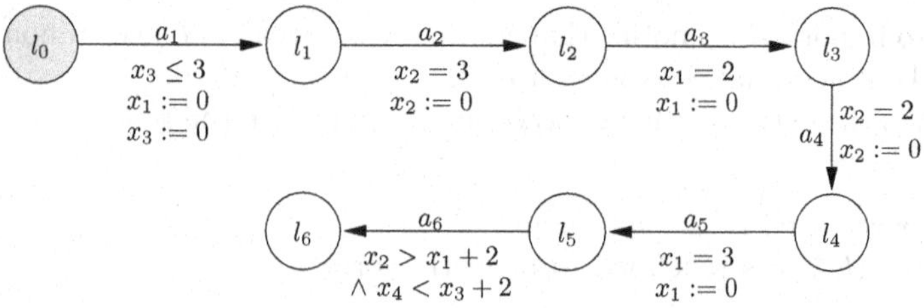

Fig. 5.36. A timed automaton with clock differences

in the description of w_5 are replaced by $x_2 - x_4 \le -3$ and $x_3 - x_1 \le \infty$, which allows to execute the action a_6 and reach the location l_6.

□

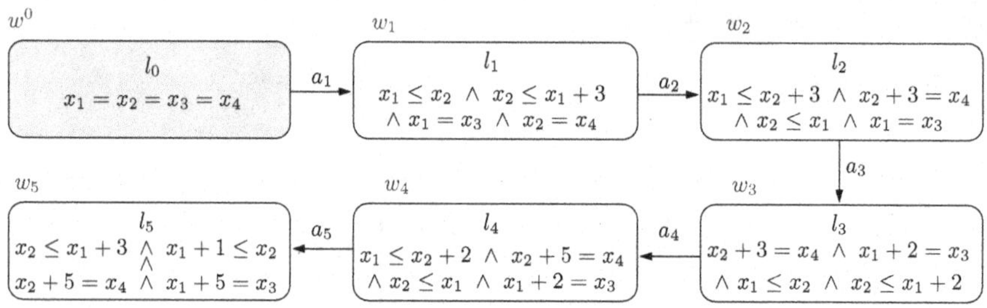

Fig. 5.37. The forward-reachability graph for the automaton of Fig. 5.36

An extrapolation abstraction for automata with diagonal constraints were described in [25–27]. Its main idea is that if an atomic constraint of $\mathcal{A}$ which compares two clocks is not satisfied by any clock valuation of a zone, then it should not be satisfied by any clock valuation of the extrapolated one. Similarly, if all the clock valuations of a zone satisfy a difference constraint, then so should also all the clock valuations of the extrapolated one.

Let D be a set of the atomic constraints which compare clocks and appear in the enabling conditions and invariants of $\mathcal{A}$. Given a zone $Z = [\![\mathfrak{cc}]\!]$ defined by a clock constraint $\mathfrak{cc} \in \mathcal{C}_{\mathcal{X}}^{\ominus}$ of such a form that none of the atoms in $\mathfrak{cc}$ can be strengthened without reducing $[\![\mathfrak{cc}]\!]$, we build its diagonal-preserving extrapolation $extr_d(Z)$ by applying the following steps:

- splitting Z in such a way that in every resulting part Z', each constraint of D either holds for each $v \in Z'$, or for none of them. The resulting set of zones is denoted $split(Z)$,
- for each zone $Z' \in split(Z)$
  - collecting all the constraints $\mathfrak{cc}' \in D$ such that
    - $[\![\mathfrak{cc}']\!] \cap Z' = \emptyset$ (i.e., none of the clock valuations in Z' satisfies $\mathfrak{cc}'$),
    - $[\![\neg\mathfrak{cc}']\!] \cap Z' = \emptyset$ (i.e., all the clock valuations of Z' satisfy $\mathfrak{cc}'$),

and defining the set

$$D_{unsat}(Z') = \{\mathfrak{cc}' \in D \mid [\![\mathfrak{cc}']\!] \cap Z' = \emptyset\} \cup \{\neg\mathfrak{cc}' \mid [\![\neg\mathfrak{cc}']\!] \cap Z' = \emptyset \ \wedge \ \mathfrak{cc}' \in D\},$$

– computing

$$ex_d(Z') = extr(Z') \cap [\![\bigwedge_{\mathfrak{cc}' \in D_{unsat}(Z')} \neg\mathfrak{cc}']\!]$$

(i.e., using all the constraints in $D_{unsat}(Z')$ to "cut" the extrapolation of the zone Z'),

- the diagonal-preserving extrapolation of Z is the union

$$extr_d(Z) = \bigcup_{Z' \in split(Z)} ex_d(Z').$$

In the simulation graph generated for the automaton $\mathcal{A}$:

- Z_0 is replaced by $extr_d(Z_0)$ (notice, however, that this equals to $extr(Z_0)$),
- $Succ_a((l, Z))$ is redefined to be of the form

$$Succ_a((l, Z)) = \{(l', Z') \mid Z' \in extr_d(((Z \cap [\![\mathfrak{cc}]\!])[X := 0]) \nearrow \cap [\![\mathcal{I}(l')]\!])\},$$

- the successor relation $\rightarrow_{\mathfrak{a}}$ of p. 138 is redefined to be of the form
  – for any $w = (l, Z) \in W$, and any $a \in A$ such that $e : l \xrightarrow{a,\mathfrak{cc},X} l' \in$ E, if $Succ_a(w) \neq \emptyset$, then for each $w' \in Succ_a(w)$ we have $w' \in W$ and $w \overset{a}{\rightarrow}_{\mathfrak{a}} w'$,

  and
- the line 9. of the algorithm in Fig. 5.29 is replaced by

  9. **for each** $a \in A$ s.t. $w' \in Succ_a(w) \wedge w' \neq \emptyset$ **do** ...

  (i.e., the equality is replaced by " $\in$ ").

Again, the graph computed this way is finite.

Other Abstractions

Besides the abstractions described above, some other solutions exist. One of them, called *convex-hull abstraction*, consists in keeping a single region (l, Z_{ch}) for each $l \in L$, where Z_{ch} is the *convex hull* of all the zones appearing in the abstract states of the form $(l, \cdot)$ generated by the algorithm, where the convex hull of two zones Z', Z'' is their smallest superset (see Fig. 5.38). This, however, allows for proving unreachability only, since some states which are not reachable in the concrete state space can be reachable in the abstracted one.

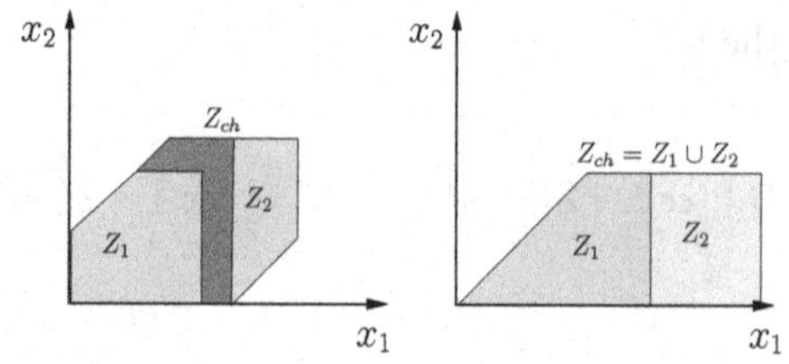

Fig. 5.38. Convex hulls for zones Z_1, Z_2

Other solutions are based on the idea of identifying a set of location-based maximal constants (i.e., counterparts of $c_{max}(\mathcal{A})$ that depend on the particular locations of the automaton), or on exploiting both the minimal and the maximal constants the clocks of $\mathcal{A}$ are compared with [22,23]. There are also many approaches aimed at reducing the memory usage while generating the model. These concern storing the valuations of *active clocks* only (a clock is considered *active* if it usefully counts time, i.e., from a time point where it is reset up to the time point where it is tested), modifications introduced to $Waiting$ and $Visited$, for example storing only some nodes sufficient for preserving reachability information [98], applying memory-saving data structures and many others [25,98].

To store and operate on abstract models usually Difference Bound Matrices [68] are used for regions of TA.

5.3 Difference Bounds Matrices

Difference Bounds Matrices are used for representing states of abstract models for timed systems. Therefore, we give a quite detailed account of them below.

Recall that each time zone $Z \in Z(n_{\mathcal{X}})$ is a (possibly unbounded) polyhedron in $\mathbb{R}_{0+}^{n_{\mathcal{X}}}$ defined by a clock constraint $\mathfrak{cc}$, i.e., $Z = [\![\mathfrak{cc}]\!]$. Notice, moreover, that $\mathfrak{cc} \in \mathcal{C}_{\mathcal{X}}^{\ominus}$ is a conjunction of a finite number of atomic clock constraints. For the sake of convenience, we assume that a zone is given by a normalised clock constraint (i.e., $\mathfrak{cc} \in \mathcal{C}_{\mathcal{X}+}^{\ominus}$, see Chap. 2). Then, we introduce the domain of *bounds*. Thus, a *bound* is an ordered pair

$$(c, r) \in (\mathbb{Z} \times \{<, \leq\}) \cup \{(\infty, <), (-\infty, <)\}.$$

The symbols $<$ and $\leq$ are totally ordered, i.e., $<$ is taken to be strictly less than $\leq$. The ordering of bounds is defined as

$$(c, r) \preceq (c', r') \text{ if either } c < c', \text{ or } c = c' \text{ and } r \leq r'.$$

Then, we introduce the notion of *Difference Bounds Matrices* [68].

Definition 5.22 (Difference Bounds Matrix). *A* difference bounds matrix *(*DBM*) in* $\mathbb{R}_{0+}^{n_\mathcal{X}}$ *is an* $(n_\mathcal{X}+1) \times (n_\mathcal{X}+1)$ *matrix of bounds, with rows and columns indexed from* 0 *to* $n_\mathcal{X}$. *The DBM*

$$D = (\mathbf{d}_{ij}),$$

where for each $i, j \in \{0, \ldots, n_\mathcal{X}\}$ $\mathbf{d}_{ij} = (d_{ij}, \sim_{ij})$, *represents the zone*

$$Z = [\![\bigwedge_{i=0}^{n_\mathcal{X}} \bigwedge_{j=0}^{n_\mathcal{X}} x_i - x_j \sim_{ij} d_{ij}]\!].$$

The zone of D *will be denoted by* $[\![D]\!]$.

It is known that for each zone Z there is a DBM D s.t. $Z = [\![D]\!]$, but there could be possibly many such D's. Thus, in order to implement the operations on zones defined in Sect. 2.1, we need to deal with *canonical* DBMs. For such DBMs all upper bounds are as "tight" as possible.

Definition 5.23 (Canonical form). *The* canonical form *of a DBM* $D = (\mathbf{d}_{ij})$ *in* $\mathbb{R}_{0+}^{n_\mathcal{X}}$, *denoted by* $cf(D)$, *is a DBM*

$$D^c = (\mathbf{d^c}_{ij})$$

in $\mathbb{R}_{0+}^{n_\mathcal{X}}$ *such that*

- $[\![D]\!] = [\![D^c]\!]$, *and*
- *for each DBM* $D' = (\mathbf{d}'_{ij})$ *in* $\mathbb{R}_{0+}^{n_\mathcal{X}}$ *with* $[\![D^c]\!] = [\![D']\!]$ *for each* $i, j \in \{0, \ldots, n_\mathcal{X}\}$ *we have* $\mathbf{d^c}_{ij} \preceq \mathbf{d}'_{ij}$.

The DBM in the canonical form is also called a canonical DBM.

Fig. 5.39 shows a graphical interpretation of bounds of the canonical form of a DBM. Let $[\![D]\!] \in Z(n_\mathcal{X})$ be a non-empty zone[25] defined by a canonical DBM matrix $D = (\mathbf{d}_{ij})$ with the elements $\mathbf{d}_{ij} = (d_{ij}, \sim_{ij})$, where $i, j \in \{1, \ldots, n_\mathcal{X}\}$, $d_{i,j} \in \mathbb{Z} \cup \{\infty\}$ and $\sim_{ij} \in \{<, \leq\}$. The zone in the picture, which is the projection of the zone $[\![D]\!]$ onto the plane given by the clocks $x_i, x_j \in \mathcal{X}$, is constrained by the dashed lines given in the picture and corresponding to the bounds in D. If for some $m, k \in \{0, i, j\}$ we have $d_{mk} = \infty$, then the corresponding line does not exist.

The canonical form of a DBM can be computed by applying the shortest-path algorithm [68]. Representation via DBMs enables very efficient tests for equality and emptiness of zones, i.e., for DBMs D, D',

$$[\![D]\!] = [\![D']\!] \text{ iff } cf(D) = cf(D'),$$

and

$$[\![D]\!] = \emptyset \text{ iff the bound } (-\infty, <) \text{ appears in } cf(D).$$

[25] Thanks to the assumption about non-emptiness of the zone we do not need to consider the bounds of the form $(-\infty, <)$ in the example.

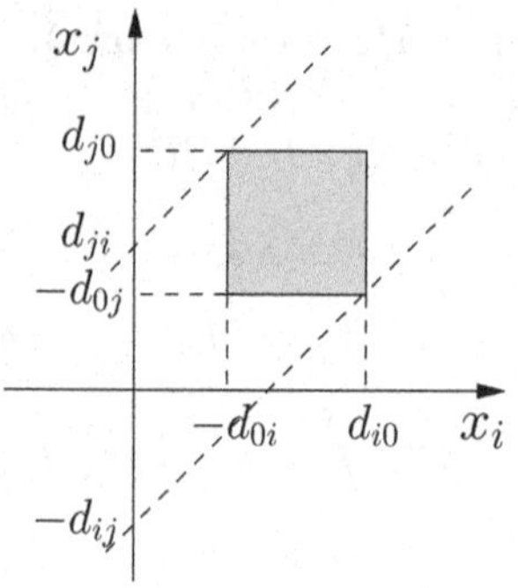

Fig. 5.39. A graphical interpretation of bounds of the canonical DBM

5.3.1 Operations on Zones Using DBMs

In order to implement some algorithms for generating abstract models for timed automata or time Petri nets, the operations of computing intersection of two zones, (immediate) time-predecessor and time-successor of a zone, clock resets, and difference of two zones have to be applied. Below, we show how to implement these operations. In general, the ideas behind the implementations have been taken from [8, 157], but our presentation differs in some details.

Intersection

Let $Z, Z' \in Z(n_{\mathcal{X}})$ be two zones, and let $D = (\mathbf{d}_{ij})$ and $D' = (\mathbf{d}'_{ij})$ be two DBMs such that $Z = [\![D]\!]$ and $Z' = [\![D']\!]$. Computing the matrix $D^{is} = (\mathbf{d^{is}}_{ij})$ such that $[\![D^{is}]\!] = Z \cap Z'$, consists in taking the lower of the bounds for each pair of clock differences, that is, for all $i, j \in \{0, \ldots, n_{\mathcal{X}}\}$,

$$\mathbf{d^{is}}_{ij} = \min(\mathbf{d}_{ij}, \mathbf{d}'_{ij}).$$

D, D' need not be canonical, and D^{is}, in general, is not canonical either. Notice that also the canonical forms of D and D' do not imply canonical form of their intersection. An example is shown in Fig. 5.40. The dashed lines correspond to the bounds in D^{is}, computed as presented above, which remain unchanged also in $cf(D^{is})$. The solid line indicates the bound computed for the intersection and resulting in a non-canonical form of the matrix D^{is}.

Clock Resets

Let $D = (\mathbf{d}_{ij})$ be a canonical DBM for a zone $Z \in Z(n_{\mathcal{X}})$, and let $X \subseteq \mathcal{X}$ be a set of clocks to be reset. A DBM $D' = (\mathbf{d}'_{ij})$ satisfying $[\![D']\!] = Z[X := 0]$ is given by

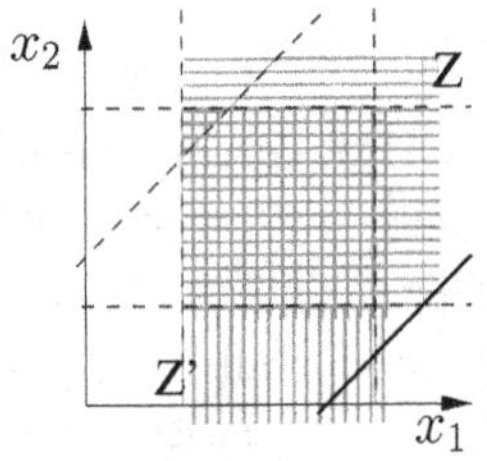

Fig. 5.40. A non-canonical DBM for intersection of two zones represented by canonical DBMs

$$\begin{array}{lll} \mathbf{d}'_{i0} = \mathbf{d}'_{0i} = (0, \leq) & \text{if} & x_i \in X \\ \mathbf{d}'_{ji} = (\infty, <) & \text{if} & x_i \in X \wedge x_j \notin X \\ \mathbf{d}'_{ij} = \mathbf{d}_{ij} & \text{otherwise.} & \end{array}$$

In general, D' does not need to be canonical.

Computing a DBM $D'' = (\mathbf{d}''_{ij})$ such that $[\![D'']\!] = [X := 0]Z$ is performed in three steps:

- Firstly, we check whether $\mathbf{d}_{0i} = (0, \leq)$ for all $x_i \in X$ (if the condition is not satisfied, then $[X := 0]Z = \emptyset$), and if so, set to 0 all the upper bounds of the values of clocks to be reset, which is done by creating a DBM $D' = (\mathbf{d}'_{ij})$ with
$$\begin{array}{ll} \mathbf{d}'_{i0} = (0, \leq) & \text{if } x_i \in X \\ \mathbf{d}'_{ij} = \mathbf{d}_{ij} & \text{otherwise.} \end{array}$$
- Secondly, we compute the canonical form for D'.
- Finally, we compute the matrix D'' by setting
$$\begin{array}{ll} \mathbf{d}''_{ij} = (\infty, <) & \text{for all } x_i \in X \text{ and } j \in \{0, \ldots, n_{\mathcal{X}}\} \text{ with } i \neq j, \text{ and} \\ \mathbf{d}''_{ij} = \mathbf{d}'_{ij} & \text{in all the other cases.} \end{array}$$

An example is shown in Fig. 5.41. The result does not need to be canonical.

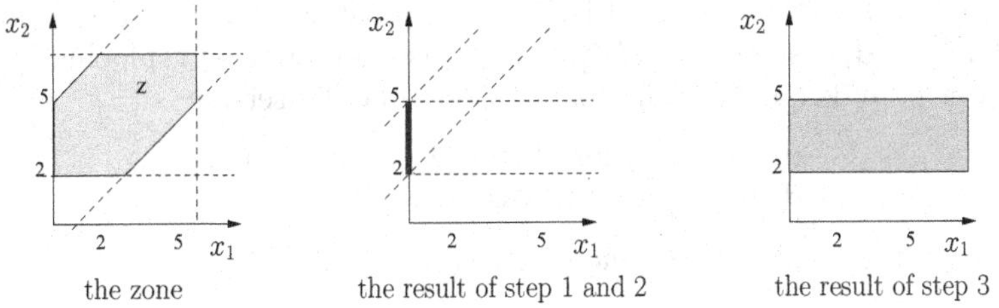

Fig. 5.41. The steps of computing $[\{x_1\} := 0]Z$

Time Successor

Let $D = (\mathbf{d}_{ij})$ be a canonical DBM, corresponding to a zone $Z \in Z(n_{\mathcal{X}})$. The DBM $D' = (\mathbf{d}'_{ij})$ such that $[\![D']\!] = Z \nearrow$ is computed by removing in D all the upper bounds on the values of clocks, i.e., setting

$$\begin{array}{ll} \mathbf{d}'_{i0} = (\infty, <) & \text{for all } 1 \leq i \leq n_{\mathcal{X}}, \text{and} \\ \mathbf{d}'_{ij} = \mathbf{d}_{ij} & \text{otherwise.} \end{array}$$

The output is canonical.

The example in Fig. 5.42 shows that the requirement of a canonical form of the matrix D, for which the time successor is computed, is necessary. The DBM $D = (\mathbf{d}_{ij})$, corresponding to the zone Z, contains the bounds $\mathbf{d}_{10} = \mathbf{d}_{20} = (5, \leq)$, $\mathbf{d}_{01} = \mathbf{d}_{02} = (-2, \leq)$, and the bounds corresponding to the dashed lines given in the picture. Notice that the zones Z' and Z'', represented by the DBMs computed by the above algorithm applied to the canonical (see part (a)) and non-canonical (part (b)) DBM representing the zone Z, are different.

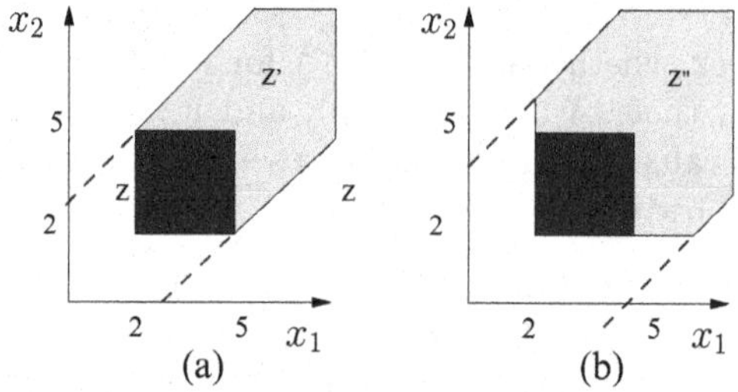

Fig. 5.42. (a) A correct time-successor of the zone represented by a canonical DBM, (b) an incorrect time-successor of the zone represented by a non-canonical DBM

Time Predecessor

Let $D = (\mathbf{d}_{ij})$ be the canonical DBM corresponding to a zone $Z \in Z(n_{\mathcal{X}})$. A DBM $D' = (\mathbf{d}'_{ij})$ such that $[\![D']\!] = Z \swarrow$, is computed by replacing by 0 all the lower bounds on the values of the clocks, i.e, by setting

$$\begin{array}{ll} \mathbf{d}'_{0i} = (0, \leq) & \text{for all } i \in \{1, \ldots, n_{\mathcal{X}}\}, \text{ and} \\ \mathbf{d}'_{ij} = \mathbf{d}_{ij} & \text{otherwise.} \end{array}$$

In general, the result is not canonical (see Fig. 5.43).

Immediate Time Predecessor

Let $Z, Z' \in Z(n_{\mathcal{X}})$. The implementation of $Z \Uparrow Z'$ is more involved. To this aim we need some auxiliary operations.

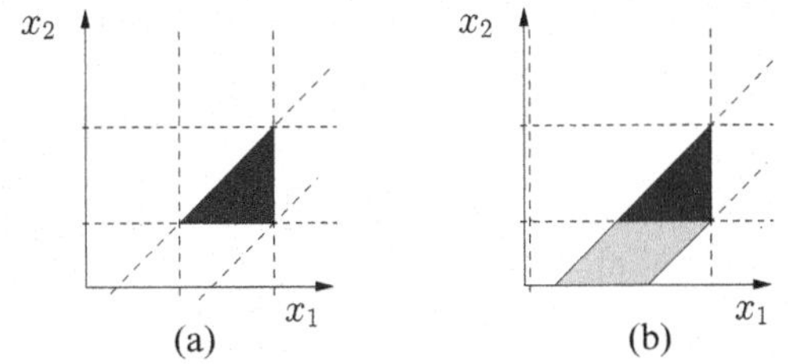

Fig. 5.43. A non-canonical result of computing $Z \swarrow$ (b) in spite of the canonical input (a). The dashed lines represent constraints appearing in DBMs

Let $D = (\mathbf{d}_{ij})$ and $D' = (\mathbf{d}'_{ij})$ be two canonical DBMs representing, respectively, the zones $Z, Z' \in Z(n_{\mathcal{X}})$. The first operation to be introduced is the closure of the zone Z, denoted by $Closure(Z)$. The matrix $D^{cl} = (\mathbf{d^{cl}}_{ij})$ such that $[\![D^{cl}]\!] = Closure(Z)$, is given by

$$\begin{array}{ll} \mathbf{d^{cl}}_{ij} = (d_{ij}, \leq) & \text{if } \mathbf{d}_{ij} = (d_{ij}, <), \text{ where } d_{ij} \neq \infty, \text{and} \\ \mathbf{d^{cl}}_{ij} = \mathbf{d}_{ij} & \text{otherwise.} \end{array}$$

The output is canonical.

The next operation is $Fill(Z)$, which consists in replacing $<$ by $\leq$ in all the lower and upper bounds of the values of clocks. The matrix $D^{fi} = (\mathbf{d^{fi}}_{ij})$ such that $[\![D^{fi}]\!] = Fill(Z)$ is given by

$$\begin{array}{ll} \mathbf{d^{fi}}_{i0} = (d_{i0}, \leq) & \text{if } \ \mathbf{d}_{i0} = (d_{i0}, <) \wedge d_{i0} \neq \infty \\ \mathbf{d^{fi}}_{0i} = (d_{0i}, \leq) & \text{if } \ \mathbf{d}_{0i} = (d_{0i}, <) \wedge d_{i0} \neq \infty \\ \mathbf{d^{fi}}_{ij} = \mathbf{d}_{ij} & \text{otherwise.} \end{array}$$

In the case of canonical input, the output of this operation is canonical as well.

Next, for two disjoint zones $Z, Z' \in Z(n_{\mathcal{X}})$ we define their *border* as follows:

$$\begin{array}{l} border(Z, Z') = \{v \in \mathbb{R}^{n_{\mathcal{X}}}_{0+} \mid \\ \quad (\, (v \in Z \text{ and } (\exists \delta > 0) \text{ s.t. } (v + \delta \in Z' \wedge (\forall 0 < \delta' < \delta)\ v + \delta' \in Z'))\,) \vee \\ \quad (\, (v \in Z' \text{ and } (\exists \delta > 0) \text{ s.t. } (v - \delta \in Z \wedge (\forall 0 < \delta' < \delta)\ v - \delta' \in Z))\,)\} \end{array}$$

(notice that the ordering of arguments is important). In case when the zones Z, Z' are not disjoint,

$$border(Z, Z') = Closure(Z \cap Z').$$

Five examples of borders of two zones are presented in Fig. 5.44.

In order to compute a DBM $D^{bd} = (\mathbf{d^{bd}}_{ij})$ such that $[D^{bd}] = border(Z, Z')$ we apply the following algorithm:

- if $Z \cap Z' \neq \emptyset$, then compute D^{bd} to get $[\![D^{bd}]\!] = Closure(Z \cap Z')$;
- otherwise:

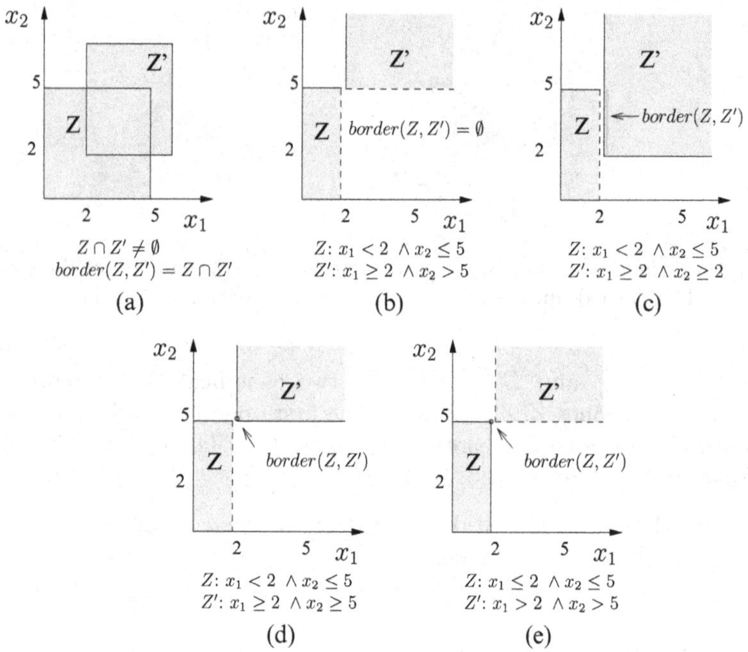

Fig. 5.44. Examples of borders of two zones

- if there exists $1 \leq i \leq n_{\mathcal{X}}$ and a constant $c \in \mathbb{Z}$ such that

$$\mathbf{d}_{i0} = (c, <) \;\wedge\; \mathbf{d}'_{0i} = (-c, \leq)$$

  then
  - if there exists $1 \leq j \leq n_{\mathcal{X}}$ with $j \neq i$ and a constant $c' \in \mathbb{Z}$ such that

$$\mathbf{d}_{j0} = (c', \leq) \;\wedge\; \mathbf{d}'_{0j} = (-c', <),$$

    then $[\![D^{bd}]\!] = \emptyset$ (see Fig. 5.44 (b));
  - otherwise compute D^{bd} to get $[\![D^{bd}]\!] = \mathit{Fill}(Z) \cap Z'$ (see Fig. 5.44 (c,d)),
- otherwise if there exists $1 \leq i \leq n_{\mathcal{X}}$ and a constant $c \in \mathbb{Z}$ such that

$$\mathbf{d}_{i0} = (c, \leq) \;\wedge\; \mathbf{d}'_{0i} = (-c', <)$$

  then compute D^{bd} to get $[\![D^{bd}]\!] = Z \cap \mathit{Fill}(Z')$ (see Fig. 5.44 (e)),
- else $[\![D^{bd}]\!] = \emptyset$.

The result, in general, is not canonical.

Finally, we can compute the DBM $D'' = (\mathbf{d}''_{ij})$, representing the zone $Z \Uparrow Z'$, from

$$[\![D'']\!] = Z \cap (border(Z, Z') \swarrow).$$

Notice that for computing this, the canonical form of the matrix representing *border*(Z, Z') is required. The matrix D'', in general, is not canonical.

The above algorithm is a slight modification of the one shown in [157].

Zone Difference

The implementation of difference of two zones $Z, Z' \in Z(n_{\mathcal{X}})$ represented by the canonical DBMs $D = (\mathbf{d}_{ij})$ and $D' = (\mathbf{d}'_{ij})$, respectively, consists in generating a partition of $Z \setminus Z'$ by successively slicing off these parts of Z which do not lie in Z'. We consider in turn each bound $\mathbf{d}'_{ij} = (d_{ij}, \sim)$ with $i \neq j$, $d_{ij} \neq \infty$ and $\sim \in \{<, \leq\}$ in D, as a potential face along which to slice Z. Slicing along this face is necessary if it "touches" any points in Z. By the operation $restrict(Z, \mathfrak{cc})$, for $\mathfrak{cc} \in \mathcal{C}^{\ominus}_{\mathcal{X}+}$, we mean restriction of the zone Z to the clock valuations which satisfy $\mathfrak{cc}$ (i.e., $restrict(Z, \mathfrak{cc}) = Z \cap [\![cc]\!]$). By $\overline{\sim}$ we mean "$\leq$" if $\sim = <$, and "$<$" otherwise. The set of DBMs representing the elements of $Z \setminus Z'$ can be generated using the algorithm shown in Fig. 5.45.

Pseudo-codes for (efficient) implementations of operations on DBMs can be found in [25, 27].

Further reading

Some overviews of the techniques of generating abstract models for timed systems can be found e.g. in [27, 159, 176].

INPUT ARGUMENTS:

a set of clocks $\mathcal{X}$ with $|\mathcal{X}| = n_\mathcal{X}$

zones Z, Z' represented by DBMs $D = (\mathbf{d}_{ij})$ and $D' = (\mathbf{d'}_{ij})$, respectively

RETURN VALUES:

zone_difference(): $2^{Z(n_\mathcal{X})}$,

restrict(): $Z(n_\mathcal{X})$

```
1.    procedure zone_difference(Z, Z', X) is
2.    begin
3.       done := false;
4.       Z \ Z' := ∅;
5.       compute the DBM D^C st [[D^C]] = Closure(Z');
6.       for each x_i, x_j ∈ X⁺ s.t. x_i ≠ x_j do
7.          if (¬done) then
8.             compute the DBM D^B s.t. [[D^B]] = Z ∩ Z';
9.             if [[D^B]] = ∅ then
10.               Z \ Z' := (Z \ Z') ∪ {[[D]]};
11.               done := true;
12.            else
13.               if [[D^B]] = Z then done := true;
14.               else
15.                  compute the DBM D^D s.t. [[D^D]] =
                     restrict([[D^C]], x_i − x_j ≤ d'_ij ∧ x_j − x_i ≤ −d'_ij); // equality
16.                  if [[D^D]] ∩ Z ≠ ∅ then
17.                     compute the DBM D^E s.t.
18.                        [[D^E]] = restrict(Z, x_j − x_i ≈ − d'_ij);
19.                     Z \ Z' := (Z \ Z') ∪ {[[D^E]]};
20.                     compute the DBM D for Z := restrict(Z, x_i − x_j ∼ d'_ij);
21.                  end if;
22.               end if;
23.            end if;
24.         end if;
25.      end do;
26.   end zone_difference;
```

Fig. 5.45. An algorithm for computing difference of zones

6

Explicit Verification

The aim of this chapter is to show how one can explicitly verify the common properties of timed systems expressible in TCTL ($TCTL_C$) and its sublogics. It turns out that most of the approaches is based on translations of the model checking problem of TCTL to the model checking problem of CTL. Thus, the first section of this chapter deals with several methods of model checking for CTL over finite Kripke models, whereas the second section shows translations.

We discuss also whether and how the model checking methods for CTL can be extended to CTL^* and modal μ-calculus. Notice that our general model checking algorithms for CTL can be directly applied to verifying timed systems, as we defined finite-state abstract models of timed automata and time Petri nets in the previous chapter.

The last two sections of this chapter overview other explicit approaches to model checking of TCTL, which are not based on a translation to CTL model checking, as well as some verification tools exploiting the above approaches.

6.1 Model Checking for CTL

There are several model checking methods for CTL. To make an easy introduction to these methods, we start with showing the simplest model checking algorithm based on a state labelling. Next, we present an automata-theoretic approach to CTL model checking.

6.1.1 State Labelling

If we do not bother about the size of the model, then the simplest approach to CTL model checking, called *state labelling*, can be used. Below, we show a deterministic algorithm, based on state labelling, for determining whether a

W. Penczek and A. Półrola: *Explicit Verification*, Studies in Computational Intelligence (SCI) **20**, 155–180 (2006)
www.springerlink.com

CTL formula φ is true at a state $s \in S$ in a finite model $M = ((S, s^0, \rightarrow), V)$, of complexity $O(|\varphi| \times (|S| + |\rightarrow|))$.[1]

The algorithm shown here is designed so that when it finishes, each state s of M is labelled with the subformulas of φ which are true at s. The algorithm operates in stages. The i-th stage handles all subformulas of φ of length i for $i \leq |\varphi|$. Thus, at the end of the last stage each state will be labelled with all subformulas of φ which are true at it. In this process we use the following equivalences (cf. p. 71):

- $\mathrm{A}(\varphi \mathrm{U} \psi) \equiv \neg(\mathrm{E}(\neg\psi \mathrm{U}(\neg\varphi \wedge \neg\psi)) \vee \mathrm{EG}(\neg\psi))$,
- $\mathrm{A}(\varphi \mathrm{R} \psi) \equiv \neg \mathrm{E}(\neg\varphi \mathrm{U} \neg\psi)$,
- $\mathrm{E}(\varphi \mathrm{R} \psi) \equiv \neg \mathrm{A}(\neg\varphi \mathrm{U} \neg\psi)$.

Thus, each of the operators of CTL can be expressed in terms of the three operators EX, EG, and EU. Because of that only six cases have to be considered, depending on whether φ is a proposition or is in one of the following forms: $\neg\psi$, $\psi_1 \wedge \psi_2$, $\mathrm{EX}\psi$, $\mathrm{E}(\psi_1 \mathrm{U} \psi_2)$, or $\mathrm{EG}\psi$. The algorithm is discussed for the last two cases only, as the others are straightforward.

To handle a formula of the form $\varphi = \mathrm{E}(\psi_1 \mathrm{U} \psi_2)$, the algorithm first finds all the states which are labelled with ψ_2 and labels them with φ. Then, it goes backwards using the relation $\rightarrow^{-1}$ and finds all the states which can be reached by a path in which each state is labelled with ψ_1. All such states are labelled with φ. This step requires time $O(|S| + |\rightarrow|)$.

Now, the case when $\varphi = \mathrm{EG}\psi$ is considered. Firstly, the graph $(S', \rightarrow')$ is constructed, where $S' = \{s \in S \mid M, s \models \psi\}$ and $\rightarrow' = \rightarrow \cap (S' \times S')$. Secondly, $(S', \rightarrow')$ is partitioned into strongly connected components[2] and those states which belong to the components of size greater than 1 or with a self-loop are selected. Finally, the algorithm goes backwards from these states using $\rightarrow^{-1}$ and finds all those states which can be reached by a path in which each state is labelled with ψ. This step also requires time $O(|S| + |\rightarrow|)$. In order to handle an arbitrary CTL formula φ, the state labelling algorithm is successively applied to the subformulas of φ, starting with the shortest and most deeply nested ones. Since each pass takes time $O(|S| + |\rightarrow|)$ and since φ has at most $|\varphi|$ different subformulas, the algorithm requires time $O(|\varphi| \times (|S| + |\rightarrow|))$.

Example 6.1. Consider the model M shown in Fig. 6.1, and the CTL formula $\varphi = \mathrm{E}(\wp_1 \mathrm{U}(\mathrm{EG}\wp_2))$. The stages of the state-labelling algorithm checking whether φ holds in M are shown in Fig. 6.2. At the beginning, the states of M are labelled with the most nested and shortest subformulas of φ. i.e., with the

---

[1] The complexity of an algorithm is often described using the so-called "big–O notation", which is a theoretical measure of how an algorithm will execute, in terms of the time or computer memory required, given the size of the problem itself (see [113]).

[2] A *strongly connected component* of a directed graph $G = (V, E)$ is a maximal subgraph $G' = (V', E')$ of G such that each two vertices in V' are connected by a path in G'.

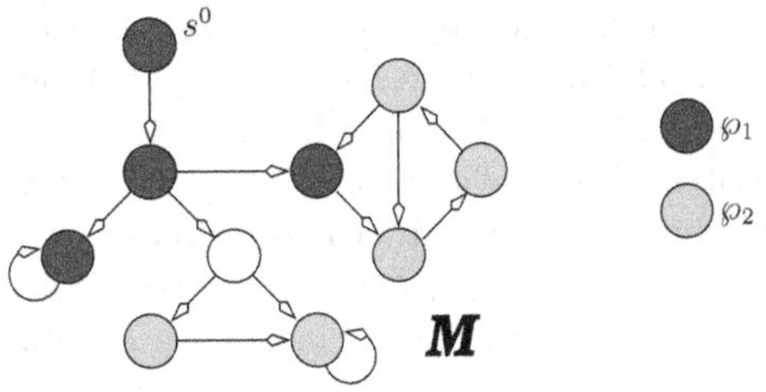

Fig. 6.1. The model M considered in Example 6.1

propositions $\wp_1$ and $\wp_2$. This is shown in part (a) of the figure. Then, the subformulas of φ of length $i = 2$ should be considered. In our example, this is only $\varphi' = \mathrm{EG}\wp_2$. In order to label the states of M with φ', we build the subgraph $(S', \rightarrow')$ whose vertices are the states labelled with $\wp_2$ (see Fig. 6.2(b)) and identify its strongly connected components (denoted SCC_i with $i = 1, 2, 3$ in Fig. 6.2(b)). Then, the algorithm labels with φ' all the states which either belong to SCC_1, SCC_3 (notice that SCC_2 is a singleton without a self-loop) or can be reached going backwards (i.e., using $\rightarrow^{-1}$) from the states in SCC_1 or SCC_3 by a path whose states are labelled with φ'. This is shown in Fig. 6.2(c)

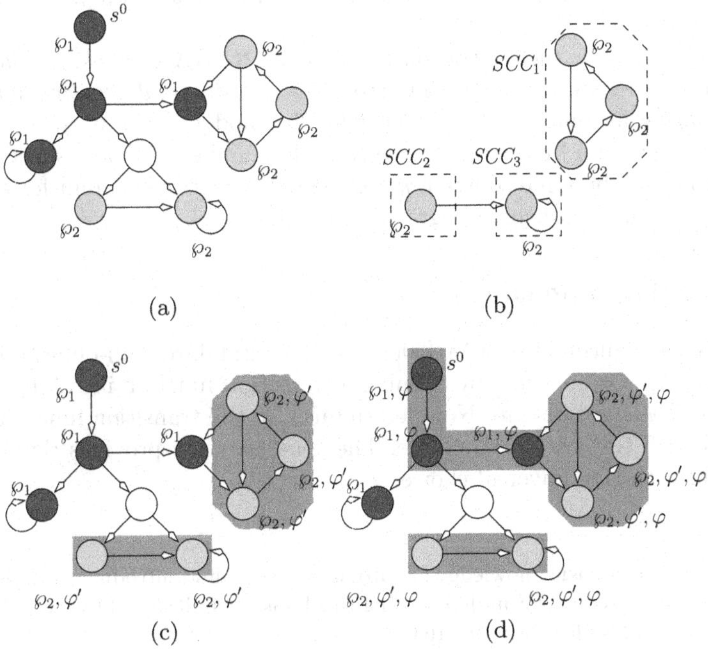

Fig. 6.2. The stages of labelling M with subformulas of $\varphi = \mathrm{E}(\wp_1 \mathrm{U}(\mathrm{EG}\wp_2))$

(the states which are labelled in this step are additionally marked with the grey background). The next stage consists in labelling the states of M with the formula φ, which is of the form $\mathrm{E}(\wp_1 \mathrm{U} \varphi')$. Thus, the algorithm labels with φ all the states which have already been marked with φ', as well as these which can be reached from them by the relation $\rightarrow^{-1}$ on a path labelled with $\wp_1$. This is shown in Fig. 6.2(d) (again, the states labelled in this step are marked with the grey background). Finally, we conclude that $M \models \varphi$, since the state s^0 is labelled with the formula φ.

□

6.1.2 Automata-Theoretic Approach

When the state labelling method is used for verifying a CTL property φ, it requires, in principle, to build a model first, and then to check the formula over this model. This can obviously make verification infeasible, especially when the size of the model for a system is prohibitive. Therefore, there are approaches which offer on-the-fly solutions, i.e., a formula is checked over a model while its construction. One of such approaches exploits automata theory[3]. Intuitively, this method consists in checking non-emptiness of the automaton[4] which is the product of two automata: one corresponding to the model and the other one obtained from a translation of the formula φ. Non-emptiness of the product automaton can be sometimes checked even before this automaton is completely built [164], which means that a part of the model is used only.

For $\varphi \in$ CTL such a translation is made to WAA (*weak alternating automata*), whereas for $\varphi \in$ CTL* to HAA (*hesitant alternating automata*). Alternating automata do not accept all the models for a given formula, but only these with a branching degree (i.e., the number of the $\rightarrow$-successors of a state) limited by some constant k. For model checking, the limit for k is taken as the maximal branching degree in the model.

Alternating Automata

Alternating automata (AA) [30] generalise standard non-deterministic automata. This is obtained by defining a transition function in AA by means of *positive boolean formulas*. Non-determinism in the transition function is associated with the existential choice. The generalisation provides the means for dealing with the universal choice.

---

[3] We assume a basic knowledge of automata theory, and introduce only the notions which are essential for understanding this book. Definitions of the complementary notions can be found in [97, 164].

[4] Non-emptiness of an automaton means that the language it accepts is non-empty, see p. 163.

For a given set Y let $\mathcal{B}^+(Y)$ be a set of *positive boolean formulas* over Y, defined as follows:

$$\beta := true \mid false \mid y \mid \beta \vee \beta \mid \beta \wedge \beta,$$

where $y \in Y$.

Notice that no negation is applied to the elements of $\mathcal{B}^+(Y)$. For a subset $Y' \subseteq Y$ we say that Y' satisfies $\beta \in \mathcal{B}^+(Y)$ iff β is satisfied when assigning *true* to the elements $y \in Y'$ and *false* to elements $y' \in Y \setminus Y'$. For example, for $Y = \{y_0, y_1, y_2, y_3\}$ the subset $Y' = \{y_0, y_1\}$ satisfies the formula $(y_0 \vee y_3) \wedge (y_1 \vee y_2)$, and it does not satisfy the formula $(y_0 \vee y_1) \wedge (y_2 \vee y_3)$.

Let $\mathbb{N}^*$ denote the set of all the finite sequences of natural numbers, and $\cdot$ denote the concatenation operation. A *tree* is a subset $\mathfrak{T} \subseteq \mathbb{N}^*$ such that if $\eta{\cdot}c \in \mathfrak{T}$ for $\eta \in \mathbb{N}^*$ and $c \in \mathbb{N}$, then also $\eta \in \mathfrak{T}$, and $\eta{\cdot}c' \in \mathfrak{T}$ for all $0 \leq c' < c$ (see Fig. 6.3).

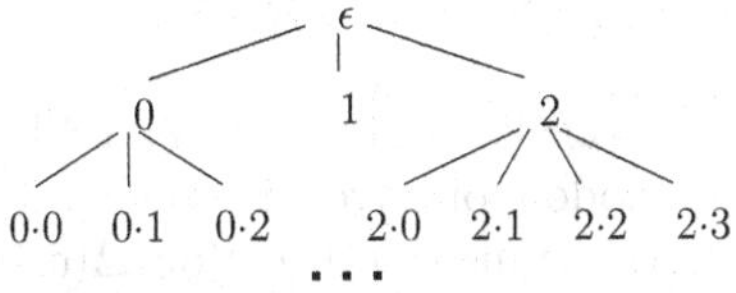

Fig. 6.3. A tree

The elements of $\mathfrak{T}$ are called *nodes*. The empty sequence ϵ is the root of $\mathfrak{T}$. For every $\eta \in \mathfrak{T}$, the nodes $\eta{\cdot}c \in \mathfrak{T}$ with $c \in \mathbb{N}$ are the *successors* of η. The number of the successors of η (denoted $d(\eta)$) is called the *degree* of η. A *leaf* is a node with no successors. A *path* π of a tree $\mathfrak{T}$ is a set $\pi \subseteq \mathfrak{T}$ such that $\epsilon \in \pi$ and for every $\eta \in \pi$, either η is a leaf or there exists a unique $c \in \mathbb{N}$ such that $\eta{\cdot}c \in \pi$. For $\mathcal{D} \subseteq \mathbb{N}$ a tree $\mathfrak{T}$ is a $\mathcal{D}$-tree if $d(\eta) \in \mathcal{D}$ for all $\eta \in \mathfrak{T}$.

Example 6.2. An example of a tree is shown in Fig. 6.3. The node 1 is a leaf so its degree is 0, i.e., $d(1) = 0$. Notice that $d(\epsilon) = d(0) = 3$, and $d(2) = 4$. One of the paths in the tree is $\{\epsilon, 0, 0{\cdot}0, \ldots\}$. The tree is a $\mathcal{D}$-tree, for some $\mathcal{D}$ with $\{0, 3, 4\} \subseteq \mathcal{D}$.

□

An $\mathbb{A}$-labelled tree, for a finite alphabet $\mathbb{A}$, is a pair $(\mathfrak{T}, V_{\mathfrak{T}})$, where $\mathfrak{T}$ is a tree, and $V_{\mathfrak{T}} : \mathfrak{T} \longrightarrow \mathbb{A}$ assigns to every node of $\mathfrak{T}$ a label in $\mathbb{A}$. In the setting used in the model checking, an $\mathbb{A}$-labelled tree with $\mathbb{A} = 2^{PV}$ (PV – a finite set propositional variables) is called a *computation tree.*

An infinite word over a set $\mathbb{A}$ is an infinite sequence $a_0 a_1 a_2 \ldots$ such that $a_i \in \mathbb{A}$ for all $i \geq 0$. Notice that an infinite word over $\mathbb{A}$ can be viewed as a $\mathbb{A}$-labelled tree in which the degree of all the nodes is 1.

Definition 6.3. *An alternating automaton over infinite trees* (ATA) *is a 6-tuple*

$$\mathcal{A} = (\mathbb{A}, \mathcal{D}, \mathfrak{Q}, \mathfrak{q}^0, \Delta, F),$$

where

- $\mathbb{A}$ *is a finite input alphabet,*
- $\mathcal{D} \subseteq \mathbb{N}$ *is a finite subset of* $\mathbb{N}$,
- $\mathfrak{Q}$ *is a finite set of states,*
- $\mathfrak{q}^0 \in S$ *is an initial state,*
- $\Delta : \mathfrak{Q} \times \mathbb{A} \times \mathcal{D} \longrightarrow \mathcal{B}^+(\mathbb{N} \times \mathfrak{Q})$ *is a partial transition function, where* $\Delta(\mathfrak{q}, a, k) \in \mathcal{B}^+(\{0, \ldots, k-1\} \times \mathfrak{Q}\})$ *if defined,*
- $F \subseteq \mathfrak{Q}$ *is an acceptance condition*[5].

Alternating automata over infinite words can be viewed as restrictions of alternating automata over infinite trees, where $\mathcal{D} = \{1\}$. Then, a transition function simplifies to a partial function:

$$\Delta : \mathfrak{Q} \times \mathbb{A} \longrightarrow \mathcal{B}^+(\mathfrak{Q}).$$

An intuition behind the transition function of an ATA is the following. If $\mathcal{A}$ is in a state $\mathfrak{q}$, and reads a node η of a tree $\mathfrak{T}$ labelled with $a \in \mathbb{A}$ which has k successors, then it applies the transition function $\Delta(\mathfrak{q}, a, k)$. This means that the automaton can send its copies, each of which moves to its own new state, to such successor states of η that, when paired with the state of the corresponding copy of the automaton, make the formula $\Delta(\mathfrak{q}, a, k)$ satisfied. This way several nodes of $\mathfrak{T}$ can be visited simultaneously, and $\mathcal{A}$ can be in more than one state at the same time (e.g., if $\Delta(\mathfrak{q}, a, k)$ contains a conjunction). This can be also viewed as running several copies of $\mathcal{A}$, each of which is in one current state.

Example 6.4. An example of alternating automaton is the following: $\mathcal{A} = (\mathbb{A}, \mathcal{D}, \mathfrak{Q}, \mathfrak{q}^0, \Delta, F)$ with $\mathbb{A} = \{a_1, a_2\}$, $\mathcal{D} = \{0, 1, 3, 4, 5\}$, $\mathfrak{Q} = (\mathfrak{q}^0, \mathfrak{q}_1, \mathfrak{q}_2)$, $F = \{\mathfrak{q}_1\}$ and the transition function given by

$$\begin{aligned}
\Delta(\mathfrak{q}^0, a_1, 3) &= (2, \mathfrak{q}_1) \vee (1, \mathfrak{q}_2), \\
\Delta(\mathfrak{q}_1, a_2, 0) &= \mathit{true}, \\
\Delta(\mathfrak{q}_1, a_2, 1) &= (0, \mathfrak{q}_1) \wedge (0, \mathfrak{q}_2), \\
\Delta(\mathfrak{q}_1, a_2, 3) &= (2, \mathfrak{q}_1) \vee (1, \mathfrak{q}_2), \\
\Delta(\mathfrak{q}_2, a_1, 4) &= ((3, \mathfrak{q}^0) \vee (1, \mathfrak{q}_1)) \wedge (2, \mathfrak{q}_1), \\
\Delta(\mathfrak{q}_2, a_2, 3) &= \mathit{true}.
\end{aligned}$$

The transition function of $\mathcal{A}$ is graphically depicted in Fig. 6.4. □

[5] An acceptance condition can be also defined as a pair of states.

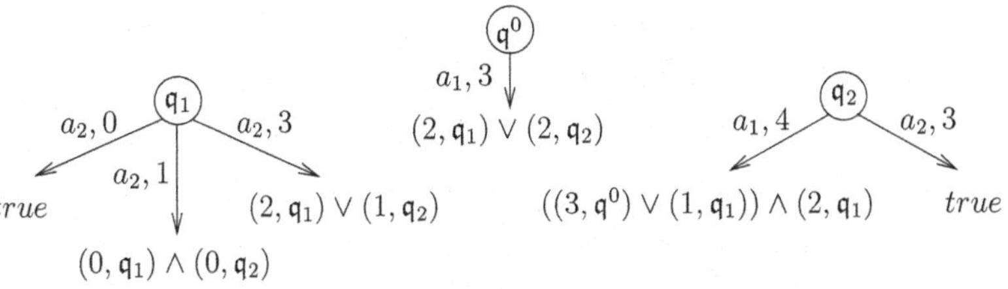

Fig. 6.4. The transition function of the automaton $\mathcal{A}$ of Example 6.4

A *run* r of an alternating automaton $\mathcal{A}$ on a leafless $\mathbb{A}$-labelled tree $(\mathfrak{T}, V_{\mathfrak{T}})$ is a tree labelled by elements of $\mathbb{N}^* \times \mathfrak{Q}$. The root of r is labelled with $(\epsilon, \mathfrak{q}^0)$ Formally, a run $r = (\mathfrak{T}_r, V_r)$ is an $\mathbb{A}_r$-labelled tree, where $\mathbb{A}_r = \mathbb{N}^* \times \mathfrak{Q}$, and $(\mathfrak{T}_r, V_r)$ satisfies the following:

1. $\epsilon \in \mathfrak{T}_r$ and $V_r(\epsilon) = (\epsilon, \mathfrak{q}^0)$.
2. Let $\eta_r \in \mathfrak{T}_r$ with $V_r(\eta_r) = (\eta, \mathfrak{q})$ and $\Delta(\mathfrak{q}, V_{\mathfrak{T}}(\eta), d(\eta)) = \beta$. Then there is a possibly empty set

$$\mathfrak{O} = \{(c_0, \mathfrak{q}_0), (c_1, \mathfrak{q}_1), \ldots, (c_n, \mathfrak{q}_n)\} \subseteq \{0, \ldots, d(\eta) - 1\} \times \mathfrak{Q},$$

   such that the following hold:
   - $\mathfrak{O}$ satisfies β, and
   - for all $0 \le i \le n$, we have $\eta_r \cdot i \in \mathfrak{T}_r$ and $V_r(\eta_r{\cdot}i) = (\eta{\cdot}c_i, \mathfrak{q}_i)$.

Example 6.5. Consider the automaton $\mathcal{A}$ of Example 6.4 and a tree $\mathfrak{T}$ a part of which is shown in Fig. 6.5(a). The labels of the nodes of $\mathfrak{T}$ (given in square brackets) are taken from the set $\mathbb{A} = \{a_1, a_2\}$. A part of a run of $\mathcal{A}$ on the tree $\mathfrak{T}$ is shown in Fig. 6.5(b).

Computing the run shown in the picture is done in the following steps:

- Initially, $\mathcal{A}$ is in its initial state, and reads the root of the tree $\mathfrak{T}$ that is labelled with the input letter a_1. Since the degree of ϵ in $\mathfrak{T}$ is 3, $\Delta(\mathfrak{q}^0, a_1, 3)$ is of our interest. As $\Delta(\mathfrak{q}^0, a_1, 3) = (2, \mathfrak{q}_1) \vee (1, \mathfrak{q}_2)$, the set $\mathfrak{O}$ can consist either of $(2, \mathfrak{q}_1)$, or $(1, \mathfrak{q}_2)$, or of both of them. Let us say that $\mathfrak{O} = \{(2, \mathfrak{q}_1)\}$. Thus, $\mathcal{A}$ moves to $\mathfrak{q}_1$, and the node 2 of $\mathfrak{T}$ is visited. This gives us the node 0 of $\mathfrak{T}_r$ with $V_r(0) = (2, \mathfrak{q}_1)$.
- Next, when the node 2 labelled with a_2 is read, we have $\mathfrak{O} = \{(0, \mathfrak{q}_1), (0, \mathfrak{q}_2)\}$ due to the value of $\Delta(\mathfrak{q}_1, a_2, 1)$. Therefore, $\mathcal{A}$ sends its copies to $\mathfrak{q}_1$ and $\mathfrak{q}_2$. The node 2·0 of $\mathfrak{T}$ is visited. As a result, the node 0 of $\mathfrak{T}_r$ has two successors with $V_r(0{\cdot}0) = (2{\cdot}0, \mathfrak{q}_1)$ and $V_r(0{\cdot}1) = (2{\cdot}0, \mathfrak{q}_2)$.
- In the next step both the copies of the automaton read the node 2·0 of $\mathfrak{T}$, which is labelled with a_2. Since $\Delta(\mathfrak{q}_2, a_2, 3) = \mathit{true}$, the set $\mathfrak{O}$ computed for this case can be empty, and therefore the node 0·1 labelled with $(2{\cdot}0, \mathfrak{q}_2)$ is a leaf of $\mathfrak{T}_r$.
  However, the node 0·0 is not a leaf, since $\Delta(\mathfrak{q}_1, a_2, 3) = (2, \mathfrak{q}_1) \vee (1, \mathfrak{q}_2)$. So we take the set $\mathfrak{O} = \{(2, \mathfrak{q}_1), (1, \mathfrak{q}_2)\}$, and therefore $\mathcal{A}$ sends its copies to

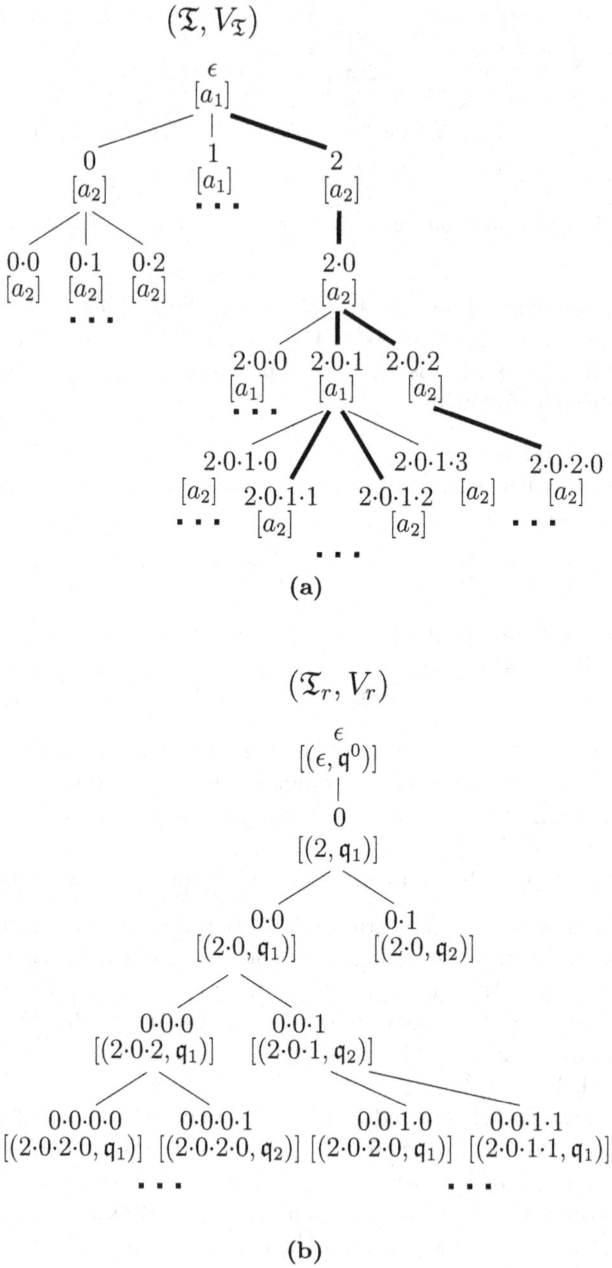

Fig. 6.5. A run $r = (\mathfrak{T}_r, V_r)$ of the automaton $\mathcal{A}$ of Example 6.4 (part (b)) on a tree $(\mathfrak{T}, V_{\mathfrak{T}})$ (part (a))

$\mathfrak{q}_1$ and $\mathfrak{q}_2$. In the tree $\mathfrak{T}$ the nodes 2·0·1 and 2·0·2 are visited. Thus, in $\mathfrak{T}_r$ we obtain two nodes: 0·0·0 and 0·0·1.

- Then, the copy of $\mathcal{A}$ which is in the state $\mathfrak{q}_1$ reads in the node 2·0·2 labelled with a_2. We obtain $\mathcal{O} = \{(0, \mathfrak{q}_1), (0, \mathfrak{q}_2)\}$, and again $\mathcal{A}$ sends its copies to $\mathfrak{q}_1$ and $\mathfrak{q}_2$. Both of them visit the node 2·0·2·0 in $\mathfrak{T}$. The node 0·0·0 of $\mathfrak{T}_r$ has therefore two successors.
- Simultaneously, the copy of $\mathcal{A}$ which is in the state $\mathfrak{q}_2$ (i.e., the one corresponding to the node 0·0·1 in $\mathfrak{T}_r$) reads the node 2·0·1 (of $\mathfrak{T}$) labelled with a_1. Taking $\mathcal{O} = \{(2, \mathfrak{q}_1), (1, \mathfrak{q}_1)\}$, the automaton $\mathcal{A}$ sends its two copies to $\mathfrak{q}_1$, and 'visits' the nodes 2·0·1·1 and 2·0·1·2 in $\mathfrak{T}$. The node 0·0·1 of $\mathfrak{T}_r$ is therefore of two successors.

Obviously, the process can be continued, since the tree $\mathfrak{T}$ is infinite.

□

A run r is *accepting* if each its infinite path satisfies the acceptance condition F, which, in our case, means that some state of F repeats infinitely often in the path. An automaton *accepts a tree* if there exists a run that accepts it. By $\mathcal{L}(\mathcal{A})$ we mean the *language* of $\mathcal{A}$, i.e., the set of all the trees accepted by an automaton.

Weak alternating automata (WAA, for short) are an important subclass of alternating automata over infinite trees. Let $\mathcal{A} = (\mathbb{A}, \mathcal{D}, \mathcal{Q}, \mathfrak{q}^0, \Delta, F)$ be an ATA. Then, for $\mathcal{A}$ to be a WAA, $F \subseteq \mathcal{Q}$ denotes the standard Büchi acceptance condition and the set of states $\mathcal{Q}$ can be partitioned into a finite number of disjoint sets $\mathcal{Q}_i$, partially ordered by $\leq$ such that for every $\mathfrak{q} \in \mathcal{Q}_k$ and $\mathfrak{q}' \in \mathcal{Q}_l$ such that $\mathfrak{q}'$ occurs in $\Delta(\mathfrak{q}, a, d)$ for some $a \in \mathbb{A}$ and $d \in \mathcal{D}$ we have $\mathcal{Q}_l \leq \mathcal{Q}_k$.

WAA are used for defining a translation from CTL. As far as CTL* is concerned we need a more powerful class of automata, called *hesitant alternating tree automaton* (HAA). Firstly, HAA have a more restricted transition structure than WAA, and secondly they use the Streett acceptance condition (see [97, 164]). Similarly the formulas of modal μ-calculus can be efficiently (i.e., in linear time[6]) translated to alternating Streett tree automata, i.e., using Streett acceptance condition. The reader is referred to [97] for more details about the above translations.

[6] Algorithms are often described as having "linear", "logarithmic", "polynomial", or "exponential" complexity, which in big-O notation is expressed as $O(n)$, $O(\log n)$, $O(n^k)$, and $O(k^n)$, where k is a constant, and n is the size of the input. Generally, for a given size of the input, linear algorithms require least resources (i.e., time or memory), exponential algorithms, the most complex, require most, whereas logarithmic and polynomial are in the middle, less to more resource-consuming (see [113]).

Translation from CTL to WAA

For a given CTL formula φ in a positive normal form and a set $\mathcal{D} \subseteq \mathbb{N}$, a WAA

$$\mathcal{A}_{\mathcal{D},\varphi} = (2^{PV(\varphi)}, \mathcal{D}, SF(\varphi), \varphi, \Delta, F)$$

can be defined such that $\mathcal{L}(\mathcal{A}_{\mathcal{D},\varphi})$ is the set of all the $\mathcal{D}$-trees satisfying φ. Below we define Δ and F of $\mathcal{A}_{\mathcal{D},\varphi}$:

- F is the set of all ER and AR formulas in $SF(\varphi)$,
- The transition function is given by

$$\begin{aligned}
\Delta(\wp, a, k) &= \begin{cases} true & \text{iff } \wp \in a, \\ false & \text{iff } \wp \notin a \end{cases} \\
\Delta(\neg\wp, a, k) &= \begin{cases} true & \text{iff } \neg\wp \in a, \\ false & \text{iff } \neg\wp \notin a, \end{cases} \\
\Delta(\varphi_1 \vee \varphi_2, a, k) &= \Delta(\varphi_1, a, k) \vee \Delta(\varphi_2, a, k), \\
\Delta(\varphi_1 \wedge \varphi_2, a, k) &= \Delta(\varphi_1, a, k) \wedge \Delta(\varphi_2, a, k), \\
\Delta(\mathrm{EX}\varphi, a, k) &= \bigvee_{c=0}^{k-1}(c, \varphi), \\
\Delta(\mathrm{AX}\varphi, a, k) &= \bigwedge_{c=0}^{k-1}(c, \varphi), \\
\Delta(\mathrm{E}(\varphi_1 \mathrm{U} \varphi_2), a, k) &= \Delta(\varphi_2, a, k) \vee (\Delta(\varphi_1, a, k) \wedge \bigvee_{c=0}^{k-1}(c, \mathrm{E}(\varphi_1 \mathrm{U} \varphi_2))), \\
\Delta(\mathrm{A}(\varphi_1 \mathrm{U} \varphi_2), a, k) &= \Delta(\varphi_2, a, k) \vee (\Delta(\varphi_1, a, k) \wedge \bigwedge_{c=0}^{k-1}(c, \mathrm{A}(\varphi_1 \mathrm{U} \varphi_2))), \\
\Delta(\mathrm{E}(\varphi_1 \mathrm{R} \varphi_2), a, k) &= \Delta(\varphi_2, a, k) \wedge (\Delta(\varphi_1, a, k) \vee \bigvee_{c=0}^{k-1}(c, \mathrm{E}(\varphi_1 \mathrm{R} \varphi_2))), \\
\Delta(\mathrm{A}(\varphi_1 \mathrm{R} \varphi_2), a, k) &= \Delta(\varphi_2, a, k) \wedge (\Delta(\varphi_1, a, k) \vee \bigwedge_{c=0}^{k-1}(c, \mathrm{A}(\varphi_1 \mathrm{R} \varphi_2))).
\end{aligned}$$

It is also useful to show the transition function for the derived operators:

$$\begin{aligned}
\Delta(\mathrm{EF}\varphi, a, k) &= \Delta(\varphi, a, k) \vee \bigvee_{c=0}^{k-1}(c, \mathrm{EF}\varphi), \\
\Delta(\mathrm{AF}\varphi, a, k) &= \Delta(\varphi, a, k) \vee \bigwedge_{c=0}^{k-1}(c, \mathrm{AF}\varphi), \\
\Delta(\mathrm{EG}\varphi, a, k) &= \Delta(\varphi, a, k) \wedge \bigvee_{c=0}^{k-1}(c, \mathrm{EG}\varphi), \\
\Delta(\mathrm{AG}\varphi, a, k) &= \Delta(\varphi, a, k) \wedge \bigwedge_{c=0}^{k-1}(c, \mathrm{AG}\varphi).
\end{aligned}$$

To see that $\mathcal{A}_{\mathcal{D},\varphi}$ is a WAA, we define a partition of $\mathcal{Q}$ into disjoint sets and the partial order over them. Each formula $\psi \in SF(\varphi)$ is a singleton set $\{\psi\}$ in the partition. The partial order is defined as follows: $\{\psi_1\} \leq \{\psi_2\}$ iff $\psi_1 \in SF(\psi_2)$.

Example 6.6. Consider a CTL formula

$$\mathrm{E}(\wp_1 \mathrm{U}(\mathrm{EG}\wp_2)),$$

which expresses that there is a path along which $\wp_1$ holds until a path starts, at which $\wp_2$ holds forever. The WAA for $\mathrm{E}(\wp_1\mathrm{U}(\mathrm{EG}\wp_2))$ is over the alphabet equal to $\{\emptyset, \{\wp_1\}, \{\wp_2\}, \{\wp_1, \wp_2\}\}$. It contains four states $\mathfrak{q}_0 = \mathrm{E}(\wp_1\mathrm{U}(\mathrm{EG}\wp_2))$ (the initial state), $\mathfrak{q}_1 = \mathrm{EG}\wp_2$, $\mathfrak{q}_2 = \wp_1$, and $\mathfrak{q}_3 = \wp_2$, whereas its acceptance condition is $F = \{\mathrm{EG}\wp_2\}$ (due to $\mathrm{EG}\wp_2$ being an abbreviation for $\mathrm{E}(false\mathrm{R}\wp_2)$, see p. 67). The transition function is computed in the following way:

- For $\mathfrak{q} = \mathfrak{q}_3 = \wp_2$ and $a \in \{\emptyset, \{\wp_1\}\}$, as well as for $\mathfrak{q} = \mathfrak{q}_2 = \wp_1$ and $a \in \{\emptyset, \{\wp_2\}\}$, we have $\Delta(\mathfrak{q}, a, k) = false$, since $\mathfrak{q} \notin a$ in any of the above cases;
- Similarly, for $\mathfrak{q} = \mathfrak{q}_3$ and $a \in \{\{\wp_2\}, \{\wp_1, \wp_2\}\}$, as well as for $\mathfrak{q} = \mathfrak{q}_2$ and $a \in \{\{\wp_1\}, \{\wp_1, \wp_2\}\}$, we have $\Delta(\mathfrak{q}, a, k) = true$,
- Due to $\mathfrak{q}_1 = \mathrm{EG}\mathfrak{q}_3$, we have
  $\Delta(\mathfrak{q}_1, a, k) = \Delta(\mathfrak{q}_3, a, k) \wedge \bigvee_{c=0}^{k-1}(c, \mathrm{EG}\mathfrak{q}_3) = \Delta(\mathfrak{q}_3, a, k) \wedge \bigvee_{c=0}^{k-1}(c, \mathfrak{q}_1)$.
  This gives us:
  - for $a \in \{\emptyset, \{\wp_1\}\}$
    $\Delta(\mathfrak{q}_1, a, k) = false \wedge \bigvee_{c=0}^{k-1}(c, \mathfrak{q}_1) = false$;
  - for $a \in \{\{\wp_2\}, \{\wp_1, \wp_2\}\}$
    $\Delta(\mathfrak{q}_1, a, k) = true \wedge \bigvee_{c=0}^{k-1}(c, \mathfrak{q}_1) = \bigvee_{c=0}^{k-1}(c, \mathfrak{q}_1)$;
- Due to $\mathfrak{q}_0 = \mathrm{E}(\mathfrak{q}_2\mathrm{U}\mathfrak{q}_1)$, we have
  $\Delta(\mathfrak{q}_0, a, k) = \Delta(\mathfrak{q}_1, a, k) \vee (\Delta(\mathfrak{q}_2, a, k) \wedge \bigvee_{c=0}^{k-1}(c, \mathfrak{q}_0))$.
  This gives us:
  - $\Delta(\mathfrak{q}_0, \emptyset, k) = false \vee (false \wedge \bigvee_{c=0}^{k-1}(c, \mathfrak{q}_0)) = false$,
  - $\Delta(\mathfrak{q}_0, \{\wp_1\}, k) = false \vee (true \wedge \bigvee_{c=0}^{k-1}(c, \mathfrak{q}_0)) = \bigvee_{c=0}^{k-1}(c, \mathfrak{q}_0)$,
  - $\Delta(\mathfrak{q}_0, \{\wp_2\}, k) = \bigvee_{c=0}^{k-1}(c, \mathfrak{q}_1) \vee (false \wedge \bigvee_{c=0}^{k-1}(c, \mathfrak{q}_0)) = \bigvee_{c=0}^{k-1}(c, \mathfrak{q}_1)$,
  - $\Delta(\mathfrak{q}_0, \{\wp_1, \wp_2\}, k) =$
    $\bigvee_{c=0}^{k-1}(c, \mathfrak{q}_1) \vee (true \wedge \bigvee_{c=0}^{k-1}(c, \mathfrak{q}_0)) = \bigvee_{c=0}^{k-1}(c, \mathfrak{q}_1) \vee \bigvee_{c=0}^{k-1}(c, \mathfrak{q}_0)$.

The above-computed transition function is summarized in the table below:

	$\Delta(\mathfrak{q}, a, k)$			
$\mathfrak{q}$	$a = \emptyset$	$a = \{\wp_1\}$	$a = \{\wp_2\}$	$a = \{\wp_1, \wp_2\}$
$\mathfrak{q}_0$	$false$	$\bigvee_{c=0}^{k-1}(c, \mathfrak{q}_0)$	$\bigvee_{c=0}^{k-1}(c, \mathfrak{q}_1)$	$\bigvee_{c=0}^{k-1}(c, \mathfrak{q}_1) \vee \bigvee_{c=0}^{k-1}(c, \mathfrak{q}_0)$
$\mathfrak{q}_1$	$false$	$false$	$\bigvee_{c=0}^{k-1}(c, \mathfrak{q}_1)$	$\bigvee_{c=0}^{k-1}(c, \mathfrak{q}_1)$
$\mathfrak{q}_2$	$false$	$true$	$false$	$true$
$\mathfrak{q}_3$	$false$	$false$	$true$	$true$

□

Example 6.7. Consider a CTL formula

$$\mathrm{AGEF}\wp,$$

which expresses that a state satisfying $\wp$ is reachable from each state. The WAA for AGEF$\wp$ is over the alphabet $\{\emptyset, \{\wp\}\}$, it contains three states $\mathfrak{q}_0 =$ AGEF$\wp$ (the initial state), $\mathfrak{q}_1 =$ EF$\wp$, and $\mathfrak{q}_2 = \wp$, whereas the acceptance condition $F = \{\mathrm{AGEF}\wp\}$. The transition function is given in the table below:

$\mathfrak{q}$	$\Delta(\mathfrak{q}, \emptyset, k)$	$\Delta(\mathfrak{q}, \{\wp\}, k)$
$\mathfrak{q}_0$	$\bigvee_{c=0}^{k-1}(c, \mathfrak{q}_1) \wedge \bigwedge_{c=0}^{k-1}(c, \mathfrak{q}_0)$	$\bigwedge_{c=0}^{k-1}(c, \mathfrak{q}_0)$
$\mathfrak{q}_1$	$\bigvee_{c=0}^{k-1}(c, \mathfrak{q}_1)$	*true*
$\mathfrak{q}_2$	*false*	*true*

□

Model Checking with WAA

Recall that the model checking problem for CTL is stated as: given a CTL formula φ and a system represented by its model M, check whether $M, s^0 \models \varphi$. Since each model corresponds to a single tree, model checking is reduced to checking the membership of that tree in the models of the formula which are accepted by the WAA $\mathcal{A}_{\mathcal{D},\varphi}$, where $\mathcal{D}$ is the minimal set containing the degrees of all states of M.

The semantics of CTL was defined over Kripke models. For the purpose of automata model checking this definition has to be extended to specific infinite trees, which are at the same time models for the formula and input trees for a formula automaton. This enables a reduction to checking the non-emptiness of automata over an alphabet of one letter only (called *1-letter non-emptiness checking*). A model $M = ((S, s^0, \to), V)$ can be viewed as a tree $(\mathfrak{T}_M, V_M)$ that corresponds to the unwinding of M from s^0 such that if M is a model for a formula, so is $(\mathfrak{T}_M, V_M)$.

Let $succ(s) = (s_0, \ldots, s_{d(s)-1})$ be an ordered list of the $\to$-successors of a state s. Then, $\mathfrak{T}_M$ and V_M are defined as follows:

1. $\epsilon \in \mathfrak{T}_M$ and $V_M(\epsilon) = s^0$,
2. For $\eta \in \mathfrak{T}_M$ with $succ(V_M(\eta)) = (s_0, \ldots, s_n)$ and for $0 \leq i \leq n$, we have $\eta{\cdot}i \in \mathfrak{T}_M$ and $V_M(\eta{\cdot}i) = s_i$.

The model checking algorithm for φ proceeds as follows:

1. Construct an alternating automaton on infinite words[7] $\mathcal{A}_{M\times\varphi} = M \times \mathcal{A}_{\mathcal{D},\varphi}$ (the *product automaton*). This automaton simulates a run of $\mathcal{A}_{\mathcal{D},\varphi}$ over

[7] Recall that this means on infinite trees such that the degree of each state is equal to 1.

$(\mathfrak{T}_M, V_M)$. Notice that the alternating automaton $\mathcal{A}_{\mathcal{D},\varphi}$ does not need to be constructed a priori, which can be seen in the next subsection, where we define the product automaton directly.

2. If $\mathcal{L}(\mathcal{A}_{M\times\varphi}) \neq \emptyset$, then the formula φ is true in the model M. Otherwise, it is false.

The efficiency of this approach follows from the fact that $\mathcal{A}_{M\times\varphi}$ can be defined over an alphabet consisting of a single symbol only, thus allowing for the 1-letter non-emptiness testing. Since the product automaton can be defined as a 1-letter alternating automaton over words, its non-emptiness can be effectively tested in linear time and in logarithmic space complexity. In general, checking for non-emptiness of a non-deterministic automaton can be reduced to checking non-emptiness of a 1-letter non-deterministic automaton, but this does not hold for alternating tree automata. Fortunately, it was shown in [30] that taking a product with the tree representing a model allows to extend this reduction to the case of our product alternating automata.

Product Automaton

Given a model $M = ((S, s^0, \rightarrow), V)$ and a CTL formula φ, a 1-letter product automaton WAA

$$\mathcal{A}_{M\times\varphi} = (\{a\}, \mathcal{Q}, \mathfrak{q}^0, \Delta, F)$$

is defined as follows:

- $\mathcal{Q} = S \times SF(\varphi)$,
- $\mathfrak{q}^0 = (s^0, \varphi)$,
- F is the set of all the pairs $(s, \mathrm{O}(\varphi_1 \mathrm{R} \varphi_2))$ with $s \in S$ and $\mathrm{O}(\varphi_1 \mathrm{R} \varphi_2) \in SF(\varphi)$, for $\mathrm{O} \in \{\mathrm{E}, \mathrm{A}\}$,
- the transition function is given by (notice that this automaton has a simplified transition function, cf. p. 160)

$$\begin{aligned}
\Delta((s,\wp),a) &= \begin{cases} \mathit{true} & \text{iff } \wp \in V(s), \\ \mathit{false} & \text{iff } \wp \notin V(s), \end{cases} \\
\Delta((s,\neg\wp),a) &= \begin{cases} \mathit{true} & \text{iff } \wp \notin V(s), \\ \mathit{false} & \text{iff } \wp \in V(s), \end{cases} \\
\Delta((s,\varphi_1 \vee \varphi_2),a) &= \Delta((s,\varphi_1),a) \vee \Delta((s,\varphi_2),a), \\
\Delta((s,\varphi_1 \wedge \varphi_2),a) &= \Delta((s,\varphi_1),a) \wedge \Delta((s,\varphi_2),a), \\
\Delta((s,\mathrm{EX}\varphi),a) &= \textstyle\bigvee_{s' \in succ(s)} (s',\varphi), \\
\Delta((s,\mathrm{AX}\varphi),a) &= \textstyle\bigwedge_{s' \in succ(s)} (s',\varphi), \\
\Delta((s,\mathrm{E}(\varphi_1 \mathrm{U} \varphi_2)),a) &= \Delta((s,\varphi_2),a) \vee \\
&\qquad (\Delta((s,\varphi_1),a) \wedge \textstyle\bigvee_{s' \in succ(s)} (s',\mathrm{E}(\varphi_1 \mathrm{U} \varphi_2))),
\end{aligned}$$

$$\begin{aligned}
\Delta((s, \mathrm{A}(\varphi_1 \mathrm{U} \varphi_2)), a) &= \Delta((s, \varphi_2), a) \vee \\
&\qquad (\Delta((s, \varphi_1), a) \wedge \textstyle\bigwedge_{s' \in succ(s)} (s', \mathrm{A}(\varphi_1 \mathrm{U} \varphi_2))), \\
\Delta((s, \mathrm{E}(\varphi_1 \mathrm{R} \varphi_2)), a) &= \Delta((s, \varphi_2), a) \wedge \\
&\qquad (\Delta((s, \varphi_1), a) \vee \textstyle\bigvee_{s' \in succ(s)} (s', \mathrm{E}(\varphi_1 \mathrm{R} \varphi_2))), \\
\Delta((s, \mathrm{A}(\varphi_1 \mathrm{R} \varphi_2)), a) &= \Delta((s, \varphi_2), a) \wedge \\
&\qquad (\Delta((s, \varphi_1), a) \vee \textstyle\bigwedge_{s' \in succ(s)} (s', \mathrm{A}(\varphi_1 \mathrm{R} \varphi_2))).
\end{aligned}$$

The function for derived operators is provided below:

$$\begin{aligned}
\Delta((s, \mathrm{EF}\varphi), a) &= \Delta((s, \varphi), a) \vee \textstyle\bigvee_{s' \in succ(s)} (s', \mathrm{EF}\varphi), \\
\Delta((s, \mathrm{AF}\varphi), a) &= \Delta((s, \varphi), a) \vee \textstyle\bigwedge_{s' \in succ(s)} (s', \mathrm{AF}\varphi), \\
\Delta((s, \mathrm{EG}\varphi), a) &= \Delta((s, \varphi), a) \wedge \textstyle\bigvee_{s' \in succ(s)} (s', \mathrm{EG}\varphi), \\
\Delta((s, \mathrm{AG}\varphi), a) &= \Delta((s, \varphi), a) \wedge \textstyle\bigwedge_{s' \in succ(s)} (s', \mathrm{AG}\varphi).
\end{aligned}$$

Example 6.8. Consider the product automaton $\mathcal{A}_{M\times\varphi}$ of the WAA for the formula

$$\varphi = \mathrm{E}(\wp_1 \mathrm{U}(\mathrm{EG}\wp_2))$$

of Example 6.6 and the model M considered in Example 6.1 and recalled in Fig. 6.6.

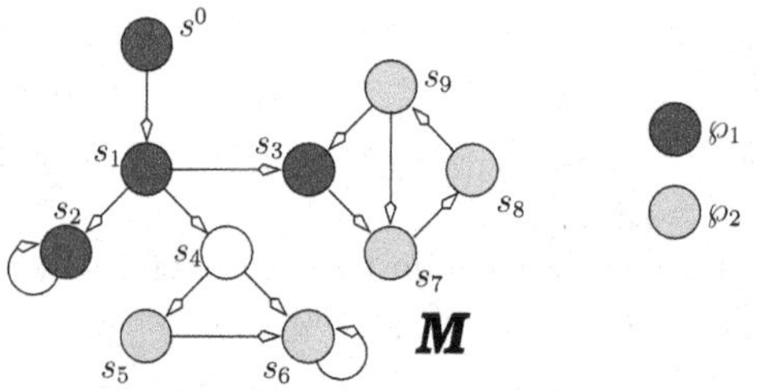

Fig. 6.6. The model M considered in Example 6.8

Let φ' denotes the subformula $\mathrm{EG}\wp_2$ of φ. The set of states of $\mathcal{A}_{M\times\varphi}$ is given by $\mathfrak{Q} = \{s^0, s_1, \ldots, s_9\} \times \{\varphi, \varphi', \wp_1, \wp_2\}$, $F = \{(s, \mathrm{EG}\wp_2) \mid s \in \{s^0, s_1, \ldots, s_9\}\}$, and the initial state $\mathfrak{q}^0$ is $(s^0, \mathrm{E}(\wp_1 \mathrm{U}(\mathrm{EG}\wp_2))$. The transition function of $\mathcal{A}_{M\times\varphi}$ is built iteratively according to the rules above. Its beginning part is graphically shown in Fig. 6.7 as an AND/OR graph[8]. To explain the method of building the graph of Fig. 6.7, consider the initial state (s^0, φ). Due to $\varphi = \mathrm{E}(\wp_1 \mathrm{U}(\mathrm{EG}\wp_2)) = \mathrm{E}(\wp_1 \mathrm{U}\varphi')$ we have

[8] This is a graph with additional nodes enabling to express existential and universal choices over successor states.

$$\Delta((s^0, \varphi), a) = \Delta((s^0, \varphi'), a) \ \vee \ (\Delta((s^0, \wp_1), a) \wedge (s_1, \varphi)),$$

since s_1 is the only successor of s^0. Then, $\Delta((s^0, \varphi'), a) = \Delta((s^0, \mathrm{EG}\wp_2), a)$ is further transformed into

$$\Delta((s^0, \varphi'), a) = \Delta((s^0, \wp_2), a) \wedge (s_1, \varphi') = \mathit{false} \wedge (s_1, \varphi'),$$

since $\wp_2$ does not hold at s^0. On the other hand, $\Delta((s^0, \wp_1), a)$ is replaced by *true*, since this proposition holds at the initial state of M. This "unwinding" of the transition function corresponds to the first level of the AND/OR graph. Computing the further levels of the graph (i.e., the values of the transition function for other states) is done in a similar way. □

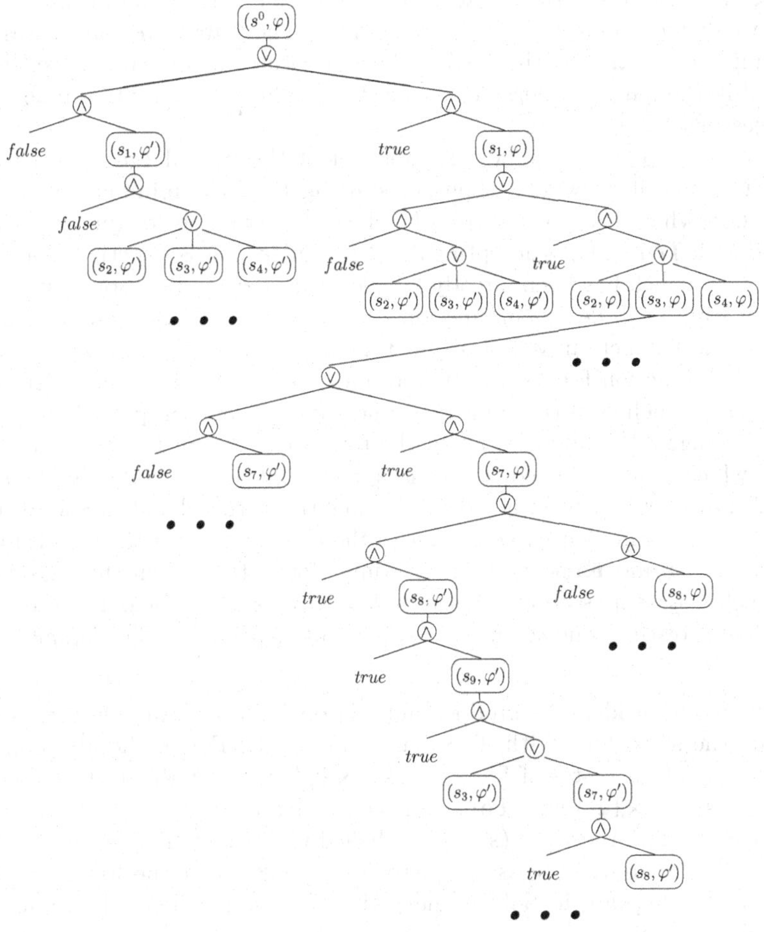

Fig. 6.7. A part of the AND/OR graph of the automaton $\mathcal{A}_{M \times \varphi}$ for the model of Fig. 6.6 and $\varphi = \mathrm{E}(\wp_1 \mathrm{U}(\mathrm{EG}\wp_2))$

Checking Non-emptiness of 1-Letter Word WAA

The 1-letter non-emptiness problem for WAA over words is of linear time complexity. Below we sketch an algorithm [97] solving this problem for a product automaton.

The algorithm labels the states $\mathcal{Q} \cup \{true, false\}$ of the OR/AND graph of the product automaton with **True** or **False**. The idea is that a state $\mathfrak{q}$ is labelled **True** iff the language accepted by the automaton, provided $\mathfrak{q}$ is an initial state, is non-empty. Therefore, the language of the product automaton is non-empty when $\mathfrak{q}^0$ is labelled with **True**.

We know (cf. p. 163) that there is a partition of $\mathcal{Q} \cup \{true, false\}$ into disjoint sets $\mathcal{Q}_i$ such that transitions from a state in $\mathcal{Q}_i$ lead either to states of the same $\mathcal{Q}_i$ or to states of a lower $\mathcal{Q}_j$ (i.e., $\mathcal{Q}_j < \mathcal{Q}_i$) [9]. Moreover, each set $\mathcal{Q}_i \subseteq \mathcal{Q}$ can be defined such that all its states share the second component. Then, we define *accepting* and *rejecting* states. The state *true* constitutes an accepting state, whereas the state *false* constitutes a rejecting state, both minimal in the partial order. Next, a set $\mathcal{Q}_i$ is defined as accepting if $\mathcal{Q}_i \subseteq F$ and rejecting if $\mathcal{Q}_i \cap F = \emptyset$.

Let $\mathcal{Q}_1 < \mathcal{Q}_2 < \ldots \mathcal{Q}_n$ be an extension of the partial order to a total order. The algorithm works in phases starting from the minimal set in the total order, which has not yet been labelled. The states belonging to $\mathcal{Q}_1$ are labelled with **True** if $\mathcal{Q}_1$ is accepting and with **False** if $\mathcal{Q}_1$ is rejecting. For each state $\mathfrak{q} \in \mathcal{Q}_i$ which has been already labelled, a transition function in which $\mathfrak{q}$ occurs is simplified, i.e., a conjunction with a conjunct **False** is simplified to **False** and a disjunction with a disjunct **True** is simplified to **True**. Thus, if a transition function for some state can be simplified to **True** or **False**, the state is then labelled in the same way, and simplification propagates further.

As the algorithm operates up to the total order, when it reaches a state $\mathfrak{q} \in \mathcal{Q}_i$ which is not labelled, it is guaranteed that all the states in the sets $\mathcal{Q}_j < \mathcal{Q}_i$ have already been labelled. The algorithm then labels $\mathfrak{q}$ and all the unlabelled states in $\Delta(\mathfrak{q}, a)$ according to the classification of $\mathcal{Q}_i$, i.e., **True** if $\mathcal{Q}_i$ is accepting and **False** if $\mathcal{Q}_i$ is rejecting. Notice that when the algorithm visits such a state $\mathfrak{q}$ as above, this state leads to a cycle or belongs to a cycle of the states of the same status, so its labelling depends on the classification of the set $\mathcal{Q}_i$.

Example 6.9. Consider the product automaton built in Example 6.8. After labelling the state *true* with **True**, and *false* with **False**, the above algorithm labels all the states of $\{s_6, s_7, s_8, s_9\} \times \{\varphi'\}$ with **True**, all the states of $\{s_1, s_2, s_3, s_4, s_5\} \times \{\varphi'\}$ with **False**, and consequently, exploting the simplified transition relation, the state (s^0, φ) is labelled with **True**. This means that φ holds at s^0. The above process is depicted in Fig. 6.8, where the states labelled in the second step are in bold frames, wheras these labelled with **True** are coloured. □

[9] Clearly, we regard *true* and *false* as states with self-loops.

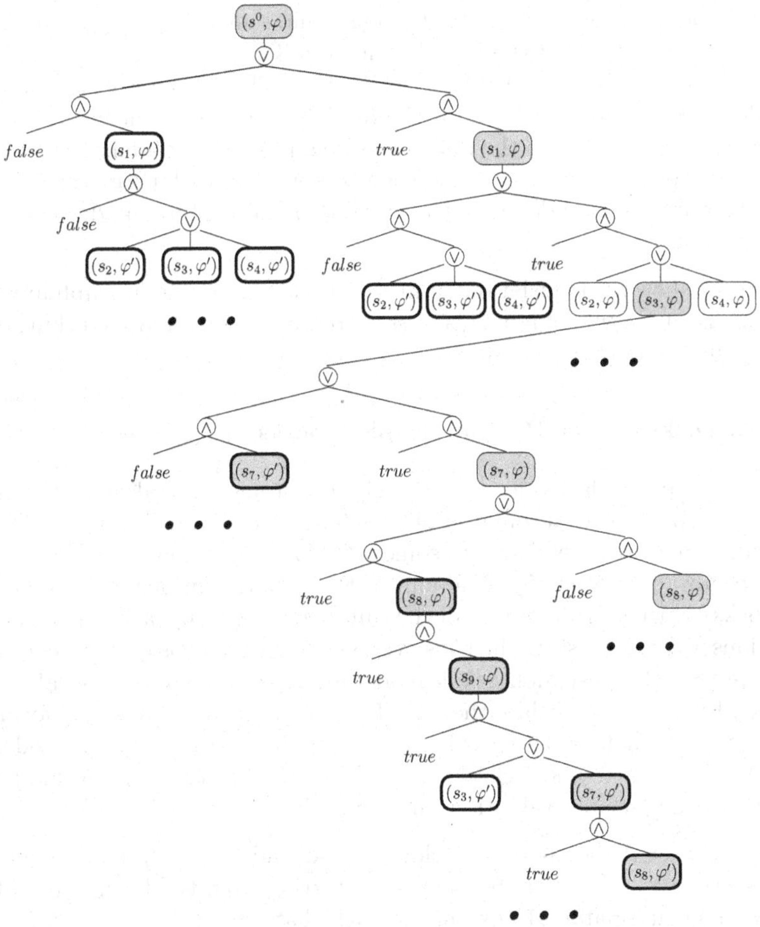

Fig. 6.8. Labelling the states while testing non-emptiness of the automaton $\mathcal{A}_{M\times\varphi}$ for the model of Fig. 6.6 and $\varphi = \mathrm{E}(\wp_1 \mathrm{U}(\mathrm{EG}\wp_2))$

6.2 Model Checking for TCTL over Timed Automata

In this section we discuss different approaches to TCTL model checking. Our focus is on timed automata only, since the approaches for time Petri nets are mainly based on existing translations to timed automata (see Chap. 3).

6.2.1 Verification Using Translation from TCTL to CTL

We start with showing a translation from TCTL to CTL model checking for timed automata which, when combined with model checking methods of the previous section, gives us model checking algorithms for TCTL. First, we define a translation for the strongly monotonic interval semantics and then discuss its adaptation for the weakly monotonic semantics and for the logic $\mathrm{TCTL}_{\mathcal{C}}$. In fact, we show the translation from TCTL to a slightly modified $\mathrm{CTL}_{-\mathrm{X}}$ (denoted $\mathrm{CTL}^{\mathrm{r}}_{-\mathrm{X}}$). Model checking for $\mathrm{CTL}^{\mathrm{r}}_{-\mathrm{X}}$ is an easy adaptation of model checking for $\mathrm{CTL}_{-\mathrm{X}}$. In general, the model checking problem for TCTL can be translated to the model checking problem for a fair[10] version of CTL [7]. However, since we have assumed that we deal with progressive timed automata only, we can define a translation to the $\mathrm{CTL}^{\mathrm{r}}_{-\mathrm{X}}$ model checking problem [159].

First, we show the translation for model checking over region graph models. Then, we discuss how to adapt this translation to perform model checking over abstract models obtained by partition refinement.

Model Checking over Region Graph Models

The idea of the translation for the strongly monotonic interval semantics is as follows. Given a timed automaton $\mathcal{A}$, a valuation function $V_{\mathcal{A}}$, and a TCTL formula φ. First, we extend $\mathcal{A}$ with some new clocks[11], actions, and transitions to obtain an automaton $\mathcal{A}_{\varphi}$. The aim of the new transitions is to reset the new clocks, which correspond to all the timing intervals appearing in φ. These transitions are used to start the runs over which subformulas of φ are checked. Then, we take the bd-region graph model for $\mathcal{A}_{\varphi}$ and augment its valuation function. Finally, we translate the TECTL formula φ to an $\mathrm{ECTL}^{\mathrm{r}}_{-\mathrm{X}}$ formula $\psi = \mathrm{cr}(\varphi)$ such that model checking of φ over the bd-region graph model of $\mathcal{A}$ can be reduced to model checking of ψ over the bd-region graph model of $\mathcal{A}_{\varphi}$ with the augmented valuation function.

Formally, let $\mathcal{X}$ be the set of clocks of $\mathcal{A}$, and $I_1, \ldots, I_r$ be a sequence of the successive intervals appearing in φ starting from the beginning of the formula. The automaton $\mathcal{A}_{\varphi}$ extends $\mathcal{A}$ such that

- the set of clocks is given by

$$\mathcal{X}' = \mathcal{X} \cup \{y_1, \ldots, y_r\},$$

- the set of actions is

$$A' = A \cup \{a_{y_1}, \ldots, a_{y_r}\},$$

- the transition relation $E' \subseteq L \times A' \times \mathcal{C}_{\mathcal{X}'} \times 2^{\mathcal{X}'} \times L$ is defined as follows:

$$E' = E \cup \{l \xrightarrow{a_{y_i}, true, \{y_i\}} l \mid l \in L,\ 1 \leq i \leq r\}.$$

---

[10] In fair CTL path quantifiers are restricted to selected subsets of paths, called *fair*.

[11] One clock is sufficient for some methods using the translation.

Example 6.10. Consider a timed automaton $\mathcal{A}$ shown in Fig. 5.17 and the formula $\varphi = \mathrm{EG}_{[2,3]}\wp_1$, where $\wp_1 \in PV$ is a proposition true at the location l_1. In order to build the automaton $\mathcal{A}_\varphi$ we extend the set of clocks of $\mathcal{A}$ by one additional clock y_1, and add a loop edge which resets this clock to each of the locations. The resulting automaton $\mathcal{A}_\varphi$ is depicted in Fig. 6.9. The new edges are marked with the arrows of white arrow-heads.

□

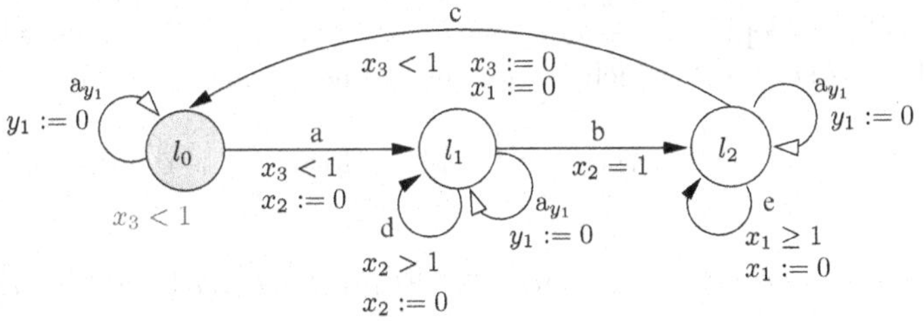

Fig. 6.9. The timed automaton $\mathcal{A}_\varphi$ augmenting that of Fig. 5.17 to verify TCTL formulas with one timing interval

Let $M_{DRG_b}(\mathcal{A}_\varphi) = (z(W, w^0, \rightarrow_{\triangleleft b}), V_a)$ be the bd-region graph model for $\mathcal{A}_\varphi$ modified such that

- the action transitions labelled with a_{y_i} are executed from the boundary regions as well,
- a region is considered as boundary in $\mathcal{A}_\varphi$ if its projection[12] on the clocks in $\mathcal{X}$ is a boundary region of $\mathcal{A}$.

Denote by $\rightarrow_{\triangleleft b_A}$ the part of $\rightarrow_{\triangleleft b}$ where transitions are labelled with elements of $A \cup \{\tau\}$, and by $\rightarrow_{\triangleleft b_{y_i}}$ the transitions that reset the clocks y_i, i.e., labelled with a_{y_i} for $1 \leq i \leq r$. Next, we extend the set of propositional variables PV to PV' and the valuation function V_a to V'_a. By $\wp_{y_i \in I_i}$ we denote a new proposition for every interval I_i appearing in φ, and by PV_φ the set of the new propositions. The proposition $\wp_{y_i \in I_i}$ is true at a state (l, Z) of $M_{DRG_b}(\mathcal{A}_\varphi)$ if $v(y_i) \in I_i$ for $v \in Z$. Let V_φ be a function labelling each state of $M_{DRG_b}(\mathcal{A}_\varphi)$ with the set of propositions from PV_φ true at that state, and labelling with $\wp_b$ each region whose projection on the clocks in $\mathcal{X}$ is a boundary region of $\mathcal{A}$. Next, set

$$PV' = PV \cup PV_\varphi \cup \{\wp_b\}$$

and define the valuation function $V'_a : W \rightarrow 2^{PV'}$ as

$$V'_a = V_a \cup V_\varphi.$$

[12] By the projection of a region R on the clocks in $\mathcal{X}$ we mean the region obtained from R by removing from the clock valuations the values of all the clocks y_i.

The model obtained from $M_{DRG_b}(\mathcal{A}_\varphi)$ by replacing V_a with V'_a is denoted by $M'_{DRG_b}(\mathcal{A}_\varphi)$.

In order to translate a TCTL formula φ to the corresponding CTL$_{-X}$ formula ψ we need to modify the language of CTL$_{-X}$ to CTL$^{\mathrm{r}}_{-X}$ by reinterpreting the operators U and R, denoted now by U_{y_i} and R_{y_i} for all $1 \leq i \leq r$, where we assume that r is the number of the intervals appearing in φ. This language is interpreted over the bd-region graph model for $\mathcal{A}_\varphi$ defined above.

Each operator U_{y_i} and R_{y_i} for $1 \leq i \leq r$ is interpreted on paths over all the transitions except for the new ones, but starting with states where the new clock y_i is set to 0, which is ensured by new transitions that reset this clock.

Formally, for $\wp \in PV'$, the set of CTL$^{\mathrm{r}}_{-X}$ formulas is defined inductively as follows:

$$\psi := \wp \mid \neg\wp \mid \psi \wedge \psi \mid \psi \vee \psi \mid \mathrm{E}(\psi \mathrm{U}_{y_i} \psi) \mid \mathrm{E}(\psi \mathrm{R}_{y_i} \psi) \mid \mathrm{A}(\psi \mathrm{R}_{y_i} \psi) \mid \mathrm{A}(\psi \mathrm{U}_{y_i} \psi).$$

A *path* in $M'_{DRG_b}(\mathcal{A}_\varphi)$ is a maximal sequence $\pi = (w_0, w_1, \ldots)$ of states such that $w_i \rightarrow_{\triangleleft b_\mathcal{A}} w_{i+1}$ for each $i \in \mathbb{N}$. Note that only $\rightarrow_{\triangleleft b_\mathcal{A}}$-steps are considered here. The relation $\models$ is defined like in Sect. 4.2 for all the CTL$^{\mathrm{r}}_{-X}$ formulas except for U_{y_i} and R_{y_i}, which is given as follows:

- $w \models \mathrm{O}(\phi \mathrm{U}_{y_i} \psi)$ iff $(\exists w' \in W)\ w \rightarrow_{\triangleleft b_{y_i}} w' \ \wedge\ w' \models \mathrm{O}(\phi \mathrm{U} \psi)$,
- $w \models \mathrm{O}(\phi \mathrm{R}_{y_i} \psi)$ iff $(\exists w' \in W)\ w \rightarrow_{\triangleleft b_{y_i}} w' \ \wedge\ w' \models \mathrm{O}(\phi \mathrm{R} \psi)$,
 for $\mathrm{O} \in \{\mathrm{E}, \mathrm{A}\}$.

Next, the TCTL formula φ is translated inductively to the CTL$^{\mathrm{r}}_{-X}$ formula $\mathrm{cr}(\varphi)$ as follows:

$$\begin{aligned}
\mathrm{cr}(\wp) &= \wp, \text{ for } \wp \in PV',\\
\mathrm{cr}(\neg\phi) &= \neg\mathrm{cr}(\phi),\\
\mathrm{cr}(\phi \vee \psi) &= \mathrm{cr}(\phi) \vee \mathrm{cr}(\psi),\\
\mathrm{cr}(\phi \wedge \psi) &= \mathrm{cr}(\phi) \wedge \mathrm{cr}(\psi),\\
\mathrm{cr}(\mathrm{O}(\phi \mathrm{U}_{I_i} \psi)) &= \mathrm{O}(\mathrm{cr}(\phi)\mathrm{U}_{y_i}(\mathrm{cr}(\psi) \wedge \wp_{y_i \in I_i} \wedge (\wp_b \vee \mathrm{cr}(\phi)))),\\
\mathrm{cr}(\mathrm{O}(\phi \mathrm{R}_{I_i} \psi)) &= \mathrm{O}(\mathrm{cr}(\phi)\mathrm{R}_{y_i}(\neg\wp_{y_i \in I_i} \vee (\mathrm{cr}(\psi) \wedge (\wp_b \vee \mathrm{cr}(\phi))))),
\end{aligned}$$

for $\mathrm{O} \in \{\mathrm{E}, \mathrm{A}\}$.

It is easy to show that the validity of the TCTL formula φ in the concrete dense model of $\mathcal{A}$ for the valuation function $V_\mathcal{A}$ (using strongly monotonic semantics) is equivalent to the validity of the corresponding CTL$^{\mathrm{r}}_{-X}$ formula $\mathrm{cr}(\varphi)$ in the bd-region graph model $M'_{DRG_b}(\mathcal{A}_\varphi)$ defined above [7].

Example 6.11. Consider the timed automaton $\mathcal{A}$ shown in Fig. 5.17, and the formula $\gamma = \mathrm{E}(\varphi \mathrm{U}_{[1,\infty)} \psi)$. The reachable part of the boundary-distinguishing

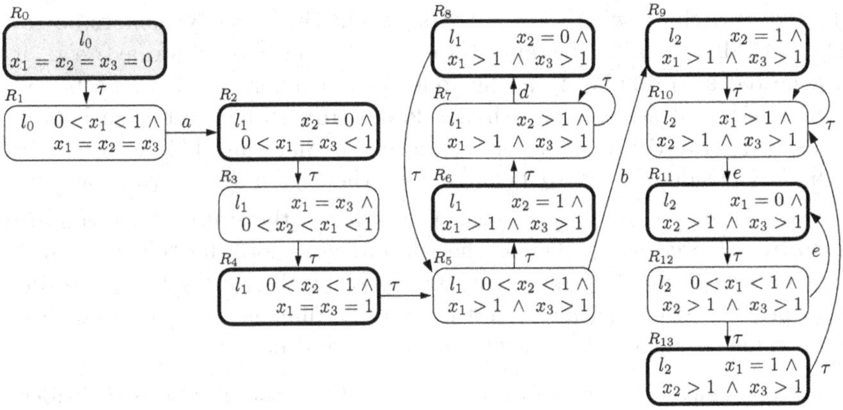

Fig. 6.10. The reachable part if the boundary-distinguishing detailed region graph for the automaton of Fig. 5.17 and $c_{max}(\mathcal{A}, \varphi) = 1$

region graph model for $\mathcal{A}$ and φ is shown in Fig. 6.10. The regions are annotated with their names (i.e., with R_i for $i = 0, \ldots, 13$), and these which are boundary are marked with bold frames.

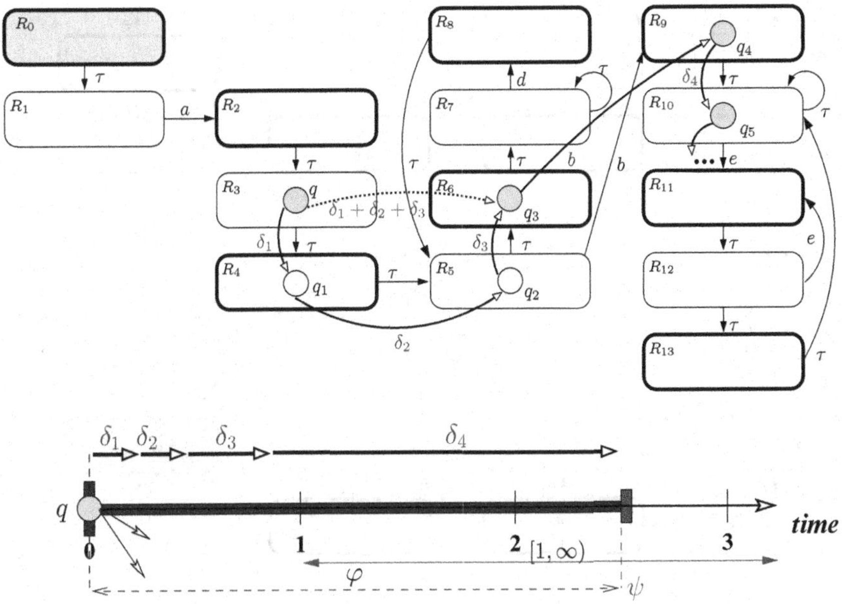

$$q \models \mathrm{E}(\varphi \mathrm{U}_{[1,\infty)} \psi)$$

Fig. 6.11. Testing whether $q \models \gamma$

In order to check whether γ (interpreted in the strongly monotonic semantics) holds in some arbitrary state q of $\mathcal{A}$ (which is performed, e.g., if γ is a nested subformula of another one, and therefore needs to be tested also in non-initial states), one should find a run starting at q such that γ holds along the dense path corresponding to that run. This is shown in Fig. 6.11. Consider a state q belonging to the region R_3. A run found is $q \xrightarrow[c]{\delta_1+\delta_2+\delta_3} q_3 \xrightarrow{b}_c q_4 \xrightarrow{\delta_4}_c q_5 \rightarrow_c \ldots$. In the figure the states of the run are marked by coloured dots, whereas the white dots denote the "intermediate" representatives of the detailed regions traversed by the run. Notice, however, that in practice searching for such a run can be difficult, since it is possible that no value of a clock of $\mathcal{A}$ shows the time passed in the run.

Next, consider the translation of γ to the $\mathrm{CTL}^{\mathrm{r}}_{-\mathrm{X}}$ formula as described above. The new clock introduced by the translation is denoted by y_1, and the resulting formula is $\mathrm{cr}(\gamma) = \mathrm{E}(\mathrm{cr}(\varphi)\mathrm{U}_{y_1}(\mathrm{cr}(\psi) \wedge \wp_{y_1 \in [1,\infty)} \wedge (\wp_b \vee \mathrm{cr}(\varphi))))$. The upper part of Fig. 6.12 shows a fragment of the detailed region graph of the

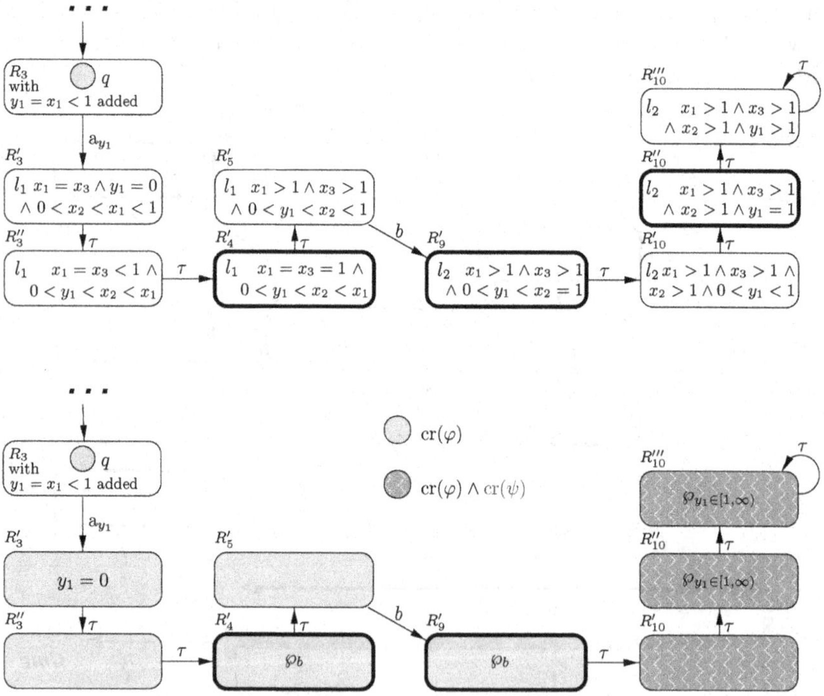

$$q \models \mathrm{E}(\mathrm{cr}(\varphi)\mathrm{U}_{y_1}(\mathrm{cr}(\psi) \wedge \wp_{y_1 \in [1,\infty)} \wedge (\wp_b \vee \mathrm{cr}(\varphi))))$$

Fig. 6.12. Testing γ translated to $\mathrm{CTL}^{\mathrm{r}}_{-\mathrm{X}}$

augmented automaton which is of our interest. The regions whose projections on the clocks in $\mathcal{X}$ are boundary regions are marked with bold frames (notice that certain boundary regions of the region graph for the augmented automaton are not marked, i.e., the value of the clock y_1 is not taken into account). The bottom part of the figure displays the same regions labelled with the new propositions $\wp_b$ and $\wp_{y_1 \in [1,\infty)}$. Notice that the region with the clock y_1 equal to zero is not labelled by $\wp_b$. The above example justifies also the requirement $(\wp_b \vee \mathrm{cr}(\varphi))$ for the region where $\mathrm{cr}(\psi) \wedge \wp_{y_1 \in [1,\infty)}$ holds: if a region R is not boundary, then its borders are open, and therefore the region of each state belonging to R contains also some time predecessors. Thus, if we require that $\mathrm{cr}(\psi)$ holds at some states of R and $\mathrm{cr}(\varphi)$ holds continuously until them, then this means that $\mathrm{cr}(\varphi)$ must hold at all the states of the region R (recall that all the states of a detailed region are equivalent w.r.t. TCTL formulas, so if $\mathrm{cr}(\psi)$ holds a state of R, then it holds at all the states of R).

□

In order to translate TCTL over the weakly monotonic semantics, a bd-region graph is replaced with a region graph in the above method. Moreover, we do not need to use the proposition $\wp_b$ for encoding the boundary regions and the last two rules of the function cr are modified as follows:

$$\begin{aligned}
\mathrm{cr}(\mathrm{O}(\phi \mathrm{U}_{I_i} \psi)) &= \mathrm{O}(\mathrm{cr}(\phi)\mathrm{U}_{y_i}(\mathrm{cr}(\psi) \wedge \wp_{y_i \in I_i})),\\
\mathrm{cr}(\mathrm{O}(\phi \mathrm{R}_{I_i} \psi)) &= \mathrm{O}(\mathrm{cr}(\phi)\mathrm{R}_{y_i}(\neg \wp_{y_i \in I_i} \vee \mathrm{cr}(\psi)))),
\end{aligned}$$

for $\mathrm{O} \in \{\mathrm{E}, \mathrm{A}\}$.

Consider now a translation for $\mathrm{TCTL}_{\mathcal{C}}$. First, we extend the language of CTL by separation predicates in the standard way. So, we do not need to use the propositions $\wp_{y_i \in I_i}$ for encoding the values of the clocks y_i anymore. In order to translate a $\mathrm{TCTL}_{\mathcal{C}}$ formula φ to the corresponding CTL formula ψ we need to modify the language of CTL to $\mathrm{CTL}^{\mathbf{z}}$ by reinterpreting the next-step operators, denoted now by X_{z_i} for all $1 \leq i \leq r$, where we assume that r is the number of of the specification clocks $z_i \in SC$ used in φ. This language is interpreted over the region graph model for $\mathcal{A}_\varphi$ defined above. Each next step operator X_{z_i} for $1 \leq i \leq r$ is interpreted only over the new transitions that reset the new clock z_i, whereas the other operators are interpreted over all the transitions except for the new ones. The relation $\models$ is defined like in Sect. 4.2 for all the $\mathrm{CTL}^{\mathbf{z}}$ formulas except for X_{z_i}, which is given as follows:

$$w \models \mathrm{X}_{z_i}\psi \text{ iff } (\exists w' \in W)\ w \rightarrow_{\triangleleft b_{z_i}} w' \ \wedge\ w' \models \psi.$$

Then, we need to translate only the reset operator in the following way:

$$\mathrm{cr}(z_i.\psi) = \mathrm{X}_{z_i}(\mathrm{cr}(\psi)) \text{ for } 1 \leq i \leq r.$$

Model Checking over (Bi)simulating Models

The above translation can be used for model checking TCTL (TACTL) over abstract models which are coarser than detailed region graphs. To this aim, bisimulating (simulating, respectively) models built for the automaton $\mathcal{A}_\varphi$ can be applied (obviously, in this case we can deal with the weakly monotonic semantics only, for which these models can be generated). However, in order to label the states of a given abstract model with the propositions of PV', they have to respect the timing intervals appearing in the tested formula φ and distinguish between zero and non-zero values of the additional clocks. Formally, each state w of the given model has to satisfy the following condition:

$$(\forall i \in \{1, \dots, r\})\{(l, v) \in w \mid v(y_i) = 0\}, \{(l, v) \in w \mid v(y_i) \in I_i\} \in \{w, \emptyset\},$$

where r denotes the number of extra clocks. Intuitively, the above means that the valuations v in the states w are consistent w.r.t. the values of the cloks y_i, which either all are equal to 0 or all belong to I_i, or none of them is equal to 0 or belongs to I_i. To achieve this, we can, for instance, build models using partition refinement algorithms, starting from an initial partition whose classes satisfy the condition given above [9, 159].

It is easy to notice that one extra clock is sufficient for the method. In this case, the initial partition has to satisfy the condition

$$(\forall i \in \{1, \dots, r\})\{(l, v) \in w \mid v(y) = 0\}, \{(l, v) \in w \mid v(y) \in I_i\} \in \{w, \emptyset\},$$

where y is the additional clock, and r denotes the number of timing intervals appearing in φ.

6.2.2 Overview of Other Approaches to TCTL Verification

Bouajjani et al. [41] defined an algorithm, which starts with building a simulation graph for a timed automaton. Then, the cycles of this graph are refined. This process is guided by a formula. If a stable cycle is found, then this means that the formula holds and at that point the verification ends.

Another solution has been suggested by Dickhofer and Wilke [67] and Henzinger et al. [96]. The idea follows the standard approach to automata-theoretic model checking for CTL. So, first an automaton accepting all the models for a TCTL formula is built and the product of this automaton with the automaton corresponding to the detailed region graph is constructed while its non-emptiness is checked [67]. The method of [96] is slightly different as the product is constructed without building the automaton for a formula first.

6.3 Selected Tools

There are many tools using the approaches considered in this and the former section. Below, we list some of them and give pointers to the literature, where more detailed descriptions can be found.

6.3.1 Tools for TPNs

Some of the existing tools for Petri nets with time are listed below:

- **CPN Tools** [134] (a replacement for **Design/CPN** [55]) – a software package for modelling and analysis of both timed and untimed Coloured Petri Nets, enabling their simulation, generating occurrence (reachability) graphs, and analysis by place invariants.
- **INA** (Integrated Net Analyser) [136] – a Petri net analysis tool, supporting place/transition nets and coloured Petri nets with time and priorities. It offers edition and simulation, as well as analysis of structural properties (i.e., liveness, safeness, boundedness etc.), computation of reachability and coverability graphs and unfolding of coloured Petri nets. Among others, INA provides verification by analysis of paths for TPNs [128] .
- **PEP (Programming Environment based on Petri nets)** [149] – a comprehensive set of modelling, compilation, simulation and verification components, linked together within a graphical user interface. The components enable designing of parallel systems, generating Petri nets from these models, simulation of high- and low-level nets, various verification algorithms (e.g., reachability and deadlock-freeness checking, partial-order based model checking), and interfaces to some other tools (like the INA package, FC2Tools [42], SMV [104] and SPIN [85]).
- **Romeo** [137] – a tool for time Petri nets analysis, which provides several methods for translating TPNs to TA [52, 103] and computation of state class graphs [73].
- **Tina** [34] is a toolbox for analysis of (time) Petri nets. It constructs state class graphs [32,33] and performs LTL or reachability verification. In addition, Tina builds atomic state class graphs [35] to be used for verification of CTL formulas. Tina includes an editor for Petri nets and time Petri nets, as well as a tool for structural analysis for nets.

6.3.2 Tools for TA

Some of the existing tools for timed automata are listed below:

- **Cospan** is a tool for verifying the behaviour of designs written in the industry standard design languages like VHDL and Verilog. Besides verifying that behaviours are correct, the tool identifies deadlocks and livelocks. It implements an automata-based approach to model checking including an on-the-fly enumerative search (using zones in the timed case), as well as symbolic search using BDDs[13]. A detailed description of timed verification can be found in [15].

[13] Binary Decision Diagrams [50].

- **Kronos** [175] is a tool which performs verification of TCTL formulas using forward or backward analysis, and behavioural analysis, which consists in building bisimulating abstract models (using partitioning), and then checking whether the abstract model of the system simulates or is equivalent to that of the specification. DBMs are used for representing zones. In order to improve on the time and memory consumption, some additional improvements like abstractions and reductions in the number of clocks [63, 64], using an on-the-fly approach [41] or binary decision diagrams [48] are implemented as well.
- **UppAal2k** [121] (a successor of UppAal) is a tool for modelling, simulation and verification of timed systems, appropriate for systems which can be described by a collection of non-deterministic processes with finite control structure and real-valued clocks, communicating through channels or shared variables. The tool consists of three parts: a description language, a simulator (a validation tool which enables examination of possible dynamic executions of a system), and a model checker, which tests invariant and bounded-liveness properties by exploring the symbolic state space of the system. i.e., reachability analysis in terms of symbolic states represented by constraints. Forward reachability analysis, deadlock detection and verification of properties expressible in a subset of TCTL are available.
- **VerICS** [65] implements partition refinement algorithms for verifying TCTL (over bisimulating or simulating models) and reachability (over pseudo-bismulating or pseudo-simulating models) for timed automata. DBMs are used to represent zones. VerICS includes an editor and simulator for time Petri nets and a translator to timed automata.

Further Reading

The interested reader is referred to several recently published books on automated verification of concurrent systems [57, 90, 115]. A chapter comparing and evaluating some selected tools can be found in [28].

7

Verification Based on Satisfiability Checking

Similarly to Chap. 6 our aim is to show how one can verify the common properties of timed systems expressible in TCTL ($\mathrm{TCTL}_{\mathcal{C}}$) and some its sublogics, but, here, using SAT-based symbolic methods. SAT-based model checking is the most recent symbolic approach that has been motivated by a dramatic increase in efficiency of SAT-solvers, i.e., algorithms solving the satisfiability problem for propositional formulas [179]. The main idea of SAT-based methods consists in translating the model checking problem for a temporal logic to the problem of satisfiability of a formula in propositional or separation logic. This formula is typically obtained by combining an encoding of the model and of the temporal property. In principle, there are two different approaches. In the first one, a model checking problem for TCTL (LTL or reachability properties) is translated to a formula in separation logic [20, 110, 145] or quantified separation logic [140] and then either solved by MathSAT[1] or translated further to propositional logic and solved by a SAT-solver. The second approach exploits a translation of the model checking problem from TCTL to CTL and then further to a propositional formula [120].

On the other hand, the approaches to SAT-based symbolic verification can be viewed as bounded (BMC) or unbounded (UMC). BMC [110, 120] applies to an existential fragment of TCTL (i.e., TECTL) on a part of the model, whereas UMC [140] is for unrestricted TCTL (or timed μ-calculus) on the whole model. However, it is possible to use the bounded approach for verifying some universal properties as well, which is shown for unreachability properties in [177].

To our knowledge, there are very few approaches to a (direct) SAT-based verification of TPNs [39], which mainly consist in describing the state space of a net in a way which is accepted by an existing symbolic tool. Other approaches (listed below) use a construction of a detailed region graph[2] for a time Petri net in many ways. A BDD-based CTL model checking is dis-

[1] MathSAT is a solver checking satisfiability of SL.

[2] This notion is inherited from timed automata.

W. Penczek and A. Pólrola: *Verification Based on Satisfiability Checking*, Studies in Computational Intelligence (SCI) **20**, 181–230 (2006)
www.springerlink.com

cussed in [111], a SAT-based approach to checking reachability is described in [118], whereas explicit TCTL verification on detailed region graphs is shown in [123,163]. Therefore, in the following sections we focus on SAT-based methods for timed automata, assuming that these provide solutions for time Petri nets when combined with translations.

In the next section, we discuss BMC for TECTL[3], for unreachability properties[4] of TA as well as for LTL. An adaptation of BMC to the existential fragment of $\mathrm{TCTL}_{\mathcal{C}}$ is an easy exercise left to the reader. Then, we look at the UMC approach. We discuss this approach for the logics $\mathrm{TCTL}_{\mathcal{C}}$ and $T\mu$. Since our algorithm is based on the symbolic model checking method by Henzinger et al. [84], it applies only to timed logics over the weakly monotonic semantics.

Next, we show how to decide formulas of separation logic (Sect. 7.4) and propositional logic (Sect. 7.5).

7.1 Bounded Model Checking Using Direct Translation to SAT

BMC for TCTL consists in translating the model checking problem of an existential TCTL formula (i.e., containing only existential quantifiers) to the problem of satisfiability of a propositional formula. This translation is based on bounded semantics satisfaction, which, instead of using possibly infinite paths, is limited to finite prefixes only. Moreover, it is known that the translation of the existential path quantifier can be restricted to finitely many computations [119].

In this section we describe how to apply BMC to TECTL. The main idea of our method consists in translating the TECTL model checking problem to the model checking problem for a branching time logic [7, 159] and then in applying BMC for this logic [119]. To this aim we start with showing a discretization of TA. Then, a translation from TCTL to $\mathrm{CTL}^{\mathrm{r}}_{-\mathrm{X}}$ on discretized region graphs models is applied, so that we can restrict ourselves to BMC for $\mathrm{CTL}^{\mathrm{r}}_{-\mathrm{X}}$. Finally, the BMC method for checking reachability and unreachability for TA is discussed.

7.1.1 Discretization of TA

We define a discretized model for a timed automaton, which is based on the discretization of [177] (a generalisation of these of [18, 76]). The idea behind this method is to represent detailed zones of a timed automaton by one or more (but finitely many) specially chosen representatives.

[3] We consider both the semantics for TCTL i.e., the time is weakly or strongly monotonic.

[4] For the time to be weakly monotonic.

Let $\mathcal{A} = (A, L, l^0, E, \mathcal{X}, \mathcal{I})$ be a diagonal-free[5] timed automaton with $n_{\mathcal{X}}$ clocks, $V_{\mathcal{A}}$ be a valuation function, and φ be a TCTL formula. As before, let $M_c(\mathcal{A}) = (C_c(\mathcal{A}), V_{\mathfrak{C}})$ be the concrete dense model for $\mathcal{A}$. We choose the discretization step

$$\Delta = 1/d,$$

where d is a fixed even number[6] not smaller than $2n_{\mathcal{X}}$. The *discretized clock space* is defined as $\mathbb{D}^{n_{\mathcal{X}}}$, where

$$\mathbb{D} = \{k\Delta \mid 0 \leq k\Delta \leq 2c_{max}(\mathcal{A}, \varphi) + 2\}.$$

This means that the clocks cannot go beyond $2c_{max}(\mathcal{A}, \varphi) + 2$, which follows from the fact that for evaluating the TCTL formula φ over diagonal-free timed automata we do not need to distinguish between clock valuations above $c_{max}(\mathcal{A}, \varphi) + 1$. Similarly, the maximal values of time delays can be restricted to $c_{max}(\mathcal{A}, \varphi) + 1$, since otherwise they would make the values of clocks greater than $c_{max}(\mathcal{A}, \varphi) + 1$. Thus, the set of values that can change a valuation in a detailed zone is defined as

$$\mathbb{E} = \{k\Delta \mid 0 \leq k\Delta \leq c_{max}(\mathcal{A}, \varphi) + 1\}.$$

To make sure that the above two definitions can be applied we will guarantee that before taking any time transition, the value of every clock does not exceed $c_{max}(\mathcal{A}, \varphi) + 1$. In what follows, by $\mathbb{E}_+$ we denote $\mathbb{E} \setminus \{0\}$.

Next, we define the set $\mathbb{U}^{n_{\mathcal{X}}}$ of valuations that are used to "properly" represent detailed zones in the discretized model, i.e., we take a subset of the valuations v of $\mathbb{D}^{n_{\mathcal{X}}}$ that preserve time delays by insisting that either the values of all the clocks in v are only even or only odd multiplications of Δ. To preserve action successors we will later use "adjust" transitions. The set $\mathbb{U}^{n_{\mathcal{X}}}$ is defined as follows:

$$\begin{aligned}\mathbb{U}^{n_{\mathcal{X}}} = \{v \in \mathbb{D}^{n_{\mathcal{X}}} \mid & (\forall x \in \mathcal{X})(\exists k \in \mathbb{N})\ v(x) = 2k\Delta \vee \\ & (\forall x \in \mathcal{X})(\exists k \in \mathbb{N})\ v(x) = (2k+1)\Delta\}.\end{aligned}$$

Example 7.1. Consider a timed automaton over two clocks x_1, x_2. Figure 7.1 shows how to discretize a part (consisting of $[0,1)^2$) of the set of its clock valuations. The discretization step chosen is $\Delta = 1/4$. The points (both the squares and the dots) represent elements of $\mathbb{D}^2$. The dots are elements of $\mathbb{U}^2$. Notice that the discretization, i.e., the dots, preserve the time-successor relation (see Fig. 7.2), if not-open regions are concerned. For representatives of open regions we need to apply a special technique, which "freezes" the valuations of clocks as soon as they become bigger than $c_{max}(\mathcal{A}, \varphi) + 1$.

---

5 A discretization for non diagonal-free timed automata has been recently defined by Zbrzezny in [178]. This new discretization was applied to checking reachability and unreachability.

6 A good choice for d is the minimal number, which equals to 2^j for some j.

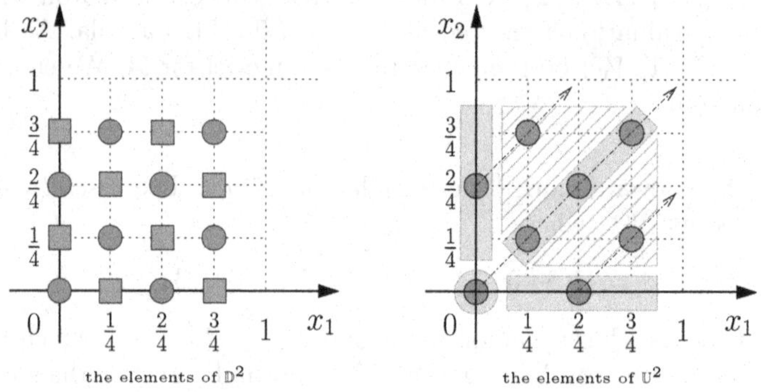

Fig. 7.1. Discretizing $[0, 1)^2$. The dots mark the elements of $\mathbb{U}^2$, while the squares belong to $\mathbb{D}^2 \setminus \mathbb{U}^2$

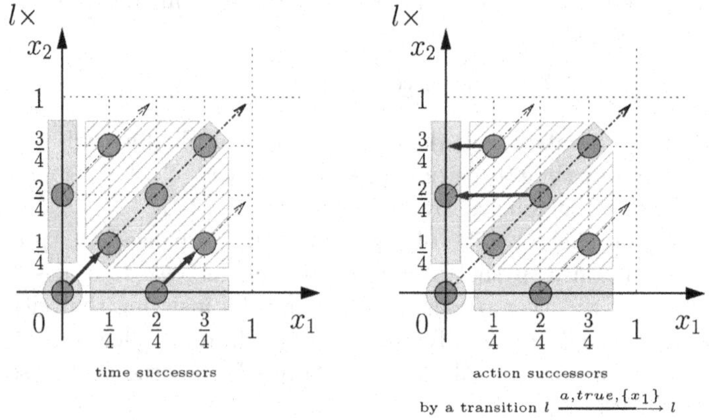

Fig. 7.2. Time- and action successor relation vs. discretization. Left: the elements of $\mathbb{U}^2$ preserve the time-successor relation. Right: the action successor relation is not preserved

Using only the squares is not possible, as there is no representation of the detailed zone $x_1 = x_2 = 0$ by a point from $\mathbb{D}^2 \setminus \mathbb{U}^2$. On the other hand, the elements of $\mathbb{D}^2$ would not preserve the time-successor relation; for example, if we consider the detailed zone $Z = [\![x_1 = 0 \;\wedge\; 0 < x_2 < 1]\!]$ and its representative $(0, 3/4)$, then there is no $\delta \in \mathbb{E}$ such that $(0 + \delta, 3/4 + \delta) \in \mathbb{D}^2$ and $(0 + \delta, 3/4 + \delta)$ is a representative of the time successor of Z (i.e., of $[\![0 < x_1 < x_2 < 1]\!]$).

Unfortunately, the elements of $\mathbb{U}^2$ do not preserve action successors. In order to solve this problem adjust transitions are used, which we discuss later. □

Discretized Model for (Un)reachability Verification

Now, we are ready to define a *discretized model* for $\mathcal{A}$ that is later used for checking reachability and unreachability of a propositional formula p, for the weakly monotonic semantics. Notice that, as we have already mentioned, reachability of p can be expressed by the CTL formula $\varphi = \mathrm{EF}p$ (or alternatively by the TCTL formula $\mathrm{EF}_{[0,\infty)}p$) for the weakly monotonic semantics. Note also that in this case $c_{max}(\mathcal{A}, \varphi)$ depends only on $\mathcal{A}$ (i.e., equals $c_{max}(\mathcal{A})$).

Definition 7.2 (Discretized (concrete) model). *The* discretized (concrete) model *for $\mathcal{A}$ is a structure*

$$DM(\mathcal{A}) = ((S_{\mathbb{D}}, s^0, \rightarrow_{\mathfrak{D}_c}), V_{\mathfrak{D}_c}),$$

where

- $S_{\mathbb{D}} = L \times \mathbb{D}^{n_{\mathcal{X}}}$,
- $s^0 = (l^0, v^0)$ *is the initial state,*
- *the labelled transition relation* $\rightarrow_{\mathfrak{D}_c} \subseteq S_{\mathbb{D}} \times (A \cup \mathbb{E} \cup \{\epsilon\}) \times S_{\mathbb{D}}$ *is defined as*
  1. $(l, v) \xrightarrow{\delta}_{\mathfrak{D}_c} (l, v')$ *iff*
     - $(\forall x \in \mathcal{X})(v(x) \leq c_{max}(\mathcal{A}, \varphi) + 1)$,
     - $v' = v + \delta$ *and*
     - $v, v' \in [\![\mathcal{I}(l)]\!]$,

     for $\delta \in \mathbb{E}$ *(*time delay transition*),*
  2. $(l, v) \xrightarrow{a}_{\mathfrak{D}_c} (l', v')$ *iff*
     - $(l, v) \xrightarrow{a}_{c} (l', v')$ *in* $C_c(\mathcal{A})$, *for* $a \in A$

     *(*action transition*),*
  3. $(l, v) \xrightarrow{\epsilon}_{\mathfrak{D}_c} (l, v')$ *iff*
     - $v' \in \mathbb{U}^{n_{\mathcal{X}}}$,
     - $(\forall x \in \mathcal{X})(v'(x) \leq c_{max}(\mathcal{A}, \varphi) + 1)$, *and*
     - $v \simeq_{C_{\mathcal{A},\varphi}} v'$

     *(*adjust transition*),*

  and
- *the valuation function* $V_{\mathfrak{D}_c} : S_{\mathbb{D}} \longrightarrow 2^{PV}$ *is given by*

$$V_{\mathfrak{D}_c}((l, v)) = V_{\mathcal{A}}(l).$$

Notice that the transitions in $DM(\mathcal{A})$ are labelled with actions of A, time delays of $\mathbb{E}$, or with the epsilon label $\epsilon \notin A \cup \mathbb{E}$. The first two types of labels correspond exactly to labels used in the concrete model. The adjust transitions are used for moving within detailed zones to a valuation in $\mathbb{U}^{n_{\mathcal{X}}}$. The reason for defining separately adjust transitions with action- and time delay transitions consists in increasing efficiency of the implementation for checking reachability in timed automata.

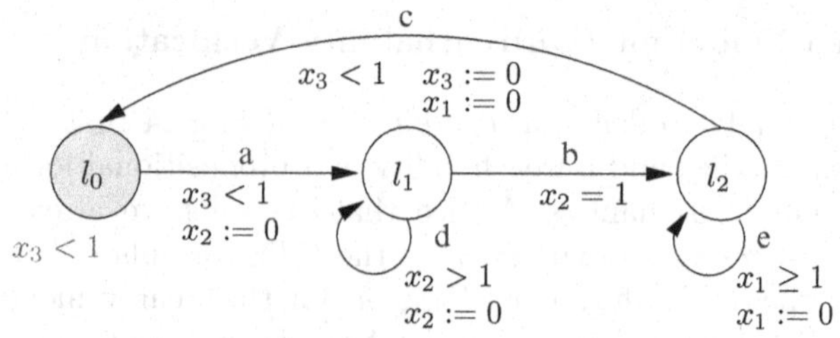

Fig. 7.3. The timed automaton used in Example 7.3

Example 7.3. Consider again the automaton $\mathcal{A}$ over three clocks x_1, x_2, x_3 used in Example 5.14 and shown in Fig. 7.3. Its detailed region graph for the weakly monotonic semantics (shown in Fig. 5.18) is recalled in Fig. 7.4. The right upper corners of the rectangles representing the regions are annotated with their names (i.e., with R_i for $i = 0, \ldots, 19$).

In order to build a discretized model for reachability verification, we chose the discretization step $\Delta = 1/8$ (notice that this is the maximal possible step for three clocks such that its denominator is a power of two, since $2n_{\mathcal{X}} = 6$). Due to $c_{max}(\mathcal{A}) = 1$, we obtain $\mathbb{D} = \{\frac{1}{8}k \mid 0 \leq \frac{1}{8}k \leq 4\}$, $\mathbb{E} = \{\frac{1}{8}k \mid 0 \leq \frac{1}{8}k \leq 2\}$ and $\mathbb{U}^3 = \{v \in \mathbb{D}^3 \mid (\forall x \in \mathcal{X})(\exists k \in \mathbb{N})\ v(x) = \frac{2k}{8} \vee (\forall x \in \mathcal{X})(\exists k \in \mathbb{N})\ v(x) = \frac{2k+1}{8}\}$.

Assume that the clock valuations for $\mathcal{A}$ are triples of the form $v = (v(x_1), v(x_2), v(x_3))$. One of the possible sequences of transitions in $DM(\mathcal{A})$ is

$$(l_0, (0,0,0)) \xrightarrow{1/8}_{\eth_c} (l_0, (\tfrac{1}{8}, \tfrac{1}{8}, \tfrac{1}{8})) \xrightarrow{a}_{\eth_c} (l_1, (\tfrac{1}{8}, 0, \tfrac{1}{8})) \xrightarrow{9/8}_{\eth_c} (l_1, (\tfrac{10}{8}, \tfrac{9}{8}, \tfrac{10}{8})) \xrightarrow{d}_{\eth_c} (l_1, (\tfrac{10}{8}, 0, \tfrac{10}{8})) \xrightarrow{9/8}_{\eth_c} (l_1, (\tfrac{19}{8}, \tfrac{9}{8}, \tfrac{19}{8})) \xrightarrow{d}_{\eth_c} (l_1, (\tfrac{19}{8}, 0, \tfrac{19}{8})) \xrightarrow{\epsilon}_{\eth_c} (l_1, (\tfrac{10}{8}, 0, \tfrac{10}{8})) \longrightarrow_{\eth_c} \ldots.$$

A symbolic presentation of this sequence is shown in Fig. 7.5. In the background, the regions of the detailed region graph are depicted. The states are denoted sequentially by $s_0, s_1, \ldots$. Similarly as in the previous example, the dots represent the elements of $L \times \mathbb{U}^3$, whereas the squares – these of $L \times (\mathbb{D}^3 \setminus \mathbb{U}^3)$. The states are placed in the regions they belong to. Notice that the time delay transitions do not necessarily correspond to the successor relation between detailed regions, e.g., the step $s_2 \xrightarrow{9/8}_{\eth_c} s_3$ is a counterpart of the sequence $R_2 \xrightarrow{\tau}_{\triangleleft} R_3 \xrightarrow{\tau}_{\triangleleft} R_4 \xrightarrow{\tau}_{\triangleleft} R_5 \xrightarrow{\tau}_{\triangleleft} R_6 \xrightarrow{\tau}_{\triangleleft} R_7$. Similarly, the step $s_4 \xrightarrow{9/8}_{\eth_c} s_5$ is a counterpart of the sequence $R_8 \xrightarrow{\tau}_{\triangleleft} R_5 \xrightarrow{\tau}_{\triangleleft} R_6 \xrightarrow{\tau}_{\triangleleft} R_7$. Notice, moreover, that the state $s_7 = (l_1, (\frac{10}{8}, 0, \frac{10}{8}))$ obtained by the adjust transition is not the only possible one – e.g., the state $(l_1, (\frac{12}{8}, 0, \frac{12}{8}))$ could have been selected as well.

□

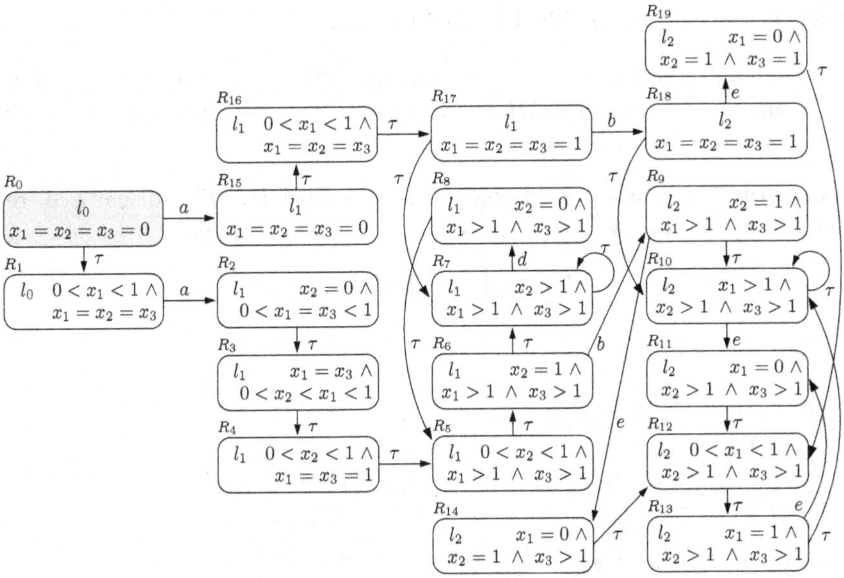

Fig. 7.4. The reachable part of the detailed region graph for the automaton of Fig. 7.3 and $c_{max}(\mathcal{A}, \varphi) = 1$

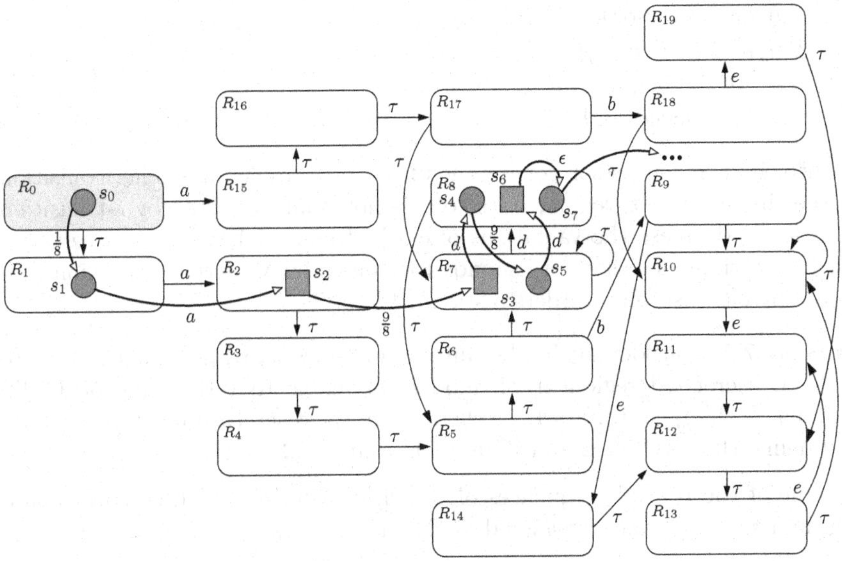

Fig. 7.5. The sequence of transitions considered in Example 7.3 in the reachable part of the discretized model for the automaton in Fig. 7.3

Discretized Model for TCTL Verification

However, for verification of more complex TCTL[7] formulas we need to put some restrictions on $DM(\mathcal{A})$ by defining the discretized region graph (discretized-rg) model.

Definition 7.4 (Discretized region graph model). *The* discretized region graph model *(*discretized rg-model*) for $\mathcal{A}$ is a structure*

$$DM_{DRG}(\mathcal{A}) = ((S, s^0, \rightarrow_{\mathfrak{d}}), V_{\mathfrak{D}}),$$

where

- $S = L \times \mathbb{U}^{n_\mathcal{X}}$,
- $s^0 = (l^0, v^0)$ *is the initial state,*
- $V_{\mathfrak{D}} = V_{\mathfrak{D}_c}|_S$, *and*
- *the transition relation* $\rightarrow_{\mathfrak{d}} \subseteq S \times (A \cup \{\tau\}) \times S$ *is given as*
  1. $(l, v) \xrightarrow{\tau}_{\mathfrak{d}} (l, v')$ *iff*
     - $(l, v) \xrightarrow{\delta}_{\mathfrak{d}_c}; \xrightarrow{\epsilon}_{\mathfrak{d}_c} (l, v')$ *for some* $\delta \in \mathbb{E}_+$, *and*
     - *if* $(l, v) \xrightarrow{\delta'}_{\mathfrak{d}_c} (l, v'') \xrightarrow{\delta''}_{\mathfrak{d}_c} (l, v')$ *with* $\delta', \delta'' \in \mathbb{E}$ *and* $(l, v'') \in S$, *then* $v \simeq_{C_{\mathcal{A},\varphi}} v''$ *or* $v' \simeq_{C_{\mathcal{A},\varphi}} v''$, *and*
     - *if* $v \simeq_{C_{\mathcal{A},\varphi}} v'$, *then* $v \simeq_{C_{\mathcal{A},\varphi}} v' + \delta''$ *for each* $\delta'' \in \mathbb{E}_+$

     *(*time successor*),*
  2. $(l, v) \xrightarrow{a}_{\mathfrak{d}} (l', v')$ *iff*
     - $(l, v) \xrightarrow{a}_{\mathfrak{d}_c}; \xrightarrow{\epsilon}_{\mathfrak{d}_c} (l', v')$

     *(*action successor*).*

Intuitively, time successor corresponds to a move by a time delay transition to the time-successor region, if its zone is not final (adjusted by ϵ-transition in order to decrease the valuations of clocks to $c_{max}(\mathcal{A}, \varphi) + 1$ if necessary), whereas action successor corresponds to a move by an action transition (adjusted by ϵ-transition in order to stay in $\mathbb{U}^{n_\mathcal{X}}$).

Example 7.5. Consider again the automaton shown in Fig. 7.3. In order to build a discretized region graph model applicable to verification of TCTL formulas with $c_{max}(\mathcal{A}, \varphi) = 1$, we chose again the discretization step $\Delta = 1/8$, and define the sets $\mathbb{D}^3$, $\mathbb{E}$ and $\mathbb{U}^3$ as in Example 7.3.

One of the possible sequences of transitions in $DM_{DRG}(\mathcal{A})$, corresponding, in a sense, to that considered in Example 7.3, is

$(l_0, (0, 0, 0)) \xrightarrow{1/8}_{\mathfrak{d}} (l_0, (\frac{1}{8}, \frac{1}{8}, \frac{1}{8})) \xrightarrow{a}_{\mathfrak{d}} (l_1, (\frac{2}{8}, 0, \frac{2}{8})) \xrightarrow{1/8}_{\mathfrak{d}} (l_1, (\frac{3}{8}, \frac{1}{8}, \frac{3}{8})) \xrightarrow{5/8}_{\mathfrak{d}} (l_1, (1, \frac{6}{8}, 1)) \xrightarrow{1/8}_{\mathfrak{d}} (l_1, (\frac{9}{8}, \frac{7}{8}, \frac{9}{8})) \xrightarrow{1/8}_{\mathfrak{d}} (l_1, (\frac{10}{8}, 1, \frac{10}{8})) \xrightarrow{1/8}_{\mathfrak{d}} (l_1, (\frac{11}{8}, \frac{9}{8}, \frac{11}{8})) \xrightarrow{d}_{\mathfrak{d}} (l_1, (\frac{10}{8}, 0, \frac{10}{8})) \xrightarrow{3/8}_{\mathfrak{d}} (l_1, (\frac{13}{8}, \frac{3}{8}, \frac{13}{8})) \xrightarrow{5/8}_{\mathfrak{d}} (l_1, (\frac{10}{8}, 1,$

[7] This is still for the weakly monotonic semantics.

$\frac{10}{8})) \xrightarrow{1/8}_{\mathfrak{d}} (l_1,(\frac{11}{8},\frac{9}{8},\frac{11}{8})) \xrightarrow{d}_{\mathfrak{d}} (l_1,(\frac{10}{8},0,\frac{10}{8})) \longrightarrow_{\mathfrak{d}} \ldots$

Another sequence is

$$(l_0,(0,0,0)) \xrightarrow{1/8}_{\mathfrak{d}} (l_0,(\tfrac{1}{8},\tfrac{1}{8},\tfrac{1}{8})) \xrightarrow{a}_{\mathfrak{d}} (l_1,(\tfrac{2}{8},0,\tfrac{2}{8})) \xrightarrow{1/8}_{\mathfrak{d}} (l_1,(\tfrac{3}{8},\tfrac{1}{8},\tfrac{3}{8})) \xrightarrow{5/8}_{\mathfrak{d}} (l_1,(1,\tfrac{6}{8},1)) \xrightarrow{1/8}_{\mathfrak{d}} (l_1,(\tfrac{9}{8},\tfrac{7}{8},\tfrac{9}{8})) \xrightarrow{1/8}_{\mathfrak{d}} (l_1,(\tfrac{10}{8},1,\tfrac{10}{8})) \xrightarrow{1/8}_{\mathfrak{d}} (l_1,(\tfrac{11}{8},\tfrac{9}{8},\tfrac{11}{8})) \xrightarrow{d}_{\mathfrak{d}} (l_1,(\tfrac{10}{8},0,\tfrac{10}{8})) \xrightarrow{3/8}_{\mathfrak{d}} (l_1,(\tfrac{13}{8},\tfrac{3}{8},\tfrac{13}{8})) \xrightarrow{5/8}_{\mathfrak{d}} (l_1,(\tfrac{10}{8},1,\tfrac{10}{8})) \xrightarrow{b}_{\mathfrak{d}} (l_1,(\tfrac{10}{8},1,\tfrac{10}{8})) \longrightarrow_{\mathfrak{d}} \ldots.$$

Notice that in both the sequences each state which results from an action successor is obtained by a combination of an action- and (non-trivial) adjust transition of $DM(\mathcal{A})$.

Both the sequences are symbolically depicted in Fig. 7.6. The states of the first one are denoted by $s'_0, s'_1, \ldots$. The second sequence corresponds to $s'_0, \ldots, s'_{10}, s'', \ldots$. Notice that in this case, unlike in $DM(\mathcal{A})$, the successor relation between pairs of states needs to correspond to that between the regions the states belong to.

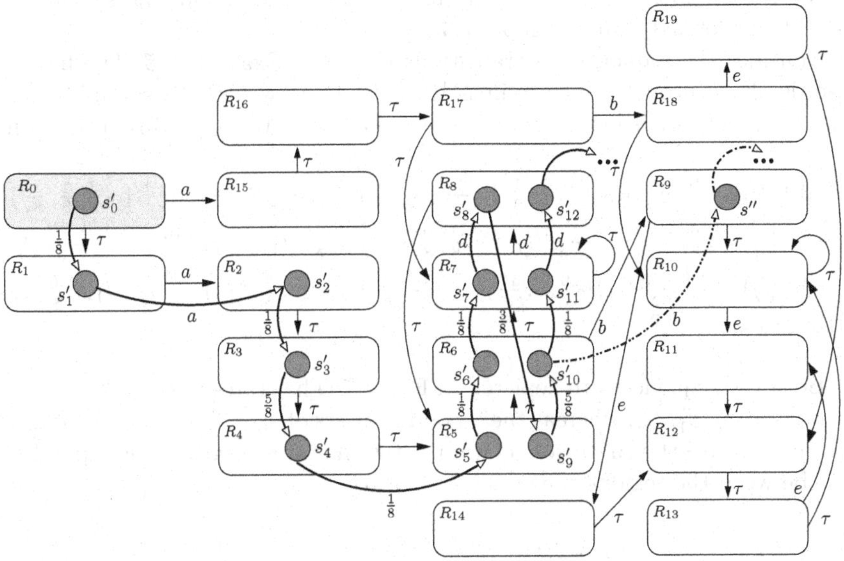

Fig. 7.6. Sequences of transitions in the reachable part of the discretized rg-model for the automaton in Fig. 7.3 and $c_{max}(\mathcal{A}, \varphi) = 1$

□

In order to define the *discretized boundary-distinguishing region graph model* (*discretized* rg_b-*model*)

$$DM_{DRG_b}(\mathcal{A}) = ((S, s^0, \rightarrow_{\mathfrak{d}}), V_{\mathfrak{D}})$$

for the strongly monotonic semantics, the definition of the action successor becomes a bit more complicated:

2. $(l, v) \xrightarrow{a}_{\mathfrak{d}} (l', v')$ iff the following conditions hold:
   - (l, v) is not boundary and
   - $((l, v) \xrightarrow{a}_{\mathfrak{d}_c}; \xrightarrow{\epsilon}_{\mathfrak{d}_c} (l', v')$ or $(l, v) \xrightarrow{\tau}_{\mathfrak{d}}; \xrightarrow{a}_{\mathfrak{d}_c}; \xrightarrow{\epsilon}_{\mathfrak{d}_c} (l', v'))$, for $a \in A$

   (*action successor*).

Intuitively, an action successor corresponds to a move by an action transition (adjusted by ϵ-transition in order to stay within $\mathbb{U}^{n_{\mathcal{X}}}$), taken from non-boundary regions, possibly preceded by the time successor step (to compensate for not executing actions from boundary regions).

Example 7.6. Consider again the automaton shown in Fig. 7.3. The boundary-distinguishing detailed region graph for this automaton and $c_{max}(\mathcal{A}, \varphi) = 1$ has been shown in Fig. 5.19. In order to build a corresponding discretized rg_b-model, we chose the discretization step $\Delta = 1/8$. Again, the sets $\mathbb{D}^3$, $\mathbb{E}$ and $\mathbb{U}^3$ are defined as in Example 7.3.

Consider the sequences of transitions shown in Example 7.5. The first of them is also a proper sequence in $DM_{DRG_b}(\mathcal{A})$. However, the second is not, since the region R_6 is boundary. Its counterpart in $DM_{DRG_b}(\mathcal{A})$ is of the form

$$(l_0, (0,0,0)) \xrightarrow{1/8}_{\mathfrak{d}} (l_0, (\tfrac{1}{8}, \tfrac{1}{8}, \tfrac{1}{8})) \xrightarrow{a}_{\mathfrak{d}} (l_1, (\tfrac{2}{8}, 0, \tfrac{2}{8})) \xrightarrow{1/8}_{\mathfrak{d}} (l_1, (\tfrac{3}{8}, \tfrac{1}{8}, \tfrac{3}{8})) \xrightarrow{5/8}_{\mathfrak{d}} (l_1, (1, \tfrac{6}{8}, 1)) \xrightarrow{1/8}_{\mathfrak{d}} (l_1, (\tfrac{9}{8}, \tfrac{7}{8}, \tfrac{9}{8})) \xrightarrow{1/8}_{\mathfrak{d}} (l_1, (\tfrac{10}{8}, 1, \tfrac{10}{8})) \xrightarrow{1/8}_{\mathfrak{d}} (l_1, (\tfrac{11}{8}, \tfrac{9}{8}, \tfrac{11}{8})) \xrightarrow{d}_{\mathfrak{d}} (l_1, (\tfrac{10}{8}, 0, \tfrac{10}{8})) \xrightarrow{3/8}_{\mathfrak{d}} (l_1, (\tfrac{13}{8}, \tfrac{3}{8}, \tfrac{13}{8})) \xrightarrow{b}_{\mathfrak{d}} (l_1, (\tfrac{10}{8}, 1, \tfrac{10}{8})) \longrightarrow_{\mathfrak{d}} \ldots.$$

Both the sequences are depicted in Fig. 7.7. The states of the first one are denoted by $s'_0, s'_1, \ldots$, whereas the second consists of $s'_0, \ldots, s'_9, s''$. Notice that also in this case the successor relation between states needs to correspond to that between the regions the states belong to.

□

Since our definition is based on the notion of the detailed region graph (Def. 5.13 (5.12)) for the strongly (weakly, respectively) monotonic semantics, it is easy to notice that $DM_{DRG_b}(\mathcal{A})$ ($DM_{DRG}(\mathcal{A})$, respectively) is its discretization, and as such, it can be used for checking the TCTL formula φ.

Basing on the translation from TCTL to $\mathrm{ECTL}^{\mathrm{r}}_{-\mathrm{X}}$ (see Sect. 6.2.1) it is sufficient to show a BMC method for $\mathrm{ECTL}^{\mathrm{r}}_{-\mathrm{X}}$ over discretized rg (rg_b) models for TA. This way we obtain a BMC method for TECTL.

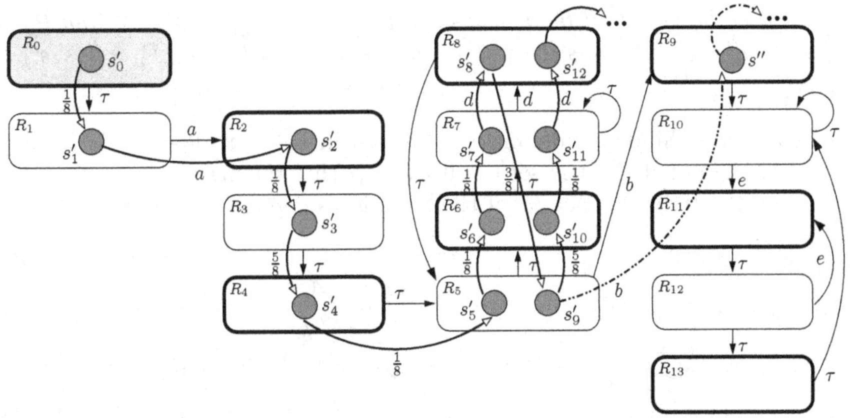

Fig. 7.7. Sequences of transitions in the reachable part of the discretized $\mathrm{rg}_{\flat}$-model for the automaton in Fig. 7.3 and $c_{max}(\mathcal{A}, \varphi) = 1$

7.1.2 Bounded Model Checking for $\mathrm{ECTL}^{\mathbf{r}}_{-\mathrm{X}}$

In this section we present a SAT-based approach to $\mathrm{ECTL}^{\mathbf{r}}_{-\mathrm{X}}$ model checking over discretized $\mathrm{rg}(\flat)$ models for timed automata, to which we refer as to models from now on. We start with giving a *bounded semantics* for $\mathrm{ECTL}^{\mathbf{r}}_{-\mathrm{X}}$ in order to define the *bounded model checking problem* and to translate it subsequently into a satisfiability problem [119].

Let φ be a TECTL formula, $\psi = \mathrm{cr}(\varphi)$, and $M = ((S, s^0, \rightarrow_{\mathfrak{d}}), V_{\mathfrak{D}})$ be a discretized $\mathrm{rg}(\flat)$ model for $\mathcal{A}_{\varphi}$ with the extended valuation function (see the text starting from p. 172). By $\rightarrow_{\mathfrak{d}\mathcal{A}}$ we denote the part of $\rightarrow_{\mathfrak{d}}$ where the transitions are labelled with elements of $A \cup \{\tau\}$, and by $\rightarrow_{\mathfrak{d}_y}$ the transitions that reset the clock y, i.e., labelled with a_y. For each state $s \in S$ by $s[y := 0]$ we denote the state $s' \in S$ such that $s \rightarrow_{\mathfrak{d}_y} s'$. Notice that in the present approach one extra clock is sufficient as we will always check subformulas over fresh finite sequences of states.

We start with some auxiliary definitions. For $k \in \mathbb{N}_+$ a *k-path* in M is finite sequence of $k+1$ states

$$\pi = (s_0, s_1, \ldots, s_k)$$

such that $(s_i, s_{i+1}) \in \rightarrow_{\mathfrak{d}\mathcal{A}}$ for each $0 \leq i < k$ (see Fig. 7.9). For a k-path $\pi = (s_0, s_1, \ldots, s_k)$, let $\pi(i) = s_i$ for each $i \leq k$. By $\Pi_k(s)$ we denote the set of all the k-paths starting at s. This is a convenient way of representing a k-bounded subtree, rooted at s, of the tree resulting from unwinding the model M from s.

Definition 7.7 (k-model). *The k-model for M is a structure*

$$M_k = ((S, s^0, P_k, P_y), V_{\mathfrak{D}}),$$

where P_k is the set of all the k-paths of M, i.e., $P_k = \bigcup_{s \in S} \Pi_k(s)$, and P_y is the set of all the pairs consisting of a state s and $s[y := 0]$, i.e., $P_y = \{(s, s') \in S \times S \mid s \rightarrow_{\mathfrak{d}_y} s'\}$.

The intuition behind this definition is shown in Fig. 7.8. The left-hand side part of the figure shows the set of all the k-paths starting at the state s, for $k = 2$. On the right, the set P_k of the 2-model is depicted.

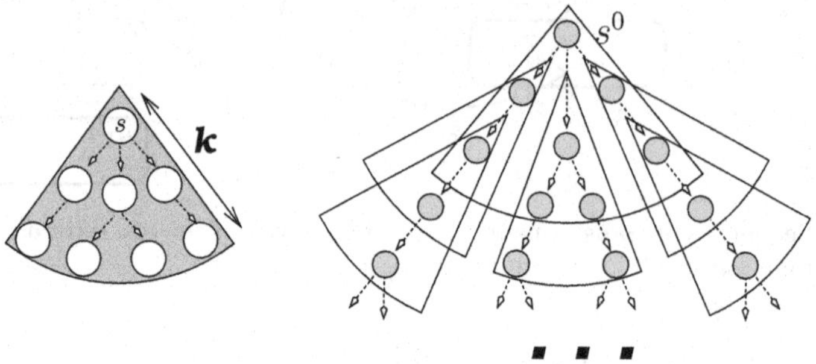

$\Pi_k(s)$ for $k = 2$ P_k of the k-model M_k for $k = 2$

Fig. 7.8. Elements of the k-model for $k = 2$

Define a function $loop : P_k \longrightarrow 2^{\mathbb{N}}$

$$loop(\pi) = \{h \mid 0 \leq h \leq k \wedge \pi(k) \rightarrow_{\mathfrak{d}_{\mathcal{A}}} \pi(h)\},$$

which returns the set of all the indices of the states for which there is a transition from the last state of π. Satisfaction of the temporal operator R_y on a k-path π in the bounded case can depend on whether or not π represents a path[8], i.e., $loop(\pi) \neq \emptyset$.

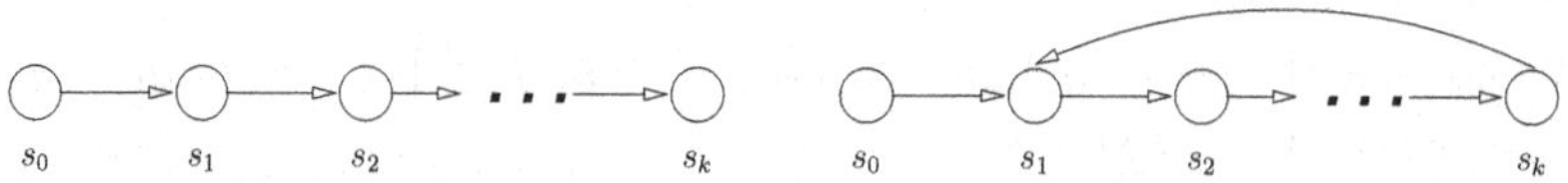

a k-path with $loop(\pi) = \emptyset$ a k-path with $loop(\pi) = \{1\}$

Fig. 7.9. Two kinds of k-paths

Next, we define a k-bounded semantics. The main reason for reformulating the semantics of the modalities in the following definition in terms of elements of k-paths rather than states and paths themselves is to restrict the semantics to a part of the model.

[8] Note that a path is infinite by definition.

Definition 7.8 (k-bounded semantics for $\mathrm{ECTL}^{\mathrm{r}}_{-\mathrm{X}}$). *Let M_k be a k-model and α, β be $\mathrm{ECTL}^{\mathrm{r}}_{-\mathrm{X}}$ subformulas of ψ.*

$$M_k, s \models \alpha$$

denotes that α is true at the state s of M_k. M_k is omitted if it is clear from the context. The relation $\models$ is defined inductively as follows:

$$\begin{array}{lll}
s \models \wp & \textit{iff} & \wp \in V_{\mathfrak{D}}(s), \text{ for } \wp \in PV', \\
s \models \neg\wp & \textit{iff} & \wp \notin V_{\mathfrak{D}}(s), \text{ for } \wp \in PV', \\
s \models \alpha \wedge \beta & \textit{iff} & s \models \alpha \;\wedge\; s \models \beta, \\
s \models \alpha \vee \beta & \textit{iff} & s \models \alpha \;\vee\; s \models \beta, \\
s \models \mathrm{E}(\alpha \mathrm{U}_y \beta) & \textit{iff} & (\exists s' \in S) \Big((s, s') \in P_y \;\wedge (\exists \pi \in \Pi_k(s')) \\
& & \quad ((\exists 0 \leq j \leq k)(\pi(j) \models \beta \wedge (\forall 0 \leq i < j)\ \pi(i) \models \alpha))\Big), \\
s \models \mathrm{E}(\alpha \mathrm{R}_y \beta) & \textit{iff} & (\exists s' \in S) \Big((s, s') \in P_y \;\wedge\; (\exists \pi \in \Pi_k(s')) \\
& & \quad (((\exists 0 \leq j \leq k)(\pi(j) \models \alpha \;\wedge\; (\forall 0 \leq i \leq j)\ \pi(i) \models \beta)) \\
& & \qquad \vee\ ((\forall 0 \leq j \leq k)(\pi(j) \models \beta \;\wedge\; loop(\pi) \neq \emptyset)))\Big).
\end{array}$$

Some examples of $\mathrm{ECTL}^{\mathrm{r}}_{-\mathrm{X}}$ formulas which hold in the state s_0, in the k-bounded semantics, are shown in Fig. 7.10.

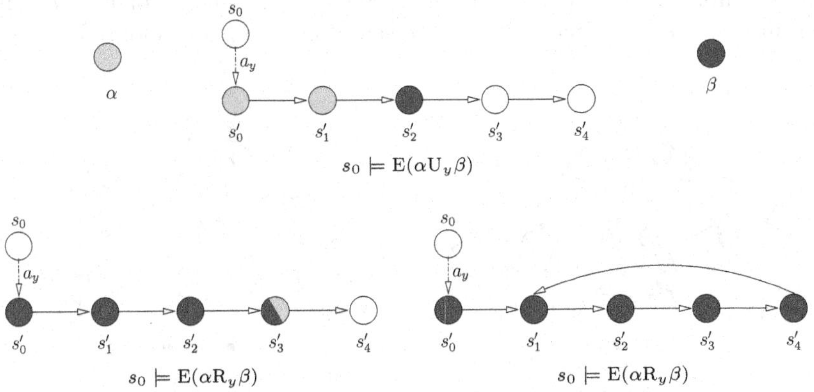

Fig. 7.10. Examples of $\mathrm{ECTL}^{\mathrm{r}}_{-\mathrm{X}}$ formulas

Next, we describe how the model checking problem ($M \models \psi$) can be reduced to the bounded model checking problem ($M_k \models \psi$). In this setting we can prove that for k equal to the number of the states of M satisfiability in the k-bounded semantics is equivalent to the unbounded one.

Theorem 7.9. *Let $M = ((S, s^0, \rightarrow_{\mathfrak{d}}), V_{\mathfrak{D}})$ be a model, ψ be an $\mathrm{ECTL}^{\mathrm{r}}_{-\mathrm{X}}$ formula, and $k = |S|$. Then, $M, s^0 \models \psi$ iff $M_k, s^0 \models \psi$.*

The rationale behind the method is that for particular examples checking satisfiability of a formula can be done on a small fragment of the model.

Next, we show how to translate the model checking problem for $\mathrm{ECTL}^{\mathrm{r}}_{-\mathrm{X}}$ on a k-model to a problem of satisfiability of some propositional formula. Our method is based on [119], but we use a variant of the operator $\mathrm{E}(\cdot\mathrm{R}\cdot)$ rather than of the less expressive $\mathrm{EG}(\cdot)$. Proofs of correctness of our approach are based on the corresponding proofs in [119, 170].

We assume the following definition of a submodel:

Definition 7.10. *Let $M_k = ((S, s^0, P_k, P_y), V_{\mathfrak{D}})$ be the k-model of M. A structure*

$$M'_k = ((S', s^0, P'_k, P'_y), V'_{\mathfrak{D}})$$

is a submodel *of M_k if*

- $P'_k \subseteq P_k$,
- $S' = States(P'_k) \cup \{s^0\}$,
- $P'_y \subseteq P_y \cap (S' \times S')$, *and*
- $V'_{\mathfrak{D}} = V_{\mathfrak{D}}|_{S'}$,

where

$$States(P'_k) = \{s \in S \mid (\exists \pi \in P'_k)(\exists i \leq k)\ \pi(i) = s\}.$$

Some examples of the sets P'_k of submodels of the k-model of Fig. 7.8 are given in Fig. 7.11. The parts of P_k corresponding to P'_k are coloured in grey.

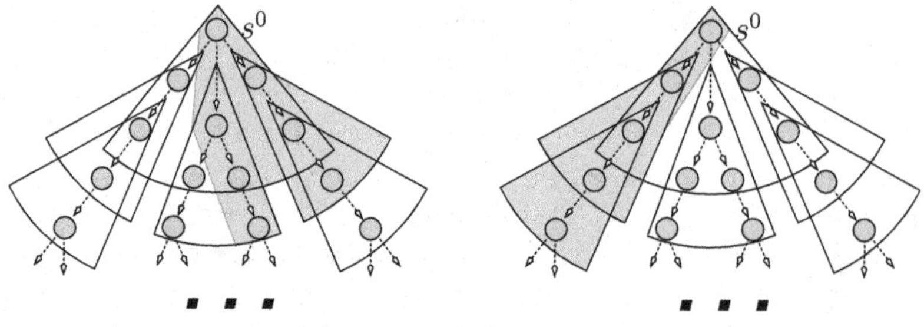

Fig. 7.11. Sets P'_k of submodels of M_k ($k = 2$)

The bounded semantics of $\mathrm{ECTL}^{\mathrm{r}}_{-\mathrm{X}}$ over submodels M'_k is defined as for M_k (see Def. 7.8). Our present aim is to give a bound on the number of k-paths in M'_k such that the validity of ψ in M_k is equivalent to the validity of ψ in M'_k. Let $\mathcal{F}_{\mathrm{ECTL}^{\mathrm{r}}_{-\mathrm{X}}}$ be a set of the formulas of $\mathrm{ECTL}^{\mathrm{r}}_{-\mathrm{X}}$.

Definition 7.11. *Define a function $f_k : \mathcal{F}_{\mathrm{ECTL}^{\mathrm{r}}_{-\mathrm{X}}} \to \mathbb{N}$ as follows:*

- $f_k(\wp) = f_k(\neg\wp) = 0$, *where* $\wp \in PV'$,
- $f_k(\alpha \vee \beta) = max\{f_k(\alpha), f_k(\beta)\}$,

- $f_k(\alpha \wedge \beta) = f_k(\alpha) + f_k(\beta)$,
- $f_k(\mathrm{E}(\alpha \mathrm{U}_y \beta)) = k \times f_k(\alpha) + f_k(\beta) + 1$,
- $f_k(\mathrm{E}(\alpha \mathrm{R}_y \beta)) = (k+1) \times f_k(\beta) + f_k(\alpha) + 1$.

The function f_k determines the number of k-paths of a submodel M'_k sufficient for checking an $\mathrm{ECTL}^{\mathbf{r}}_{-\mathrm{X}}$ formula. An intuitive explanation for the above formulas can be derived from the picture in Fig. 7.12. Notice that for checking formulas of the form $\mathrm{E}(\alpha \mathrm{U}_y \beta)$ we need to use k paths for checking α's on the first k states of the current path and one more for checking β in its last state. For formulas of the form $\mathrm{E}(\alpha \mathrm{R}_y \beta)$ the intuition behind the function f_k is similar (we need k paths for checking β's on the first k states, and two paths to check α and β on the state $k+1$ of the current path; see Fig. 7.13).

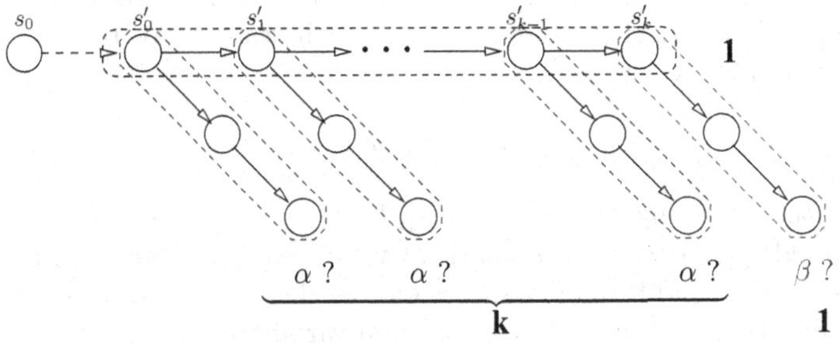

Fig. 7.12. An intuition behind the number of k-paths in $f_k(\mathrm{E}(\alpha \mathrm{U}_y \beta))$

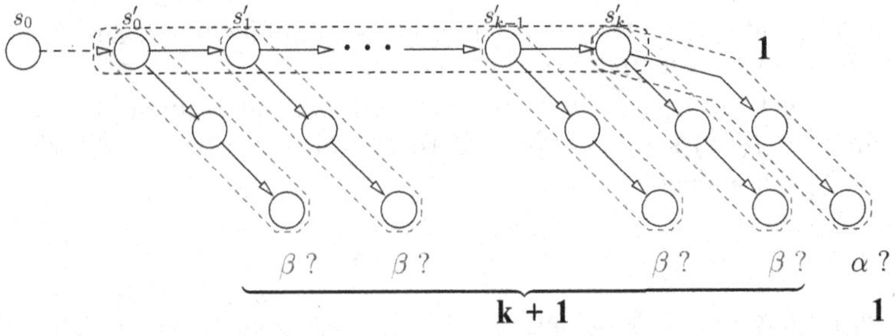

Fig. 7.13. An intuition behind the number of k-paths in $f_k(\mathrm{E}(\alpha \mathrm{R}_y \beta))$

Now, we are ready to check φ over M_k. The main idea is that this can be translated to checking the satisfiability of the propositional formula

$$[M, \psi]_k = [M^{\psi, s^0}]_k \wedge [\psi]_{M_k},$$

where the first conjunct represents (a part of) the model under consideration and the second a number of constraints that must be satisfied on M_k for ψ

to be satisfied. Once this translation is defined, checking satisfiability of an $\mathrm{ECTL}^{\mathrm{r}}_{-\mathrm{X}}$ formula can be done by means of a SAT-checker. Although from a theoretical point of view the complexity of this operation is not lower, in practice the efficiency of modern SAT-checkers makes the process worthwhile in many instances.

We now give details of this translation. We begin with the encoding of the transitions in the model under consideration. Since the set of states S of our model is finite, every element of S can be encoded as a bit vector $(s[1], \ldots, s[\mathfrak{l}_b])$ of length $\mathfrak{l}_b$ depending on the number of locations in L, the size of the set $\mathbb{D}$ and the number of clocks. The bit vector consists of two parts, the first of which is used to encode the location of a state of the automaton, whereas the second encodes the timed part of that state (i.e., the clock valuation)[9]. We do not give more details of this encoding here. The interested reader is referred to [119, 120, 177]. Each state s can be represented by a valuation of a vector

$$w = (w[1], \ldots, w[\mathfrak{l}_b])$$

(called a *global state variable*), where $w[i]$, for $i = 1, \ldots, \mathfrak{l}_b$, is a propositional variable (called *state variable*). Notice that we distinguish between states of S encoded as sequences of 0's and 1's (we refer to these as valuations of w) and their representations in terms of propositional variables $w[i]$. A finite sequence

$$(w_0, \ldots, w_k)$$

of global state variables is called a *symbolic k-path*.

In general we shall need to consider not just one but a number of symbolic k-paths. This number depends on the formula ψ under investigation, and it is returned as the value $f_k(\psi)$ of the function f_k.

To construct $[M, \psi]_k$, we first define a propositional formula $[M^{\psi,s^0}]_k$ that constrains the $f_k(\psi)$ symbolic k-paths to be valid k-paths of M_k. The j-th symbolic k-path is denoted as

$$(w_{0,j}, \ldots, w_{k,j}),$$

where $w_{i,j}$ are global state variables for $1 \leq j \leq f_k(\psi)$, $0 \leq i \leq k$. Let PV_s be a set of state variables, $\mathcal{F}$ be a set of propositional formulas over PV_s, and let $lit : \{0,1\} \times PV_s \to \mathcal{F}$ be a function defined as follows:

- $lit(0, \wp) = \neg\wp$ and $lit(1, \wp) = \wp$.

Furthermore, let w, w' be global state variables. We define the following propositional formulas:

[9] If the system considered consists of n automata, each part of the vector can be divided into n subvectors, each of which encodes respectively the location and the valuation of the local clocks for the i-th component, for $i = 1, \ldots, n$.

- $I_s(w) := \bigwedge_{i=1}^{l_b} lit(s[i], w[i])$
is a formula over w, which is true for the valuation s, (encodes the state s of the model M, i.e., $s[i] = 1$ is encoded by $w[i]$, and $s[i] = 0$ is encoded by $\neg w[i]$),
- $\wp(w)$
is a formula over w, which is true for a valuation s_w of w iff $\wp \in V_{\mathfrak{D}}(s_w)$, where $\wp \in PV'$ (encodes the proposition $\wp$),
- $H(w, w') := \bigwedge_{i=1}^{l_b} (w[i] \Leftrightarrow w'[i])$
is a formula over w, w', which is true for two valuations, s_w of w and $s_{w'}$ of w' iff $s_w = s_{w'}$ (equality of the two state encodings),
- $\mathcal{R}(w, w')$
is a formula over w, w', which is true for two valuations, s_w of w and $s_{w'}$ of w', iff $s_w \rightarrow_{\mathfrak{d}_{\mathcal{A}}} s_{w'}$ (encodes the transition relation of the paths),
- $R_x(w, w')$
is a formula over w, w', which is true for two valuations, s_w of w and $s_{w'}$ of w', iff the state s'_w is like s_w with the clock $x \in \mathcal{X}'$ reset. Notice that in the case of $x = y$ the formula is true iff $s_w \rightarrow_{\mathfrak{d}_y} s_{w'}$ (encodes the transitions resetting the clock y),
- $L_j(k, h) := \mathcal{R}(w_{k,j}, w_{h,j})$
is a formula over $w_{k,j}, w_{h,j}$, which is true for two valuations, s_k of $w_{k,j}$ and w_h of $w_{h,j}$, iff $s_k \rightarrow_{\mathfrak{d}_{\mathcal{A}}} s_h$ (encodes a backward loop from the k-th state to the h-th state in the symbolic k-path j, for $0 \leq h \leq k$).

The translation of $[M^{\psi,s^0}]_k$, representing the transitions in the k-model is given by the following definition:

Definition 7.12 (Encoding of Transition Relation). *Let $M_k = ((S, s^0, P_k, P_y), V_{\mathfrak{D}})$ be the k-model of M, and ψ be an $\mathrm{ECTL}^{\mathrm{r}}_{-\mathrm{X}}$ formula. The propositional formula $[M^{\psi,s^0}]_k$ is defined as follows:*

$$[M^{\psi,s^0}]_k := I_{s^0}(w_{0,0}) \wedge \bigwedge_{j=1}^{f_k(\psi)} \bigwedge_{i=0}^{k-1} \mathcal{R}(w_{i,j}, w_{i+1,j}),$$

where $w_{0,0}$, and $w_{i,j}$ for $0 \leq i \leq k$ and $1 \leq j \leq f_k(\psi)$ are global state variables. $[M^{\psi,s^0}]_k$ constrains the $f_k(\psi)$ symbolic k-paths to be valid k-paths in M_k.

The next step of our algorithm is to translate an $\mathrm{ECTL}^{\mathrm{r}}_{-\mathrm{X}}$ formula ψ into a propositional formula. We use $[\alpha]_k^{[m,n]}$ to denote the translation of an $\mathrm{ECTL}^{\mathrm{r}}_{-\mathrm{X}}$ subformula α of ψ at $w_{m,n}$ to a propositional formula, where $w_{m,n}$ is a global state variable with $(m, n) \in \{(0, 0)\} \cup \{0, ..., m\} \times \{1, ..., f_k(\psi)\}$. Note that the index n denotes the number of a symbolic path, whereas the index m the position at that path.

$$
\begin{array}{lll}
[\wp]_k^{[m,n]} & := & \wp(w_{m,n}), \text{ for } \wp \in PV', \\
[\neg\wp]_k^{[m,n]} & := & \neg\wp(w_{m,n}), \text{ for } \wp \in PV', \\
[\alpha \wedge \beta]_k^{[m,n]} & := & [\alpha]_k^{[m,n]} \wedge [\beta]_k^{[m,n]}, \\
[\alpha \vee \beta]_k^{[m,n]} & := & [\alpha]_k^{[m,n]} \vee [\beta]_k^{[m,n]}, \\
[\mathrm{E}(\alpha \mathrm{U}_y \beta)]_k^{[m,n]} & := & \bigvee_{1 \le i \le f_k(\psi)} \Big(R_y(w_{m,n}, w_{0,i}) \wedge \\
 & & \qquad \bigvee_{j=0}^{k} \big([\beta]_k^{[j,i]} \wedge \bigwedge_{l=0}^{j-1} [\alpha]_k^{[l,i]}\big)\Big), \\
[\mathrm{E}(\alpha \mathrm{R}_y \beta)]_k^{[m,n]} & := & \bigvee_{1 \le i \le f_k(\psi)} \Big(R_y(w_{m,n}, w_{0,i}) \wedge \\
 & & \qquad \big(\bigvee_{j=0}^{k} \big([\alpha]_k^{[j,i]} \wedge \bigwedge_{l=0}^{j} [\beta]_k^{[l,i]}\big) \vee \\
 & & \qquad \bigwedge_{j=0}^{k} [\beta]_k^{[j,i]} \wedge \bigvee_{l=0}^{k} L_i(k,l)\big)\Big).
\end{array}
$$

Some intuitions behind the translations are presented in Fig. 7.14. For simplicity, the numbers in the subscripts of $[\alpha]_k^{[\cdot]}$, $[\beta]_k^{[\cdot]}$ are omitted.

Given the translations above, we can now check ψ over M_k by checking satisfiability of the propositional formula

$$[M^{\psi,s^0}]_k \wedge [\psi]_{M_k},$$

where $[\psi]_{M_k} = [\psi]_k^{[0,0]}$. The translation presented above can be shown to be correct and complete. This can be done in a way similar to [119].

Theorem 7.13. *Let M be a model, M_k be a k-model of M, and ψ be an* $\mathrm{ECTL}^{\mathbf{r}}_{-\mathrm{X}}$ *formula. Then, $M \models_k \psi$ iff $[\psi]_{M_k} \wedge [M^{\psi,s^0}]_k$ is satisfiable.*

We have all ingredients in place to give the algorithm for BMC of TECTL. Its pseudo-code is presented in Fig. 7.16.

Since we have defined the translation cr from $\mathrm{TCTL}_{\mathcal{C}}$ to $\mathrm{CTL}^{\mathbf{z}}$, the above method could be adapted to the $\mathrm{TECTL}_{\mathcal{C}}$ formulas. This requires to redefine the k-bounded semantics (an intuition behind this is shown in Fig. 7.17), to extend the function f_k to deal with X_{z_i} by

$$f_k(\mathrm{X}_{z_i}\alpha) = f_k(\alpha) + 1,$$

define the translation for $\mathrm{X}_{z_i}\alpha$ (see below and Fig. 7.15):

$$[\mathrm{X}_{z_i}\alpha]_k^{[m,n]} := \bigvee_{1 \le i \le f_k(\psi)} \Big(R_{z_i}(w_{m,n}, w_{0,i}) \wedge [\alpha]_k^{[0,i]}\Big),$$

and the translations for EU and ER by modifying the translations for EU_y and ER_y changing $\mathcal{R}_y$ to H. The rest of the details is left as an exercise to the reader.

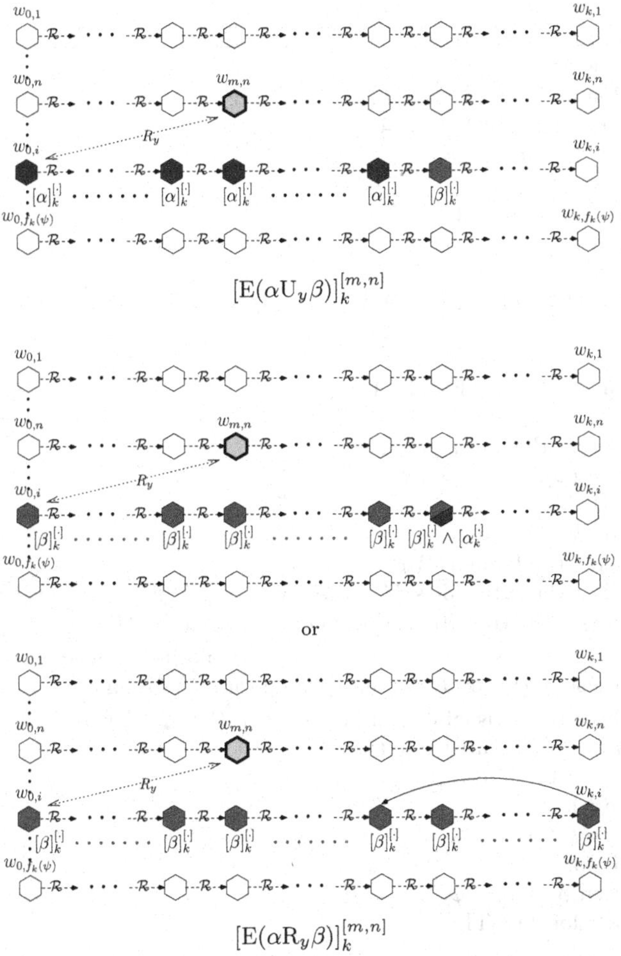

Fig. 7.14. An intuition behind the translations of TCTL formulas into $\mathrm{ECTL}^{\mathbf{r}}_{-X}$ formulas

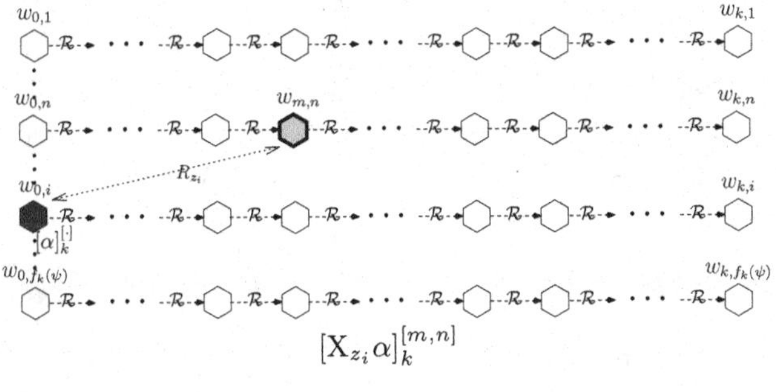

Fig. 7.15. An intuition behind the translation of $X_{z_i}\alpha$

INPUT ARGUMENTS:

  a timed automaton $\mathcal{A} = (A, L, l^0, E, \mathcal{X}, \mathcal{I})$ of $n_{\mathcal{X}}$ clocks
  a valuation function $V_{\mathcal{A}}$ for $\mathcal{A}$
  a TECTL formula φ

GLOBAL VARIABLES:

  ψ: an $\mathrm{ECTL}^{\mathrm{r}}_{-\mathrm{X}}$ formula
  k: $\mathbb{N}_+$

RETURN VALUES:

  BMC_for_TECTL(): $\{M_c(\mathcal{A}) \models \varphi, M_c(\mathcal{A}) \not\models \varphi\}$

```
function BMC_for_TECTL($\mathcal{A}$, $V_{\mathcal{A}}$, $\varphi$) is
begin
   construct the automaton $\mathcal{A}_\varphi$;
   $\psi := \mathrm{cr}(\varphi)$;
   $k := 1$;
   while $k \leq |M|$ do
     select the $k$-model $M_k$;
     select the submodels $M'_k$ of $M_k$ with $|P'_k| \leq f_k(\psi)$ and $|P'_y| \leq f_k(\psi)$;
     encode the transition relation of all $k$-paths of $M_k$
                                   by a propositional formula $[M^{\psi,s^0}]_k$;
     translate $\psi$ over all $M'_k$ into a propositional formula $[\psi]_{M_k}$;
     check the satisfiability of $[M,\psi]_k := [M^{\psi,s^0}]_k \wedge [\psi]_{M_k}$;
     if $[M,\psi]_k$ is satisfiable then
       return $M_c(\mathcal{A}) \models \varphi$;
     end if;
     $k := k + 1$;
   end do;
   return $M_c(\mathcal{A}) \not\models \varphi$;
end BMC_for_TECTL;
```

Fig. 7.16. A BMC algorithm for TECTL

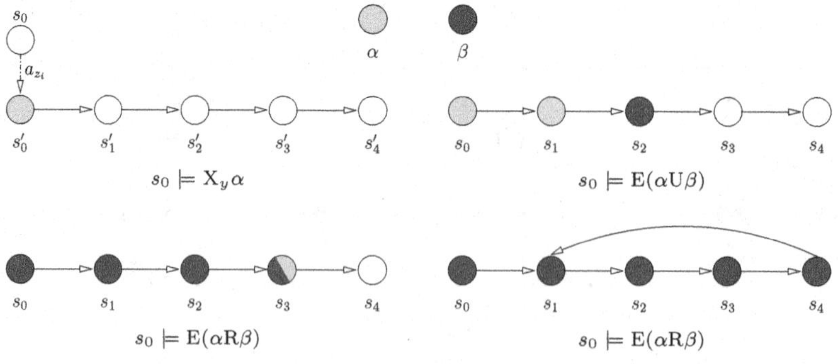

Fig. 7.17. Examples of $\mathrm{ECTL}^{\mathrm{r}}_{-\mathrm{X}}$ formulas

7.1.3 Checking Reachability with BMC

Reachability of a propositional formula p in a timed automaton $\mathcal{A}$ can be specified by the ECTL formula EFp. So, in principle, reachability can be verified using the above approach over $DM_{DRG}(\mathcal{A})$ or $DM_{DRG_b}(\mathcal{A})$ depending on the semantics considered.

It turns out, however, that a slight change in the technique can dramatically influence efficiency of the method in this case. We explain the idea in detail for the weakly monotonic semantics. Again, an adaptation of this method for the alternative semantics is left as an exercise to the reader.

First of all, we consider k-paths over $DM(\mathcal{A})$, rather than over $DM_{DRG}(\mathcal{A})$ or $DM_{DRG_b}(\mathcal{A})$, which means that adjust transitions are not combined with action- and time successor transitions, and the delay transition relation is transitive. Secondly, we use the notion of a *special k-path*[10] which satisfies the following conditions:

- It begins with the initial state.
- The first transition is a time delay one.
- Each time delay transition is directly followed by an action one.
- Each action transition is directly followed by an adjust one.
- Each adjust transition is directly followed by a time delay one.
- The above three rules do not apply only to the last transition of a special k-path.

An example of a special k-path is shown in Fig. 7.18.

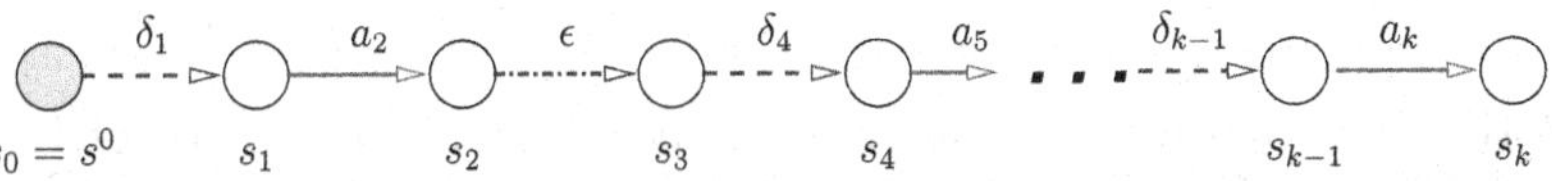

Fig. 7.18. A special k-path

Obviously, it is sufficient to use only one symbolic path to encode all the special k-paths. Thus, the reachability problem is translated to conjunction of the encoding of the symbolic path and the encoding of the propositional property p at the last state of that path. If this conjunction is satisfiable, then p is reachable.

7.1.4 Checking Unreachability with BMC

Unreachability of a propositional formula p in a timed automaton $\mathcal{A}$ means that the ACTL formula AG$\neg p$ holds in $\mathcal{A}$. Again, for verification, we could

[10] This means that our bound on $k = |S|$ in Theorem 7.9 should be extended to $3|S|$ in order to ensure that all the states of S can appear in a path.

check the ECTL formula EFp over $DM_{DRG}(\mathcal{A})$ or $DM_{DRG_b}(\mathcal{A})$, but, in this case, we have to prove that this formula does not hold in the model. This is, obviously, one of the major problems with BMC, as in the worst case the algorithm needs to reach the upper bound for k, i.e., $3|S| + 1$ in this case.

There is, however, another approach to checking unreachability, which in many cases (the method is not complete) gives striking results. The idea is to use a SAT-solver to find a minimal (possible) k such that if $\neg p$ holds at all the k-paths, then it means that p is unreachable. The method described below finds such a k, if each path at which p holds only at the final state is finite. To this aim, a *free special k-path* is defined. It satisfies all the conditions on a special k-path except for the first one, i.e., it does not need to start at the initial state.

In addition we require that for a special k-path the following conditions hold:

- p holds only at the last state if the last transition is an action one,
- p holds only at the last two states if the last transition is an adjust one,
- p holds only at the last three states if the last transition is a time delay one.

An example of a free special k-path is depicted in Fig. 7.19.

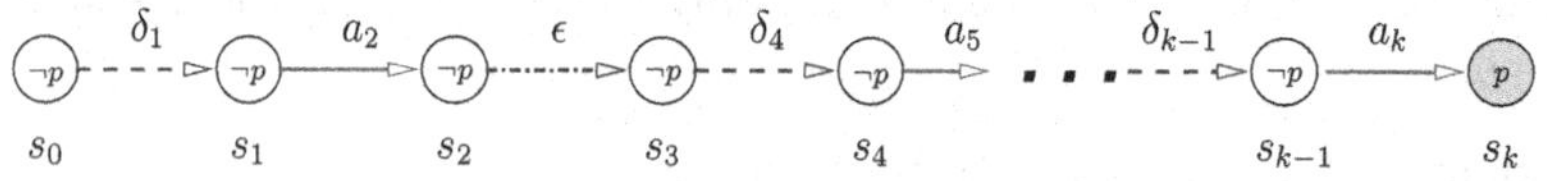

Fig. 7.19. A free special k-path

Notice that if p holds in our model, then it holds at a path of length restricted by the length of a longest special k-path. So, using one symbolic k-path, we encode all the free special k-paths in order to find the length of a longest one satisfying the above three conditions. If the above encoding is unsatisfiable for some k (denoted by k_0), then it means that we have found a longest free path. It is easy to see that we can look for such a k by running the algorithm for the values of k satisfying $k \; mod \; 3 = 2$ only, since only the transitions from s_{k-1} to s_k with k satisfying the above condition can change the location (which influences reachability). Then, when we find k_0 for which the encoding is unsatisfiable, we can run the check for reachability of p up to $k = k_0 - 3$. If the reachability algorithm does not find the formula satisfiable for such a k, then it means that p is indeed unreachable.

Unfortunately, the above method is not complete, it fails when there are loops in the unreachable part of the state space involving states satisfying $\neg p$, from which a state satisfying p is reachable. A solution to make the method complete by encoding that a free special path is loop-free [38], i.e., no state repeats at the path, turns out to be sometimes ineffective in practice [177].

7.2 Bounded Model Checking via Translation to SL

There is another approach to BMC for timed automata, which consists in translating the model checking problem to the question of satisfiability of some SL formula. It seems that such an approach could be applied to TACTL, but since it has been only defined for LTL [20], and TCTL and LTL are not comparable w.r.t. expressiveness, it is interesting to discuss it.

In the presentation below, we follow the method of [20] defined for the LTL properties of networks of timed automata $\mathfrak{A} = \{\mathcal{A}_\mathfrak{i} \mid \mathfrak{i} \in \mathfrak{I}\}$, where $\mathcal{A}_\mathfrak{i} = (A_\mathfrak{i}, L_\mathfrak{i}, l_\mathfrak{i}^0, E_\mathfrak{i}, \mathcal{X}_\mathfrak{i}, \mathcal{I}_\mathfrak{i})$, $\mathfrak{I} = \{\mathfrak{i}_1, \ldots, \mathfrak{i}_{n_\mathfrak{I}}\}$, over the weakly monotonic semantics. Let $\mathcal{A} = (A, L, l^0, E, \mathcal{X}, \mathcal{I})$ denote the product of the component automata $\{\mathcal{A}_\mathfrak{i} \mid \mathfrak{i} \in \mathfrak{I}\}$. We use the notations concerning boolean encodings defined in the previous section, but in addition we need more boolean and real variables.

- **Boolean variables**
  - For each label $a \in A$, the boolean variable $\underline{a}$ is introduced which holds iff $\mathcal{A}$ executes a transition labelled a,
  - For each transition $e \in E_\mathfrak{i}$ we use the boolean variable $\underline{e}$, which holds iff e is executed,
  - The boolean variable e_δ is used to denote that the time elapses by some $\delta > 0$,
  - The boolean variable $e_{null}^\mathfrak{i}$ is used to denote that $\mathcal{A}_\mathfrak{i}$ is idle.
- **Real variables**
  - The clocks $\mathcal{X}$ of $\mathcal{A}$ are represented by real variables in the encoding,
  - The real variable z (called *absolute time reference*), whose negation, i.e., $-z$, is used to denote the time elapsed from the beginning of the execution, i.e., from the start of the system. This means that $z = 0$ at the start of the system and then z is continuously decreasing.
  - For each clock x of $\mathcal{A}$ we define the variable ox, whose negated value is equal to the absolute time when the clock was last reset. So, the value of x is obtained as $-z - (-ox) = ox - z$.

We use r' to denote the value of the real variable r after a transition of the automaton. The fact that a transition resets a clock is encoded by the constraint $ox' = z'$, which says that after executing a transition the absolute time is equal to the absolute time when the clock was last reset.

7.2.1 Encoding k-Paths

As before k-paths are encoded as sequences of encoded transitions, but now for each component $\mathcal{A}_\mathfrak{i}$ separately, and then formulas are added, which disallow execution of transitions labelled by different actions in the same step. So, it is crucial to this method to encode transitions in each component automaton. This encoding for the component $\mathfrak{i}$ is shown below. Notice that unlike in the BMC encoding for TACTL, where state variables were used to encode concrete state, here state variables w, w' are used to encode locations only. The time components of the states are described by SL formulas. Moreover, we are dealing with one symbolic k-path only. The subscript $j \in \mathbb{N}$ of a state variable w_j denotes its position on the path.

The formulas used to encode the component $\mathfrak{i}$ are as follows:

- $I_{l_{\mathfrak{i}}^0}(w_0) \wedge \bigwedge_{x \in \mathcal{X}_{\mathfrak{i}}} (ox = z)$
  encodes the initial location and the value of each clock set to 0,
- $\bigwedge_{l \in L_{\mathfrak{i}}} (I_l(w) \Rightarrow \mathcal{I}(l))$
  encodes the invariants of all the locations,
- for $e \in E_{\mathfrak{i}}$ and $e = l \xrightarrow{a,\mathfrak{cc},X} l'$
  $\underline{e} \Rightarrow (I_l(w) \wedge \underline{a} \wedge \mathfrak{cc} \wedge I_{l'}(w') \wedge \bigwedge_{x \in X} (x' = z') \wedge \bigwedge_{x \in \mathcal{X}_{\mathfrak{i}} \setminus X} (x' = x) \wedge (z' = z))$
  encodes execution of the transition e, so change of the locations from l to l', execution of a, resetting the clocks of $X \subseteq \mathcal{X}_{\mathfrak{i}}$, and the requirements that the other clocks do not change their values, the guard $\mathfrak{cc}$ holds, and no time elapses which makes $z = z'$,
- $e_\delta^i \Rightarrow ((z' - z < 0) \wedge \bigvee_{l \in L_{\mathfrak{i}}} I_l(w) \wedge H(w, w') \wedge \bigwedge_{x \in \mathcal{X}_{\mathfrak{i}}} (x' = x) \wedge \bigwedge_{a \in A_{\mathfrak{i}}} (\neg \underline{a}))$
  encodes δ-time transitions, so the requirements that time elapse must be greater than 0, the location does not change, the values of clocks do not change (no resetting of clocks), no action transition can occur at the same time,
- $e_{null}^{\mathfrak{i}} \Rightarrow ((z' = z) \wedge \bigvee_{l \in L_{\mathfrak{i}}} I_l(w) \wedge H(w, w') \wedge \bigwedge_{x \in \mathcal{X}_{\mathfrak{i}}} (x' = x) \wedge \bigwedge_{a \in A_{\mathfrak{i}}} (\neg \underline{a}))$
  encodes the null transition, so time elapse is equal to 0, the location does not change, the values of clocks do not change (no resetting of clocks), no action transition can occur at the same time,
- $(\bigvee_{e \in E_{\mathfrak{i}}} \underline{e}) \vee e_\delta \vee e_{null}^{\mathfrak{i}}$
  encodes that at least one transition occurs, i.e., either an action transition, or a δ-time transition, or the null transition,
- $\bigwedge_{a,b \in A_{\mathfrak{i}}, a \neq b} (\neg \underline{a} \vee \neg \underline{b})$,
  encodes that no two different actions can be executed simultaneously,
- $\bigwedge_{e,e' \in E_{\mathfrak{i}},\, e \neq e'} (\neg \underline{e} \vee \neg \underline{e'})$
  encodes that no two different action transitions can be executed simultaneously.

A k-path $(w_0, \ldots, w_k)$ is called an *h-loop* $(0 \leq h \leq k)$ if there is a transition from the k-th to h-th state of the k-path. An h-loop of a k-path $w_0, \ldots, w_k$ is encoded by imposing that each propositional variable has the same value at w_k and w_h. For each clock we require that $ox^{(k)} - z^{(k)} = ox^{(h)} - z^{(h)}$, where the superscripts (k), (h) mean the values at the states w_k and w_h, respectively.

7.2.2 Product Encoding

The encoding for the product automaton $\mathcal{A} = \mathcal{A}_{\mathfrak{i}_1} \| \ldots \| \mathcal{A}_{\mathfrak{i}_{n_\mathfrak{I}}}$ is defined as conjunction of the encodings for each $\mathcal{A}_{\mathfrak{i}}$ together with the following formula

$$\bigwedge_{\mathfrak{i},\mathfrak{i}'\in\mathfrak{I},\mathfrak{i}\neq\mathfrak{i}'} \quad \bigwedge_{a\in A_{\mathfrak{i}}\setminus A_{\mathfrak{i}'},\, b\in A_{\mathfrak{i}'}\setminus A_{\mathfrak{i}}} (\neg\underline{a}\vee\neg\underline{b})$$

which prevents two different local actions to occur simultaneously.

For a given model M and $k > 0$, let $[M]_k$ denote the SL formula representing all the k-paths of M, and $L(k,h)$ be the encoding of the h-loop condition. The BMC problem is formulated as

$$M_k \models \mathrm{E}\varphi,$$

where φ is an LTL path formula, stating that there is a k-path in M starting at the initial state satisfying φ. This is again equivalent to the satisfiability of the SL formula

$$[M]_k \wedge [\varphi]_k,$$

where $[\varphi]_k$ is defined as

$$[\varphi]_k^0 \vee \bigvee_{h=0}^{k} (L(k,h) \wedge {}_h[\varphi]_k^0),$$

with $[\varphi]_k^0$ and ${}_h[\varphi]_k^0$ denoting translations of φ on k-paths. Recall that unlike in the BMC encoding for TACTL we are dealing with one symbolic k-path, so only one superscript is used with $[\varphi]_k$ to denote the position at this k-path. The translation of φ on a k-path is defined below for two cases, where either there is no loop ($[\varphi]_k^0$) or there is a loop from k to h (${}_h[\varphi]_k^0$)).

7.2.3 Encoding of LTL Formulas

First, we show a translation of φ on the m-th position of a k-path ($m \leq k$), which is not a loop, and then for a loop from k to h.

No loop case

The translation for the case when the k-path is not a loop looks as follows:

$$\begin{array}{lll}
[\wp]_k^{[m]} & := & \wp(w_m), \\
[\neg\wp]_k^{[m]} & := & \neg\wp(w_m), \\
[\alpha\wedge\beta]_k^{[m]} & := & [\alpha]_k^{[m]} \wedge [\beta]_k^{[m]}, \\
[\alpha\vee\beta]_k^{[m]} & := & [\alpha]_k^{[m]} \vee [\beta]_k^{[m]}, \\
[\mathrm{X}\alpha]_k^{[m]} & := & [\alpha]_k^{[m+1]} \text{ if } m < k \text{ and } \mathit{false} \text{ otherwise}, \\
[\alpha\mathrm{U}\beta]_k^{[m]} & := & \bigvee_{j=m}^{k} \left([\beta]_k^{[j]} \wedge \bigwedge_{i=m}^{j-1} [\alpha]_k^{[i]} \right), \\
[\alpha\mathrm{R}\beta]_k^{[m]} & := & \bigvee_{j=m}^{k} \left([\alpha]_k^{[j]} \wedge \bigwedge_{i=m}^{j} [\beta]_k^{[i]} \right).
\end{array}$$

Intuitions behind the last two cases of the translation are illustrated in Fig. 7.20. For simplicity, the numbers in the subscripts in $[\alpha]_k^{[\cdot]}$, $[\beta]_k^{[\cdot]}$ are omitted.

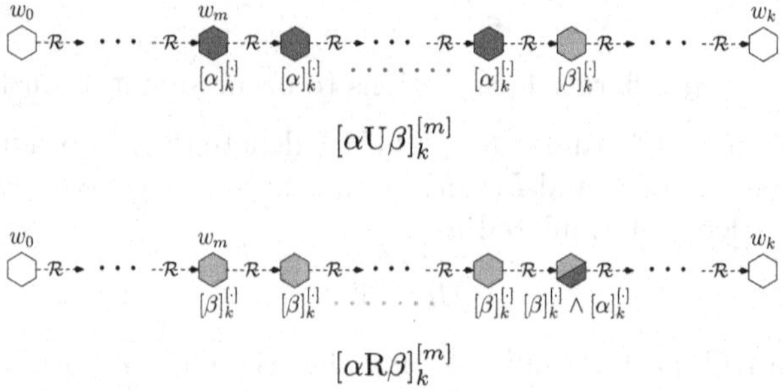

Fig. 7.20. An intuition behind the translations of $\varphi = \alpha \mathrm{U} \beta$ and $\varphi = \alpha \mathrm{R} \beta$ when the k-path is not a loop

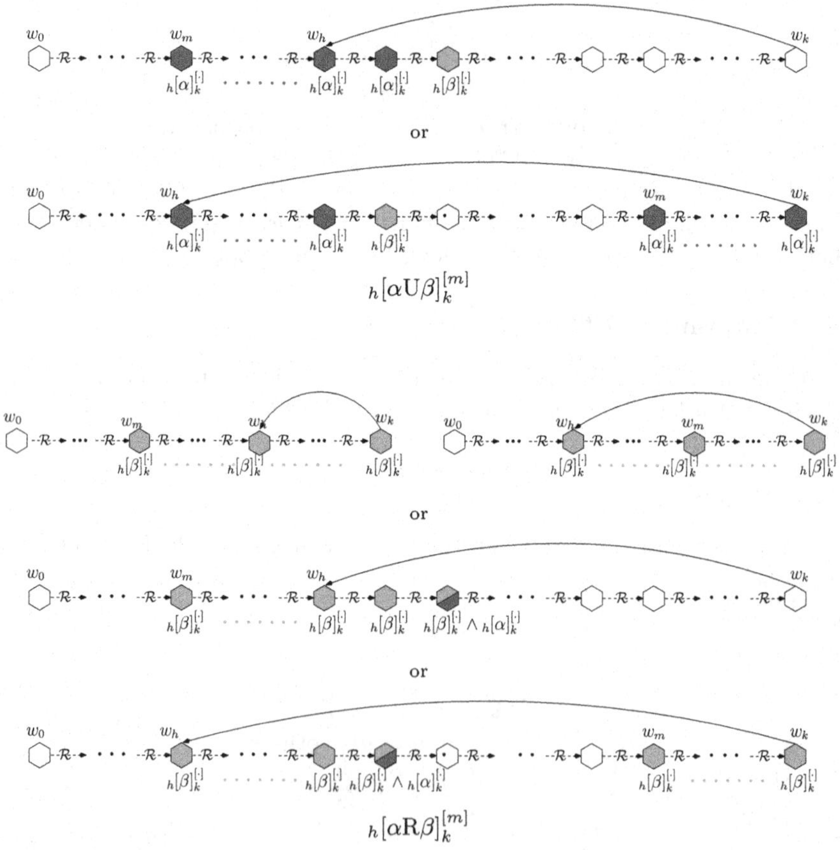

Fig. 7.21. An intuition behind the translations of $\varphi = \alpha \mathrm{U} \beta$ and $\varphi = \alpha \mathrm{R} \beta$ when the k-path is a loop

Loop case

In the case when a k-path is a loop from k to h, the translation is given by the following formulas:

$$
\begin{array}{lll}
{}_h[\wp]_k^{[m]} & := & \wp(w_m), \\
{}_h[\neg\wp]_k^{[m]} & := & \neg\wp(w_m), \\
{}_h[\alpha \wedge \beta]_k^{[m]} & := & {}_h[\alpha]_k^{[m]} \wedge {}_h[\beta]_k^{[m]}, \\
{}_h[\alpha \vee \beta]_k^{[m]} & := & {}_h[\alpha]_k^{[m]} \vee {}_h[\beta]_k^{[m]}, \\
{}_h[\mathrm{X}\alpha]_k^{[m]} & := & {}_h[\alpha]_k^{[m+1]} \text{ if } m < k \text{ and } {}_h[\alpha]_k^{[h]} \text{ otherwise}, \\
{}_h[\alpha \mathrm{U} \beta]_k^{[m]} & := & \bigvee_{j=m}^{k} \left({}_h[\beta]_k^{[j]} \wedge \bigwedge_{i=m}^{j-1} {}_h[\alpha]_k^{[i]} \right) \vee \\
 & & \bigvee_{j=h}^{m-1} \left({}_h[\beta]_k^{[j]} \wedge \bigwedge_{i=m}^{k} {}_h[\alpha]_k^{[i]} \wedge \bigwedge_{i=h}^{j-1} {}_h[\alpha]_k^{[i]} \right), \\
{}_h[\alpha \mathrm{R} \beta]_k^{[m]} & := & \bigwedge_{j=min(m,h)}^{k} {}_h[\beta]_k^{[j]} \vee \bigvee_{j=m}^{k} \left({}_h[\alpha]_k^{[j]} \wedge \bigwedge_{i=m}^{j} {}_h[\beta]_k^{[i]} \right) \\
 & & \vee \bigvee_{j=h}^{m-1} \left({}_h[\alpha]_k^{[j]} \wedge \bigwedge_{i=m}^{k} {}_h[\beta]_k^{[i]} \wedge \bigwedge_{i=h}^{j} {}_h[\beta]_k^{[i]} \right)
\end{array}
$$

Intuitions behind some parts of the above translation are presented in Fig. 7.21. Again, for simplicity, the numbers in the subscripts in $[\alpha]_k^{[\cdot]}$, $[\beta]_k^{[\cdot]}$ are omitted.

7.2.4 Optimizations of the Encoding

Different optimizations to the above encoding have been defined. Firstly, several mathematical formulas can be added, which encode that some constraints imply others. By making this information explicit in the encoding, a SAT-solver saves time on avoiding branching on the truth value of mathematical constraints. Secondly, parallelism of operations can be exploited by encoding their parallel execution, which can result in making counterexamples shorter. Moreover, symmetry reduction can be encoded. The detailed discussion of the above optimizations can be found in [20].

7.3 Unbounded Model Checking for $T\mu$

There are several model checking methods based on the symbolic encoding (presented below) of the model checking problem for $T\mu$. In what follows we discuss a combination of the approach by Bryant [140] and McMillan [105]. So, the translation from QSL to QPL is combined with the translation from QPL to propositional logic.

Given a timed automaton $\mathcal{A}$, a valuation function $V_{\mathcal{A}} : L \longrightarrow 2^{PV}$, and a TCTL$_C$ or $T\mu$ formula φ. In order to check whether the formula φ holds in the initial state of the dense concrete model $M_c(\mathcal{A})$, we first find a *characteristic (state) predicate* (i.e., an SL formula) ψ, which characterises all the states of the model $M_c(\mathcal{A})$ where φ holds. Formally, define

$$[\![\psi]\!] = \{s \in S \mid M_c(\mathcal{A}), s \models \varphi\}.$$

Then, we need to check whether the initial state is characterised by ψ, which means to check whether $s^0 \in [\![\psi]\!]$. This, for example, can be solved by testing whether the formula

$$enc_{\mathrm{SL}}(s^0) \wedge \psi$$

is satisfiable, where $enc_{\mathrm{SL}}(s^0)$ denotes an SL encoding of the state s^0.

The model checking algorithm given by Henzinger et al. [84] generates a characteristic predicate $[\![\varphi]\!]$ for a TCTL$_{\mathcal{C}}$ formula φ in two steps. Firstly, the formula φ is translated into an equivalent $T\mu$ formula φ'. Secondly, one applies the algorithm p, shown in Fig. 7.22, which computes a characteristic predicate for φ' by a successive approximation of fixpoints.

Before providing the above-mentioned algorithm for computing the characteristic state predicate, we introduce some additional operations. Assume that the (Q)SL formulas considered below are >-normalised, and $\psi + \delta$ denotes the formula obtained by adding δ to all the clock variables (except for x_0) occurring in ψ. The model checking algorithm uses the following three operators on SL formulas:

- **Time elapse**:

  $$\psi_1 \rightsquigarrow \psi_2$$

  encodes the set of states from which the state set $[\![\psi_2]\!]$ is reachable by allowing time to elapse, while staying in the state set $[\![\psi_1]\!]$ at all the time points in between. It is given by the formula

  $$\psi_1 \rightsquigarrow \psi_2 := \exists\delta(\delta \geq x_0 \wedge \psi_2 + \delta \wedge \forall\delta'(x_0 \leq \delta' \leq \delta \Rightarrow \psi_1 + \delta')).$$

  Note that δ is a fresh real time variable. The formula encoding the time elapse is not in QSL as it includes expressions involving the sum of real variables, but it can be transformed to QSL by using, instead of δ and δ', variables $\overline{\delta}$ and $\overline{\delta'}$ representing their negations. The corresponding transformation is as follows:

  $$\psi_1 \rightsquigarrow \psi_2 := \exists\overline{\delta}(\overline{\delta} \leq x_0 \wedge \psi_2[x_0 \leftarrow \overline{\delta}] \wedge \neg\exists\overline{\delta'}(\overline{\delta'} \leq x_0 \wedge \overline{\delta} \leq \overline{\delta'} \wedge \neg\psi_1[x_0 \leftarrow \overline{\delta'}]).$$

  The above formula can be understood as decreasing the absolute time reference, and this way increasing the value of any clock variable which is relative to the absolute time. Note that the substitution $\psi_2[x_i \leftarrow x_i + (-\overline{\delta}), 1 \leq i \leq n]$ (i.e., $\psi_2 + \delta$) can be computed as $\psi_2[x_0 \leftarrow \overline{\delta}]$, since subtracting from the value of each clock $\overline{\delta}$ is equivalent to setting x_0 to $\overline{\delta}$. Intuitively, this corresponds to the clock z previously used in Sect. 7.2.
- **Assignment**:

  $$\psi[\mathbf{A}],$$

  where $\mathbf{A}$ is sequence of assignments $\alpha_1 \leftarrow \beta_1, \ldots, \alpha_n \leftarrow \beta_n$, denotes the formula obtained by substituting in ψ each formula α_i with β_i, for all $1 \leq i \leq n$.

- **Weakest precondition**:
$$pre_{\mathcal{A}}(\psi)$$
denotes the weakest precondition of ψ with respect to the timed automaton $\mathcal{A}$, i.e., characterises the set of states from which a state satisfying ψ can be obtained either by doing nothing or by executing some action. It is given by
$$pre_{\mathcal{A}}(\psi) := \mathcal{I}_{\mathcal{A}} \wedge (\psi \vee \bigvee_{e \in E} pre_e(\mathcal{I}_{\mathcal{A}} \wedge \psi)),$$
with
$$pre_e(\alpha) := \phi \wedge \alpha[\mathbf{A}],$$
where ϕ is conjunction of the formula $guard(e)$ and the boolean encoding of the source location $enc_{PL}(source(e))$. $\mathbf{A}$ is a sequence of assignments in which all the clock variables in $reset(e)$ are assigned 0 and the encoding of the location variable is assigned the encoding of the target location. Moreover,
$$\mathcal{I}_{\mathcal{A}} := \bigwedge_{l \in L} (enc_{PL}(l) \Rightarrow \mathcal{I}(l)),$$
with $enc_{PL}(l)$ denoting a boolean encoding of l. Intuitively, $pre_e(\alpha)$ characterises the set of states from which a state satisfying α can be obtained either by doing nothing[11] or by executing the action labelling e.

The model checking algorithm computes the characteristic state predicate $[\![\varphi]\!]$ for a given $T\mu$ formula φ. This is defined by the recursive function p operating on the structure of the formula. A pseudo-code of this function is presented in Fig. 7.22. We assume that if a formula is covered by several entries of the instruction `case`, then the instructions of the first matching case are executed only.

The main components of the algorithm **p** require quantifier elimination in the time elapse operation, substitution of state variables in an assignment, and the decision procedure to check containment in the fixed-point computation. In order to define a symbolic model checker that represents sets of states as SL formulas, these operations are defined as operations in QSL. So, $p((p(\psi_1) \vee p(\psi_2)) \rightsquigarrow pre_{\mathcal{A}}(p(\psi_2)))$ uses a method for eliminating (existential) quantifiers over real variables from a QSL formula. This is discussed in Sect. 7.3.1.

Notice that checking containment of the set of states encoded by ϕ_{new} in the set ϕ_{old} (ϕ_{old} in ϕ_{new}), is obtained by deciding the validity of $\phi_{new} \Rightarrow \phi_{old}$ ($\phi_{old} \Rightarrow \phi_{new}$, respectively) or, alternatively, satisfiability of $\neg(\phi_{new} \Rightarrow \phi_{old})$ ($\neg(\phi_{old} \Rightarrow \phi_{new})$, respectively). Procedures for deciding separation formulas [19, 110, 151] are described in Sect. 7.4.

One thing that remains to be discussed is an implementation of substitution of a clock variable. For a clock variable x_i, the substitution $[x_i \leftarrow d]$ is

[11] The time cannot elapse then.

INPUT ARGUMENTS:
   a $T\mu$ formula ψ
GLOBAL VARIABLES:
   ψ', ψ_1, ψ_2, ϕ_{old}, ϕ_{new}: $T\mu$ formulas
RETURN VALUES:
   $p()$:   an SL formula

```
1.    function p(ψ) is
2.    begin
3.        case ψ is
4.            when an SL formula   => return 𝓘_𝒜 ∧ ψ;
5.            when ¬ψ'              => return 𝓘_𝒜 ∧ ¬p(ψ);
6.            when ψ₁ ∧ ψ₂          => return p(ψ₁) ∧ p(ψ₂);
7.            when ψ₁ ∨ ψ₂          => return p(ψ₁) ∨ p(ψ₂);
8.            when ψ₁ ▷ ψ₂          => return p((p(ψ₁) ∨ p(ψ₂)) ⤳ pre_𝒜(p(ψ₂)));
9.            when z.ψ'             => return p(ψ')[z ← 0];
10.           when μZ.ψ'            => φ_new := false;
11.               repeat
12.                   φ_old := φ_new;
13.                   φ_new := p(ψ'[Z ← φ_old]);
14.               until (φ_new ⇒ φ_old);
15.               return p(φ_old);
16.           when νZ.ψ'            => φ_new := true;
17.               repeat
18.                   φ_old := φ_new;
19.                   φ_new := p(ψ'[Z ← φ_old]);
20.               until (φ_old ⇒ φ_new);
21.               return p(φ_old);
22.       end case;
23.   end p;
```

Fig. 7.22. An algorithm computing the characteristic state predicate for a $T\mu$ formula

performed by replacing all the boolean variables corresponding to formulas involving x_i with variables corresponding to the substituting formulas, using fresh variables if necessary.

Next, we discuss the main ideas behind an implementation of the above-discussed operations using a boolean encoding of QSL. Firstly, quantification of real variables is replaced by quantification of boolean variables. Secondly, SL formulas are represented by boolean formulas and model checking is implemented as operations in QPL.

7.3.1 From Real to Boolean Quantification

Below we show a method for eliminating existential quantifiers over real variables from a QSL formula. This is applicable to the formulas of the form $\exists x_a.\psi$, where x_a, with $a \in \mathbb{N}$, is a real time (clock) variable, and ψ is a $>$-normalised SL formula. The formula $\exists x_a.\psi$ is transformed to an equivalent QSL formula, with quantifiers over boolean variables only, in the following three steps:

1. Each separation predicate $x_i \sim x_j + c$ in ψ (where $a \in \{i,j\}$) is encoded by the boolean variable $\wp_{i,j}^{\sim,c}$. The separation predicates that are negations of each other are represented by boolean literals[12] that are negations of each other. The resulting formula is denoted by ψ_{bool}^a.
2. In order to disallow satisfying boolean assignments that do not have corresponding assignments to the real-valued variables, *transitivity constraints* are added to some pairs of boolean literals that encode predicates having at least one real variable in common. A transitivity constraint for x_a can be of the following types:
   - $\wp_{i,a}^{\sim_1,c_1} \wedge \wp_{a,j}^{\sim_2,c_2} \Rightarrow (x_i \sim x_j + c_1 + c_2)$,
     where if $\sim_1 = \sim_2$, then $\sim := \sim_1$, otherwise we have to write the above constraint for both the $\sim := \sim_1$ and $\sim := \sim_2$ ($\sim \in \{>, \geq\}$),
   - $\wp_{i,j}^{\sim_1,c_1} \Rightarrow \wp_{i,j}^{\sim_2,c_2}$,
     where $c_1 > c_2$ and $a \in \{i,j\}$,
   - $\wp_{i,j}^{>,c_1} \Rightarrow \wp_{i,j}^{\geq,c_1}$,
     where $a \in \{i,j\}$.

   After generating all the transitivity constraints for x_a, we conjoin them to get the formula ψ_{cons}^a.
3. The resulting formula is the conjunction of the above formulas preceded by the existential quantifiers over all the boolean variables added in the step 1.

This is shown in detail in Example 7.14.

Example 7.14. Let $\alpha_a = \exists x_a.\psi$, where $\psi = x_a \leq x_0 \wedge x_1 \geq x_a \wedge x_2 \leq x_a$. After transforming ψ to the $>$-normalised version we get

$$\psi = x_0 \geq x_a \wedge x_1 \geq x_a \wedge x_a \geq x_2.$$

Define the boolean variables $\wp_{0,a}^{\geq,0}, \wp_{1,a}^{\geq,0}, \wp_{a,2}^{\geq,0}$ corresponding respectively to the above SL formulas occurring in the conjunction. Then,

$$\psi_{bool}^a = \wp_{0,a}^{\geq,0} \wedge \wp_{1,a}^{\geq,0} \wedge \wp_{a,2}^{\geq,0}.$$

The formula ψ_{cons}^a is the conjunction of the following constraints:

- $\wp_{0,a}^{\geq,0} \wedge \wp_{a,2}^{\geq,0} \Rightarrow x_0 \geq x_2$,

---

[12] By a *literal* we mean both a propositional variable and its negation.

- $\wp_{1,a}^{\geq,0} \wedge \wp_{a,2}^{\geq,0} \Rightarrow x_1 \geq x_2$.

Note that the transitivity constraints are built only for the pairs of boolean variables in which a occurs at the second and at the first position in the list of the subscripts. Thus, we do not have a transitivity constraint for $\wp_{0,a}^{\geq,0} \wedge \wp_{1,a}^{\geq,0}$, where a occurs at the second position only. Note that the transitivity constraints of the last two types do not need to be added (there are no $\wp_{i,j}^{\sim_1,c_1}, \wp_{i,j}^{\sim_2,c_2}$ with $c_1 > c_2$ or with $c_1 = c_2$, $\sim_1=$ "$>$" and $\sim_2=$ "$\geq$").

Then, we obtain the formula

$$\alpha_{bool} = \exists \wp_{0,a}^{\geq,0}, \wp_{1,a}^{\geq,0}, \wp_{a,2}^{\geq,0}.[\psi_{bool}^{a} \wedge \psi_{cons}^{a}],$$

which evaluates to $x_0 \geq x_2 \wedge x_1 \geq x_2$. This is easily seen when, according to the semantics of $\exists$, we replace the quantified variables with *true*.

□

7.3.2 From QSL Formulas Without Real-Time Quantification to Equivalent Boolean Formulas

Since the formula obtained after the above translation is in QSL, but it contains only existential quantifiers over boolean variables, we can eliminate the quantifiers, translate the SL formula to a boolean formula (see Sect. 7.4) and use a SAT-solver to decide the resulting boolean formula. Existential quantifiers can be eliminated using a translation to BDDs, or using a method described in Sect. 7.6, where we show a translation from QPL to boolean formulas. This translation is for universal quantifiers, but each existentially quantified QPL formula $\exists \wp.\psi$ can be replaced by the equivalent universally quantified QPL formula $\neg\forall\wp.\neg\psi$. Alternatively, one could use a QBF-solver [122] to decide QPL formulas. Several optimizations to the above method are applied in [140].

7.4 Deciding Separation Logic (MATH-SAT)

There are several methods of deciding SL formulas (see [19, 110, 151, 154]). In what follows we show in detail the method of [151], which consists in translating the problem of satisfiability of an SL formula to the problem of satisfiability of a propositional formula. Then, we briefly discuss the other existing approaches.

7.4.1 From SL to Propositional Logic

We start with the general idea of the method and an instructive example. So, assume that φ is a $>$-normalised SL formula. The decision procedure consists of the following three stages:

1. **Propositional encoding.** Deriving a propositional formula φ' from φ such that all the φ predicates are encoded by fresh propositional variables.
2. **Building the inequality graph G_φ.** The vertices of the graph are defined by the SL variables in φ and the edges correspond to the propositional variables and their duals.
3. **Transitivity constraints.** Adding to the conjunction of the propositional variables all the transitivity constraints for every simple cycle in G_φ.

Next, we give details of the above method.

Propositional Encoding

This step involves replacing all predicates with fresh propositional variables. So, each predicate $x_i \sim x_j + c$, where $\sim \in \{>, \geq\}$, is replaced by the proposition $\wp_{i,j}^{\sim,c}$. Notice that by such a translation we loose all the transitivity constraints between predicates. The simplest example illustrating this is when we translate $x_i > x_j$ and $x_j > x_k$. Then, the constraint $x_i > x_k$ is lost. In order to compensate for the lost we use a graph-theoretic approach for deriving propositional variables corresponding to the transitivity constraints.

Building the Inequality Graph

Let $G_\varphi = (\mathcal{X}, \mathcal{E})$ be a weighted directed multigraph, where each edge $e \in \mathcal{E}$ is a four-tuple

$$(x_i, x_j, \sim, c),$$

where x_i is the source node, x_j is the target node, $\sim \in \{>, \geq\}$ – the type of edge, and c – the weight. For each edge e, its components are denoted by *source*(e), *target*(e), *iqsgn*(e) and *weight*(e), respectively. Moreover, the dual edge of $e = (x_i, x_j, \sim, c)$ is defined as

$$\overline{e} = (x_j, x_i, \overline{\sim}, -c),$$

where $\overline{>} :=$ "$\geq$" and $\overline{\geq} :=$ "$>$". Notice that for the edge $e = (x_i, x_j, \sim, c)$ corresponding to $\wp_{i,j}^{\sim,c}$ the dual edge of e corresponds to the negation of $\wp_{i,j}^{\sim,c}$.

The graph $G_\varphi = (\mathcal{X}, \mathcal{E})$ is constructed as follows:

- $\mathcal{X}$ is a set of all the real variables in φ,
- $\mathcal{E}$ is a set of edges $e = (x_i, x_j, \sim, c)$ and their duals $\overline{e}$, for each predicate $x_i \sim x_j + c$ appearing in φ.

Transitivity Constraints

The idea is to add new propositions encoding the transitivity constraints imposed by separation predicates (see [132, 143]). We start with analyzing a simple cycle of size 2 in G_φ, i.e., a cycle composed of two edges. Let $\mathfrak{cc} = x_1 \sim x_2 + c$ and $\mathfrak{cc}' = x_2 \sim' x_1 + c'$ be two clock constraints in φ. Notice that $x_1 - x_2 \sim c$ and $x_2 - x_1 \sim' c'$, so $0 \geq c + c'$. Thus, if $c + c' > 0$, then obviously $\mathfrak{cc} \wedge \mathfrak{cc}'$ is unsatisfiable. Additionally, if $c + c' = 0$ and at least one of $\sim, \sim'$ is equal to "$>$", then $\mathfrak{cc} \wedge \mathfrak{cc}'$ is also unsatisfiable. The constraints in the other direction can be inferred by applying the above constraints to the duals of $\mathfrak{cc}$ and $\mathfrak{cc}'$. So, if $c + c' < 0$, then obviously $(\neg\mathfrak{cc}) \wedge (\neg\mathfrak{cc}')$ is unsatisfiable. Additionally, if $c + c' = 0$ and at least one of $\sim, \sim'$ is equal to "$>$", then $(\neg\mathfrak{cc}) \wedge (\neg\mathfrak{cc}')$ is also unsatisfiable.

In order to generalize the above analysis on more complex cycles composed of more than two edges we need to define some auxiliary notions.

Definition 7.15. *A directed path in G_φ of length m from x to x' is a sequence of edges $(e_1, \ldots, e_m)$ in G_φ such that*

- $x = source(e_1)$,
- $x' = target(e_m)$,
- $target(e_i) = source(e_{i+1})$ *for each* $1 \leq i \leq m - 1$.

The notations source(), target(), weight() and iqsgn() are extended to the path ξ such that

- $source(\xi) = source(e_1)$,
- $target(\xi) = target(e_m)$,
- $weight(\xi) = \Sigma_{i=1}^{m} weight(e_i)$, *and*
- $iqsgn(\xi) = \begin{cases} \text{"}\geq\text{"} & \text{iff } iqsgn(e_i) = \text{"}\geq\text{" for each } 1 \leq i \leq m, \\ \text{"}>\text{"} & \text{iff } iqsgn(e_i) = \text{"}>\text{" for each } 1 \leq i \leq m, \\ \text{"}\gg\text{"} & \text{otherwise.} \end{cases}$

Below, we give the transitivity constraints of a *simple cycle* $\mathcal{C}$, i.e., a directed path ξ such that $source(\xi) = target(\xi)$ and each sub-cycle in ξ is iterated only once[13]. They are given by the following conditions:

- if $iqsgn(\mathcal{C}) =$ "$\geq$", then apply **R1**, **R2**,
- if $iqsgn(\mathcal{C}) =$ "$>$", then apply **R3**, **R4**,
- if $iqsgn(\mathcal{C}) =$ "$\gg$", then apply **R2**, **R3**,

where the rules **R1**, **R2**, **R3**, **R4** are defined as follows (we identify the edges with the corresponding constraints):

> **R1.** if $weight(\mathcal{C}) > 0$, then $\bigwedge_{e_i \in \mathcal{C}} e_i = false$,
> **R2.** if $weight(\mathcal{C}) \leq 0$, then $\bigvee_{e_i \in \mathcal{C}} e_i = true$,

[13] It is clear that iterations over cycles do not add transitivity constraints.

R3. if $weight(\mathcal{C}) \geq 0$, then $\bigwedge_{e_i \in \mathcal{C}} e_i = false$,
R4. if $weight(\mathcal{C}) < 0$, then $\bigvee_{e_i \in \mathcal{C}} e_i = true$.

The above rules express additional (implied) conditions that should be satisfied, which in turn requires to add some additional constraints to the translation of the original SL formula. Therefore, the rules should be understood as follows. The rule **R1** (**R3**) requires to generate a constraint, which when satisfied implies that $\bigwedge_{e_i \in \mathcal{C}} e_i = false$. So, we take the formula $\bigvee_{e_i \in \mathcal{C}} \neg e_i$. The rule **R2** (**R4**) requires to generate a constraint, which when satisfied implies that $\bigvee_{e_i \in \mathcal{C}} e_i = true$. So, we take the formula $\bigwedge_{e_i \in \mathcal{C}} \neg e_i$. This is shown in Example 7.16 for **R3**.

It turns out that we can concentrate on simple cycles only, as if there is an assignment $\mathbf{A}_{\mathcal{C}}$ to a non-simple cycle $\mathcal{C}$ which does not satisfy it, then there is a simple cycle $\mathcal{C}'$, which is a subgraph of $\mathcal{C}$, such that $\mathbf{A}_{\mathcal{C}}$ does not satisfy $\mathcal{C}'$ either. Thus, the last step of our translation consists in adding to the conjunction of the propositional variables corresponding to the separation predicates all the transitivity constraints for every simple cycle in G_φ.

Example 7.16. Consider the formula

$$\varphi = x_1 > x_2 - 1 \ \vee\ \neg(x_3 > x_2 - 2) \ \vee\ \neg(x_1 \geq x_3 - 3).$$

Notice that $\neg(x_3 > x_2 - 2)$ is equivalent to $x_2 \geq x_3 + 2$, whereas $\neg(x_1 \geq x_3 - 3)$ to $x_3 > x_1 + 3$. After the step 1 (see p. 213) we have

$$\varphi' = \wp_{1,2}^{>,-1} \vee \neg\wp_{3,2}^{>,-2} \vee \neg\wp_{1,3}^{\geq,-3}.$$

The variable dual to $\wp_{3,2}^{>,-2}$ is $\wp_{2,3}^{\geq,2}$. The other dual variables are $\wp_{2,1}^{\geq,1}$ and $\wp_{3,1}^{>,3}$. The graph G_φ (see Fig. 7.23) contains only one simple cycle $\mathcal{C}$ such its weight, i.e., $weight(\mathcal{C})$, is equal to 4, $iqsgn(\mathcal{C}) =$ "$\gg$", and consisting of the vertices x_1, x_2, x_3; and the dual of this cycle. Therefore, following **R3**, we add to φ' the transitivity constraint $\neg\wp_{1,2}^{>,-1} \vee \neg(\neg\wp_{3,2}^{>,-2}) \vee \neg(\neg\wp_{1,3}^{\geq,-3})$. Note that applying the rules to the dual cycle gives exactly the same constraints.

□

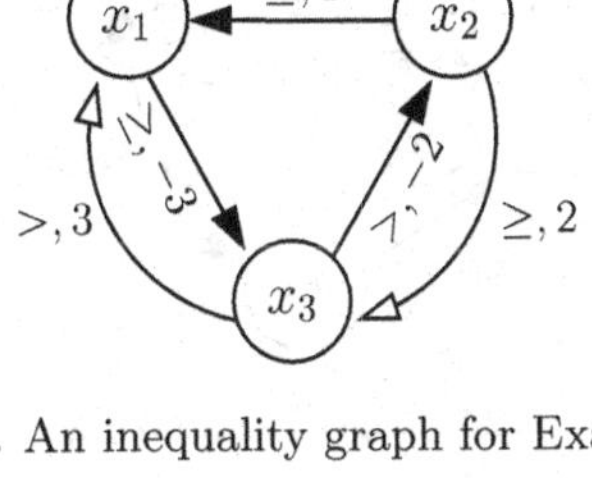

Fig. 7.23. An inequality graph for Example 7.16

Complexity and Optimization

The complexity of enumerating constraints for all the simple cycles is linear in the number of cycles. But, there may be an exponential number of such cycles (see Fig. 7.24).

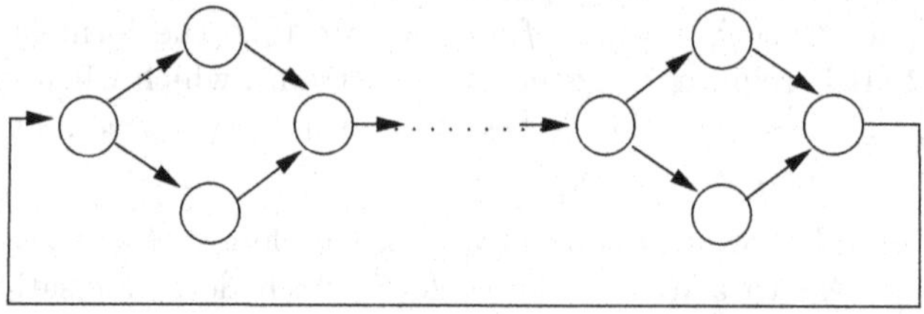

Fig. 7.24. The structure of a graph with exponentially many simple cycles

The optimization suggested in [151] consists in adding to the graph G_φ new edges (called *chords*) connecting nodes of the cycles of size 4 and more which were originally not connected by a single edge. Thanks to this construction one can then consider only simple cycles of size 2 and 3. Below, we give a short description of the above method.

Definition 7.17. *Let C be a simple cycle in G_φ and x_i and x_j be two non-adjacent nodes in C. Denote the path from x_i to x_j in C by $\xi_{i,j}$. A* chord *e from x_i to x_j is called $\xi_{i,j}$-accumulating if the following two requirements are satisfied:*

- *$weight(e) = weight(\xi_{i,j})$, and*
- *$iqsgn(e)$ = " $\geq$ " if $iqsgn(\xi_{i,j})$ = " $\geq$ " or if ($iqsgn(\xi_{i,j})$ = " $\gg$ " and $iqsgn(\xi_{j,i})$ = " $>$ "). Otherwise, $iqsgn(e)$ = " $>$ ".*

Example 7.18. The edge from x_1 to x_3 is $\xi_{1,3}$-accumulating chord in the graph in Fig. 7.25. The edge from x_3 to x_1 is the dual one.

□

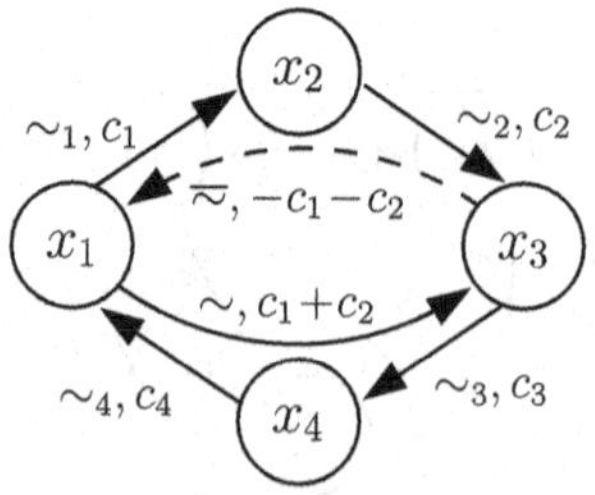

Fig. 7.25. An accumulating chord

Definition 7.19. *The graph G_φ is called* chordal *if all the simple cycles in G_φ of size greater or equal to 4 contain an accumulating chord.*

It turns out that for each simple cycle $\mathcal{C}$ and an assignment $\mathbf{A}_\mathcal{C}$, if $\mathbf{A}_\mathcal{C}$ does not satisfy $\mathcal{C}$, then there is a simple cycle $\mathcal{C}'$ of size (at most) 3 such that $\mathbf{A}_\mathcal{C}$ does not satisfy $\mathcal{C}'$. Then, one can concentrate only on cycles of size 2 and 3.

The graph construction phase of p. 213 is extended by a step of making the graph chordal.

- Make the graph chordal:

```
1.     unmarked := X;
2.     while (∃xᵢ ∈ unmarked) do
3.         mark the vertex xᵢ; unmarked := unmarked \ {xᵢ};
4.         for each (xⱼ, xᵢ, ∼₁, c₁), (xᵢ, xₖ, ∼₂, c₂) ∈ ℰ
                    s.t. (xⱼ, xₖ ∈ unmarked ∧ j ≠ k) do
5.             add (xⱼ, xₖ, ∼₁, c₁ + c₂) and its dual to ℰ;
6.             if ∼₁ ≠ ∼₂ then
7.                 add (xⱼ, xₖ, ∼₂, c₁ + c₂) and its dual to ℰ;
8.             end if;
9.         end do;
10.    end do;
```

In the worst case, the process of making the graph chordal can add an exponential number of edges, but in many cases, the chordal method can reduce complexity. For example this happens when all weights are equal to 0.

7.4.2 Other Approaches to Deciding SL

A quite different method was presented in [19]. The whole algorithm is decomposed into 5 layers L_0–L_5. The idea is to refine the process of finding a satisfying assignment from layer to layer. If at some layer the formula appears unsatisfiable, then it is indeed unsatisfiable. Otherwise, it has to be considered at the next layer. The layers are shortly described below:

L_0 considers only propositional connectives in an SL formula φ. Mathematical atoms are abstracted into propositional literals. The process is realized by a DPLL-like procedure[14], which is discussed in Sect. 7.5.

L_1 considers also equalities, performing equality propagation, building equality-driven clusters of variables and detecting equality-driven unsatisfiabilities.

L_2 handles also inequalities of the form $x_i \sim x_j + c$ for unrestricted relational $\sim$, by a variant of the Bellman–Ford minimal path algorithm.

---

[14] DPLL is a SAT-solver algorithm. Its name comes from the first letters of the names of its authors (Davis, Putnam, Logemann and Loveland).

L_3 takes into account also general inequalities (i.e., variables may have coefficients) using the standard simplex algorithm.
L_4 considers also negated equalities.

Another algorithm was given in [110]. The key idea of this approach is that a conjunction of clock constraints can be represented by a DBM. The Floyd–Warshall algorithm is used to normalize the constraints, to find contradiction in the set of numerical clauses, and to extract a solution from a consistent set of inequalities in polynomial time. Whereas typical SAT-solvers perform both syntactic transformations of formulas that may simplify clauses and remove variables as well as search in the state space of valuations for the remaining variables, their solver gives priority to the former and resorts to search only when simplifications are impossible. It is based on Davis–Putnam procedure.

7.5 Deciding Propositional Formulas (SAT)

This section aims at explaining the main principles followed by propositional SAT-solvers, i.e., algorithms testing satisfiability of propositional formulas. Our presentation is based on [38, 109].

Assume we are given a propositional formula φ. The aim of a SAT-solver is to find a satisfying assignment for φ if it exists, or return "unsatisfiable" otherwise. It is well known that the problem of establishing whether a formula is satisfiable or not (known as a SAT-problem) is NP-complete. Therefore, in general, one cannot expect that a SAT-solver will return a result in a polynomial time. We should be aware of the fact that a SAT-solver is a heuristics only, but it can be very "clever". Modern SAT-solvers can decide formulas composed of hundreds of thousands of propositional variables in a reasonable time.

Typically, SAT-solvers accept formulas in conjunctive normal form (CNF), i.e., a conjunction of clauses, where a *clause* is a disjunction of literals. Such a form is quite useful for checking satisfiability, as any valuation, which makes at least one literal of each clause satisfied, makes the whole formula satisfied.

Every propositional formula φ can be translated to a CNF formula in two ways. Either the resulting CNF formula preserves only satisfiability of φ or it is logically equivalent to φ. Clearly, the former translation is much easier than the latter and it is used for checking satisfiability of φ, what is the main subject of this section. If we need to operate further on φ, then an equivalence-preserving translation is necessary. In Sect. 7.6 we show how to use it for translating a fragment of QPL to PL.

We start with some basic definitions and formalizations. Then, we describe how to construct a satisfiability-preserving CNF formula for φ.

Consider a set of propositional variables PV extended by two constants $true$ and $false$ with the standard meaning. A disjunction of zero literals is taken to mean the constant *false*. A conjunction of zero clauses is taken to mean the constant *true*. An *assignment* $\mathbf{A}$ is a partial function from PV to $\{true, false\}$. The domain of an assignment $\mathbf{A}$ is denoted by $dom(\mathbf{A})$. An assignment is said to be *total* (*φ-total*) when its domain is PV ($PV(\varphi)$, respectively). A φ-total assignment $\mathbf{A}$ is said to be *satisfying* for φ when $\mathbf{A} \models \varphi$,[15] i.e., the value of φ given by $\mathbf{A}$ is *true*. It is convenient to generalize an assignment $\mathbf{A}$ to the formulas over the propositions of $dom(\mathbf{A})$ such that $\mathbf{A}(\varphi) = true$ iff $\mathbf{A} \models \varphi$.

Next, we equate also an assignment $\mathbf{A}$ with the conjunction of a set of literals, specifically the set containing $\neg\wp$ for all $\wp \in dom(\mathbf{A})$ such that $\mathbf{A}(\wp) = false$, and $\wp$ for all $\wp \in dom(\mathbf{A})$ such that $\mathbf{A}(\wp) = true$. For example for $\mathbf{A} = \{(\wp_1, true), (\wp_2, false)\}$, we get the conjunction of the literals of the set $\{\wp_1, \neg\wp_2\}$. In the following we show a polynomial algorithm that, given a propositional formula φ, constructs a CNF formula which is unsatisfiable exactly when φ is valid. The procedure works as follows. First of all, for every subformula ψ of the formula φ, we introduce a distinct propositional variable l_ψ. Furthermore, if ψ is a propositional variable from PV, then

$$l_\psi = \psi.$$

Next, we assign the formula $toCNF(\psi)$ to every subformula ψ of φ according to the following rules:

- if ψ is a propositional variable, then $toCNF(\psi) = true$,
- if $\psi = \neg\phi$, then
  $toCNF(\psi) = toCNF(\phi) \wedge (l_\psi \vee l_\phi) \wedge (\neg l_\psi \vee \neg l_\phi)$,
- if $\psi = \phi_1 \vee \phi_2$, then
  $toCNF(\psi) = toCNF(\phi_1) \wedge toCNF(\phi_2) \wedge (l_\psi \vee \neg l_{\phi_1}) \wedge (l_\psi \vee \neg l_{\phi_2}) \wedge (\neg l_\psi \vee l_{\phi_1} \vee l_{\phi_2})$,
- if $\psi = \phi_1 \wedge \phi_2$, then
  $toCNF(\psi) = toCNF(\phi_1) \wedge toCNF(\phi_2) \wedge (\neg l_\psi \vee l_{\phi_1}) \wedge (\neg l_\psi \vee l_{\phi_2}) \wedge (l_\psi \vee \neg l_{\phi_1} \vee \neg l_{\phi_2})$,

and for derived boolean operators

- if $\psi = \phi_1 \Rightarrow \phi_2$ then
  $toCNF(\psi) = toCNF(\phi_1) \wedge toCNF(\phi_2) \wedge (l_\psi \vee l_{\phi_1}) \wedge (l_\psi \vee \neg l_{\phi_2}) \wedge (\neg l_\psi \vee \neg l_{\phi_1} \vee l_{\phi_2})$,
- if $\psi = \phi_1 \Leftrightarrow \phi_2$ then
  $toCNF(\psi) = toCNF(\phi_1) \wedge toCNF(\phi_2) \wedge (l_{\phi_1} \vee \neg l_{\phi_2} \vee \neg l_\psi) \wedge (\neg l_{\phi_1} \vee l_{\phi_2} \vee \neg l_\psi) \wedge (l_{\phi_1} \vee l_{\phi_2} \vee l_\psi) \wedge (\neg l_{\phi_1} \vee \neg l_{\phi_2} \vee l_\psi)$.

It can be shown [105] that the formula φ is valid when the CNF formula

$$toCNF(\varphi) \wedge \neg l_\varphi$$

[15] A φ-total assignment coincides with a valuation function, which motivates the notion $\models$.

is unsatisfiable. This follows from the fact that for any assignment $\mathbf{A}$ with $dom(\mathbf{A}) = PV(\varphi)$ there is a unique satisfying assignment $\mathbf{A}'$ of $toCNF(\varphi)$ consistent with $\mathbf{A}$ such that $\mathbf{A}'(l_\varphi) = \mathbf{A}(\varphi)$.

Notice also that the CNF formula

$$toCNF(\varphi) \wedge l_\varphi$$

preserves satisfiability, but is not equivalent to φ. The reason is that $toCNF(\varphi) \wedge l_\varphi$ is defined over an extended set of propositional variables and there exist assignments which make φ satisfied and $toCNF(\varphi) \wedge l_\varphi$ not satisfied.

Example 7.20. Let $\varphi = \wp_1 \wedge \wp_2$. Then,

$$toCNF(\varphi) \wedge l_\varphi = (\neg l_\varphi \vee \wp_1) \wedge (\neg l_\varphi \vee \wp_2) \wedge (l_\varphi \vee \neg\wp_1 \vee \neg\wp_2) \wedge l_\varphi.$$

Consider the assignment $\mathbf{A} = \{\wp_1, \wp_2, \neg l_\varphi\}$. We have $\mathbf{A}(\varphi) = true$ and $\mathbf{A}(toCNF(\varphi) \wedge l_\varphi) = false$.

□

There are several approaches used to check satisfiability of propositional formulas. They can be based on Stålmarck's method [141], use methods of soft computing (Monte Carlo, evolutionary algorithms) or exploit the theory of resolution [71]. Here, we discuss the algorithm proposed by Davis and Putnam [62] and later improved by Davis, Logemann and Loveland [61], known as DPLL. The solution is based on a backtracking search algorithm through the space of possible assignments of a CNF formula. The algorithm uses the methods of *boolean constraint propagation* (*BCP*), *conflict-based learning* (*CBL*), and *variable selection* (*VS*).

A template of the generic SAT algorithm is given in Fig. 7.26. Below, we explain in detail the procedures it uses and the ideas behind them.

7.5.1 Boolean Constrain Propagation

The idea behind the algorithm is to identify assignments that are necessary for satisfiability of the CNF formula and to efficiently propagate each variable's assignment found this way.

The algorithm starts with identifying the *unit clauses* of the CNF formula, i.e., the clauses of one unassigned literal only and the other literals evaluating to $false$. Obviously, a clause composed of one unassigned literal is the unit clause. All the unassigned literals in the unit clauses are assigned *true* by the algorithm. This is obviously necessary for any potential satisfying assignment. Then, the algorithm computes all the immediate implications of the assignment just made by iteratively assigning the unassigned literals of the newly created unit clauses. Consider the formula

$$\wp_1 \wedge (\neg\wp_1 \vee \neg\wp_2).$$

INPUT ARGUMENTS:
  a propositional formula φ
GLOBAL VARIABLES:
  d: $\mathbb{N}$
  χ: a set of clauses over $PV(\varphi)$
  $\mathbf{A}$: $2^{PV(\varphi)}$
RETURN VALUES:
  SAT(): $2^{PV(\varphi)}$
  decide(): {DECISION, ALL-ASSIGNED}
  deduce(): {OK, CONFLICT}
  diagnose(): Integer $\times 2^{PV(\varphi)}$

```
1.   function SAT(φ) is
2.   begin
3.       d := 0; χ := clauses(φ); A := ∅;
4.       deduce(d);
5.       while true do
6.         d := d + 1;
7.         if decide(d) = ALL-ASSIGNED then return A; end if;
8.         if deduce(d) = CONFLICT then
9.           (dₗ, cₗ) := diagnose(d);
10.          d := dₗ;
11.          if d = 0 then return ∅; end if;
12.          erase(d);
13.          χ := χ ∪ {cₗ};
14.        end if;
15.      end do;
16.  end SAT;
```

Fig. 7.26. A generic SAT algorithm

The first (implied) decision is $(\wp_1, true)$. This in turn implies $(\wp_2, false)$, which for our formula produces the satisfying assignment and means that the algorithm stops.

When no more unit clauses can be found, the algorithm selects an unassigned literal and assigns it a boolean value, which could be either *true* or *false*. Then, again the algorithm applies the unit clause rule. The above is realized by the procedures **decide** and **deduce**. By a *decision variable* we mean a propositional variable which could have been assigned both the values at the moment of the decision, i.e., its assignment was not implied by other assignments. Every assigned variable $\wp$ is given the *decision depth*, which is equal to the number of the decision variables assigned so far, i.e., before $\wp$ was assigned.

Such a procedure can likely make a clause unsatisfiable. This would for example happen for the formula

$$\wp_1 \wedge (\wp_2 \vee \wp_3) \wedge (\wp_2 \vee \neg\wp_3).$$

After assigning $(\wp_1, true)$ and selecting $\wp_2$ and its value as $false$ and then assigning $(\wp_3, true)$, the clause $\wp_2 \vee \neg\wp_3$ becomes unsatisfiable. This is called a *conflict*, as having the variables $\wp_1$ and $\wp_2$ assigned as above there is no assignment of $\wp_3$ leading to a satisfying total assignment. Then, clearly the decision $(\wp_2, true)$ must be changed and the implications of the new decision must be recomputed. Notice that backtracking implicitly prunes parts of the search tree, which is of size 2^n in case n unassigned literals remain in a point of starting backtracking. When a conflict occurs, the technique of *conflict based-learning* (see below) is used to deduce a new clause (called a *conflict clause*), which is added to the working set if clauses, denoted by χ in our algorithm in Fig. 7.26. This mechanism prevents similar conflicts from reoccurring, as the algorithm backtracks immediately if such an assignment is repeated.

7.5.2 Conflict-Based Learning

We explain the CBL mechanism of deriving conflict clauses using a more complicated example. Let

$$\varphi = c_1 \wedge c_2 \wedge c_3 \wedge c_4,$$

where

- $c_1 = (\neg\wp_1)$,
- $c_2 = (\wp_1 \vee \wp_4 \vee \neg\wp_5)$,
- $c_3 = (\neg\wp_2 \vee \wp_3)$, and
- $c_4 = (\wp_4 \vee \wp_5)$.

Notice that at the start of the procedure the assignment of $\wp_1$ is implied, which follows from the unit clause rule. Thus, we have $\mathbf{A} = \{\neg\wp_1\}$. Now, no more unit clauses exist. So, the algorithm decides an assignment for another unassigned variable, say $\mathbf{A}(\wp_2) = true$ (it could have decided as well an assignment for $\wp_3$, $\wp_4$ or $\wp_5$). This implies the assignment of $\wp_3$, namely $\mathbf{A}(\wp_3) = true$, so that the clause $(\neg\wp_2 \vee \wp_3)$ is satisfied. Next, the assignment $\mathbf{A}(\wp_4) = false$ is decided, but notice that this implies both $\wp_5$ (because of the clause $(\wp_4 \vee \wp_5)$) and $\neg\wp_5$ (because of the clause $(\wp_1 \vee \wp_4 \vee \neg\wp_5)$). So, whatever assignment comes first, i.e., either $\mathbf{A}(\wp_5) = true$ or $\mathbf{A}(\wp_5) = false$, we have a conflict.

When such a conflict is identified, the procedure `diagnose` is responsible for finding the assignments that are directly responsible for the conflict and could have been potentially selected differently in order to find a satisfying assignment. In our example these are $\{\neg\wp_1, \neg\wp_4\}$, which we explain below. The conjunction of these assignments gives a sufficient condition for the conflict to occur. Consequently, the negation of this conjunction must be satisfied if our formula is satisfiable. Notice that $\neg\wp_1$ clearly cannot be changed as it is equal to the clause c_1 of one literal only. Thus, we generate the new clause

$$c_5 = (\wp_1 \vee \wp_4),$$

remove $\wp_1$ (in order to optimize on c_5), and add the clause $\wp_4$ to the working set of clauses of φ. The above process is performed by the procedure `diagnose`, which returns an (optimised) learned clause c_l that corresponds to an assignment found, and in addition computes the decision level d_l to which the search has to backtrack. The procedure `erase(`d`)` (called for $d{=}d_l$) cancels all the assignments of the variables that were assigned at the decision levels greater than d and leaves the others intact.

In order to find a conflict clause we use the *implication graph*, which records the unit clause propagation process implied by the decision assignment. Each node in this graph corresponds to a variable assignment, so is represented by a literal. The incoming directed edges $(l_1, l_j), \ldots, (l_i, l_j)$ labelled by a clause c represent the fact that

$$c = (\neg l_1 \vee \ldots \vee \neg l_i \vee l_j),$$

which means that l_j has been implied by the unit clause rule. Thus, the vertices without incoming edges correspond to decision assignments, while the others correspond to implied assignments. A graph represents a conflict if it contains two vertices l and $\neg l$ for some literal l. The conjunction of the roots of the graph backwards reachable from the above two vertices (excluding these which belong to one-literal clauses) represents the assignment responsible for the conflict.

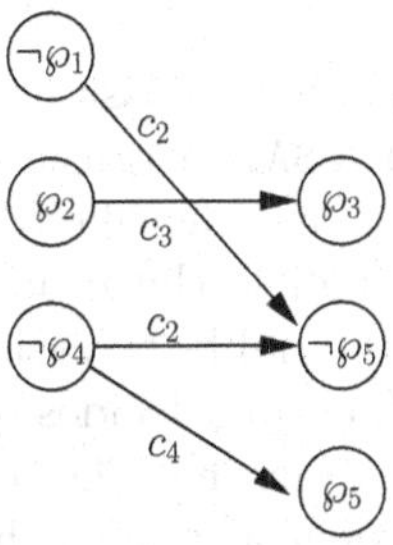

Fig. 7.27. An implication graph for φ

The implication graph for the above example is shown in Fig. 7.27. Its roots are $\neg\wp_1$, $\wp_2$, and $\neg\wp_4$, and the edges between literals are labelled by the clauses that induced the assignments. So, the edge from $\wp_2$ to $\wp_3$ is labelled by the clause c_3, whereas the edges from $\neg\wp_1$ and $\neg\wp_4$ to $\neg\wp_5$ are labelled by c_2. The edge from $\neg\wp_4$ to $\wp_5$ is labelled by c_4. The conflict in the graph is represented by two contradictory vertices, namely $\wp_5$ and $\neg\wp_5$. The search shows that the above vertices are reachable only from the roots $\neg\wp_1$ and $\neg\wp_4$. Therefore, the conjunction of them, i.e., $\neg\wp_1 \wedge \neg\wp_4$, is a sufficient condition for a conflict. Thus, the negation of this conjunction must be satisfied if our formula is satisfiable.

Next, the algorithm backtracks and withdraws from the assignment of $\neg\wp_4$, returning to $\mathbf{A} = \{\neg\wp_1, \wp_2, \wp_3\}$. Notice that the learned conflict clause c_5 implies $\mathbf{A}(\wp_4) = true$. Then, the algorithm decides an assignment for $\wp_5$, say $\mathbf{A}(\wp_5) = true$. This is a φ-total assignment, which does not produce a conflict. Thus, a satisfying assignment that is found is

$$\mathbf{A} = \{\neg\wp_1, \wp_2, \wp_3, \wp_4, \wp_5\}.$$

It is important to mention that the original Davis–Putnam algorithm backtracks one step at a time (i.e., $d_l = d-1$), whereas modern SAT-solvers exploit *non-chronological backtracking search* strategies (i.e., $d_l = d - j$, for $j \leq d$), which allows them to skip many irrelevant assignments.

Still several details of the algorithm in Fig. 7.26 remain to be discussed. This is done in the next section.

7.5.3 Variable Selection (VS)

At each decision level d in the search, a propositional variable $\wp_d$ is assigned either *true* or *false*. This assignment is selected by the procedure decide(d). Many heuristics may serve this aim. For example the authors of [109] suggest the following procedure: *"At each node in the decision tree evaluate the number of clauses directly satisfied by each assignment to each variable. Choose the variable and the assignment that directly satisfies the largest number of clauses."*

If all the variables have been already assigned and no conflict occurs (indicated by ALL–ASSIGNED), then SAT returns a satisfying assignment **A**. Otherwise, the implied assignments are identified by the procedure deduce(d). If this terminates without a conflict, the procedure SAT increases the decision level and looks for a new variable to assign. Otherwise, the procedure diagnose(d) analyses the conflict and decides on the next step. If $\wp_d$ was assigned only one boolean value so far, the other one is decided and the process is repeated. If the other assignment also fails, then it means that the value of $\wp_d$ is not responsible for the conflict. Then, the procedure diagnose(d) identifies the assignment responsible for the conflict and computes the decision level d_l which SAT must backtrack to. The procedure will backtrack $d - d_l$ times, each time erasing (the procedure erase(d)) the current decision and its implied assignments in line 12.

7.6 From a Fragment of QPL to PL

For defining a translation from a fragment of QPL to propositional logic, we need to know how to compute a CNF formula which is equivalent to a given propositional formula φ. In order to do this we use a version of the algorithm *toCNF* [105], which is known is a *cube reduction*. We refer the reader to [53, 72], where alternative solutions can be found. We present the algorithm `equCNF`, which is a slight modification of the `SAT` algorithm, but in fact it could be presented in a general way, abstracting away from the specific realization of SAT.

Assume that φ is an input formula. Initially, the algorithm `equCNF` (see Fig. 7.28) builds a satisfying assignment for the formula

$$toCNF(\varphi) \wedge \neg l_{\varphi},$$

i.e., the assignment which falsifies φ. If such one is found, instead of terminating, the algorithm constructs a new clause that is in conflict with the current assignment (i.e., it rules out the satisfying assignment). This new clause is called a *blocking clause* and it has to meet the following properties:

- it contains only input variables, i.e., the propositional variables of $PV(\varphi)$,
- it is false in the current assignment $\mathbf{A}$,
- it is implied by $toCNF(\varphi) \wedge l_{\varphi}$.

In the algorithm, each time a satisfying assignment is obtained, a blocking clause is generated by the algorithm `blocking_clause` and added to the working set of clauses χ (line 9) and then in line 10 to the formula ψ. This clause rules out a set of cases where φ is false. Thus, on termination, when there is no satisfying assignment for the current set of clauses, ψ is a conjunction of the blocking clauses and precisely characterises φ.

A blocking clause could in principle be generated using the conflict-based learning procedure. However, we require the blocking clause to contain only input variables, i.e., literals of $PV(\varphi)$. To this aim one could use an (alternative) implication graph [105], in which all the roots are input literals. However, here we show another method introduced by Szreter [152, 153]. Before we go into details of this method, let us consider the simplest approach to generating blocking clauses. Assume that the procedure `SAT(`ψ`)` returned a ψ-total satisfying assignment $\mathbf{A}$ of ψ. Then, a blocking clause c_b could be defined as the disjunction of literals l corresponding to $PV(\varphi)$ such that $\wp$ ($\neg\wp$) is an element of c_b iff $\mathbf{A}(\wp) = false$ ($\mathbf{A}(\wp) = true$, respectively). Formally,

$$c_b = \bigvee_{\wp \in PV(\varphi)} N(\wp),$$

where $N(\wp) = \wp$ iff $\mathbf{A}(\wp) = false$ and $N(\wp) = \neg\wp$ iff $\mathbf{A}(\wp) = true$. This is clearly very simple, but highly inefficient. It is easy to notice that this

---

INPUT ARGUMENTS:
 a propositional formula φ
GLOBAL VARIABLES:
 χ: a set of clauses over $2^{PV(toCNF(\varphi))}$
 $\mathbf{A}$: $2^{PV(toCNF(\varphi))}$
RETURN VALUES:
 equCNF(): the formulas in CNF over $2^{PV(\varphi)}$
 SAT(): $2^{PV(\varphi)}$
 blocking-clause(): the clauses over $2^{PV(\varphi)}$

```
function equCNF(φ) is
begin
    χ := ∅;
    ψ := toCNF(φ) ∧ l_¬φ;
    while true do
        A := SAT(ψ);
        if A = ∅ then return χ; end if;
        c_b := blocking_clause(A);
        χ := χ ∪ {c_b};
        ψ := ψ ∧ c_b;
    end do;
end equCNF;
```

Fig. 7.28. A generic `equCNF` algorithm

way we could likely generate exponentially many blocking clauses, each one of the length equal to the number of propositions of φ. So, the question is how to generate shorter blocking clauses, which would block more than one assignment at the same time.

The idea is to represent the formula φ by a directed acyclic graph (*dag*), where the leaves are its literals, the other nodes correspond to the boolean operators, while the edges point to the arguments of these operators. Rather than formalizing the above intuitive description, we show in Fig. 7.29 a dag representing the formula

$$\varphi = \wp_1 \vee \psi,$$

where

$$\psi = \wp_2 \wedge \wp_3 \wedge (\wp_4 \Leftrightarrow \wp_5).$$

Then, for a given assignment $\mathbf{A}$, this dag of φ is searched with the DFS algorithm in order to identify a minimal number of its leaves (i.e., subformulas of φ being literals), which under the assignment $\mathbf{A}$ make already φ evaluate to $false$. The search is guided by properties of the boolean operators. That is, if a node corresponding to $\wedge$ is considered and one of its descendants is $false$,

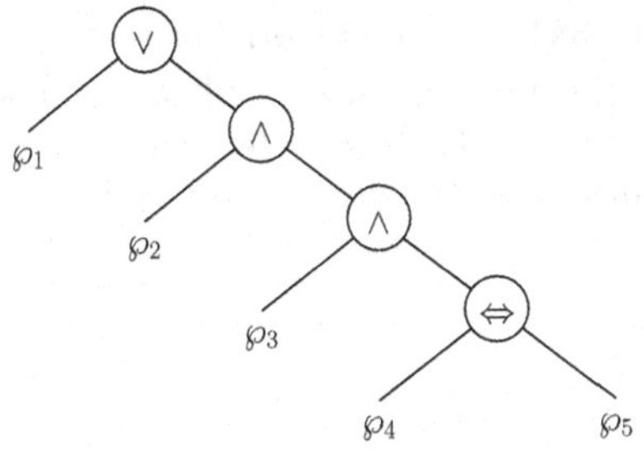

Fig. 7.29. A dag for the formula $\varphi = \wp_1 \vee (\wp_2 \wedge \wp_3 \wedge (\wp_4 \Leftrightarrow \wp_5))$

then the whole conjunction is *false* irrespectively on the other descendant. Similarly, if a node corresponding to $\vee$ is considered and one of its descendants is *true*, then the whole disjunction is *true* irrespectively on the other descendant. Again, rather than formalizing the above algorithm, we show how it works on a simple formula:

Example 7.21. Consider again the formula

$$\varphi = \wp_1 \vee \psi,$$

where

$$\psi = \wp_2 \wedge \wp_3 \wedge (\wp_4 \Leftrightarrow \wp_5).$$

For the assignment

$$\mathbf{A} = \{\neg\wp_1, \neg\wp_2, \neg\wp_3, \wp_4, \wp_5\}$$

found by the algorithm `SAT(`φ`)`, the algorithm searching through the dag of φ identifies the blocking clause $\wp_1 \vee \wp_2$ (see Fig. 7.30). This follows from the fact that φ evaluates to *false* when $\wp_1$ is *false* and ψ is *false*, but for ψ to evaluate to *false* it is sufficient when $\wp_2$ is *false*.

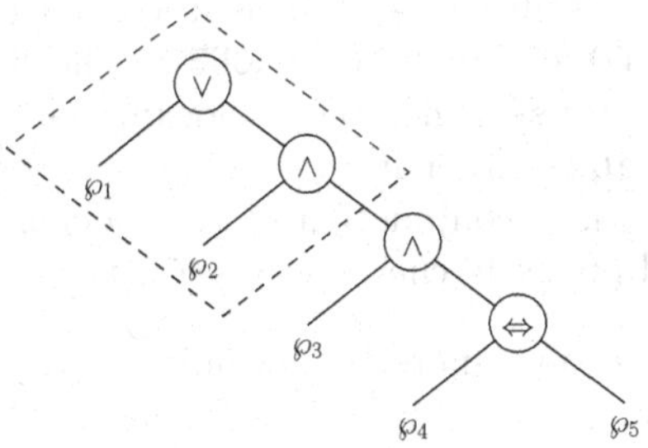

Fig. 7.30. Identifying the blocking clause $\wp_1 \vee \wp_2$

For the assignment

$$\mathbf{A} = \{\neg\wp_1, \wp_2, \neg\wp_3, \neg\wp_4, \wp_5\}$$

found by the algorithm `SAT(`φ`)`, the algorithm searching through the dag of φ identifies the blocking clause $\wp_1 \vee \wp_3$ (see Fig. 7.31). Similarly, this follows from the fact that φ evaluates to *false* when $\wp_1$ is *false* and ψ is *false*, but for ψ to evaluate to *false* it is sufficient when $\wp_3$ is *false*.

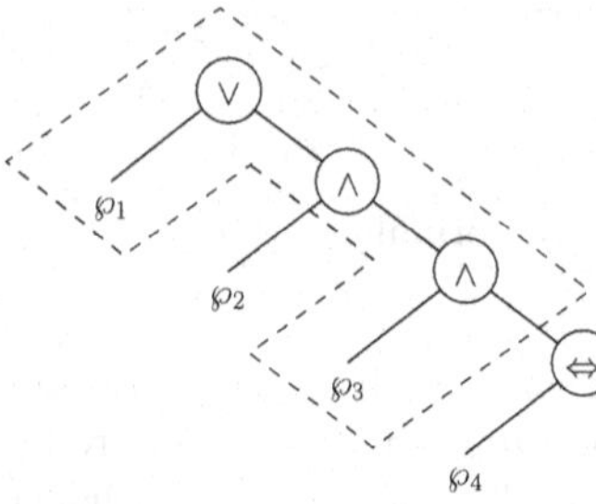

Fig. 7.31. Identifying the blocking clause $\wp_1 \vee \wp_3$

We do not analyse the other satisfying assignments and the resulting blocking clauses, but show the final result.

$$equCNF(\varphi) = (\wp_1 \vee \wp_2) \wedge (\wp_1 \vee \wp_3) \wedge (\wp_1 \vee \neg\wp_4 \vee \wp_5) \wedge (\wp_1 \vee \wp_4 \vee \neg\wp_5).$$

□

Now our aim is to compute a propositional formula equivalent to a given QPL formula containing only universal quantifiers.

We will use the notation $\forall\overline{\wp}.\varphi$, where $\overline{\wp} = (\wp[1], \ldots, \wp[m])$ is a vector of propositional variables, to denote $\forall\wp[1].\forall\wp[2]\ldots\forall\wp[m].\varphi$. What is important here, is that for a given QPL formula $\forall\overline{\wp}.\varphi$, we construct a CNF formula equivalent to it by using a version the algorithm `forall` [105]. This algorithm constructs a formula ψ equivalent to $\forall\overline{\wp}.\varphi$ and eliminates the quantified variables on-the-fly. This is sufficient since ψ is in conjunctive normal form. The algorithm `forall(`$\varphi, \overline{\wp}$`)` differs from `equCNF` in the line 8 only, where the procedure `blocking_clause` generates a blocking clause and deprives it of the propositional variables either from $\overline{\wp}$ or the negation of these. On termination, the formula ψ is a conjunction of the blocking clauses without the propositions of $\overline{\wp}$ and precisely characterises $\forall\overline{\wp}.\varphi$.

Example 7.22. Consider once more the formula $\varphi = \wp_1 \vee \psi$, where $\psi = \wp_2 \wedge \wp_3 \wedge (\wp_4 \Leftrightarrow \wp_5)$. Then,

$$forall(\varphi, (\wp_1)) = (\wp_2) \wedge (\wp_3) \wedge (\neg\wp_4 \vee \wp_5) \wedge (\wp_5 \vee \neg\wp_6).$$

□

7.7 Remaining Symbolic Approaches - Overview

A standard symbolic approach to representation of state spaces and model checking of untimed systems is based on Binary Decision Diagrams (BDDs) [50]. A similar approach is to apply BDDs to encoding discretizations of TA using the so-called Numeric Decision Diagrams (NDDs) [18]. Discretizations of TA can be also implemented using propositional formulas (see Sect. 7.1).

Another approach follows the solution suggested by Henzinger et al. [84], where the characteristic function of a set of states is a formula in separation logic (SL). SL formulas can be represented using Difference Decision Diagrams (DDDs) [107,108]. A DDD is a data structure using separation predicates with the ordering of predicates induced by the ordering of the clocks. A similar approach is taken in [140], where a translation from quantified SL to quantified boolean logic [151] is exploited.

One can use also Clock Difference Diagrams (CDDs) to symbolically represent unions of regions [24, 99]. Each node of a CDD is labelled with the difference of clock variables, whereas the outgoing edges with intervals bounding this difference. Alternatively, model checkers are based on data structures called Clock Restriction Diagrams (CRD) [168]. CRD is like CDD except for the fact that for each node the outgoing edges are labelled with an upper bound rather than with the interval of the corresponding difference of clock variables.

7.8 Selected Tools

There are many tools using the approaches considered so far. Below, we list some of them and give pointers to the literature, where more detailed descriptions can be found.

- **HyTech** [82] is an automatic tool for the analysis of embedded systems. The tool computes the condition under which a linear hybrid system satisfies a temporal requirement. Hybrid systems are specified as collections of automata with discrete and continuous components, and temporal requirements are verified by symbolic model checking. If the verification fails, a diagnostic error trace is generated. Real-time requirements are specified in the logic TCTL and its modification - ICTL (*Integrator Computation Tree Logic*), used to specify safety, liveness, time-bounded and duration requirements of hybrid automata. Verification is performed by a successive approximation of the set of states satisfying the formula to be checked, by iterating boolean operations and weakest-precondition operations on regions (see [14]).
- **NuSMV** [56] (an extension of SMV [104]) is a symbolic model checker which combines BDD-based model checking and SAT-based model checking. The tool offers an analysis of invariants (on-the-fly for reachability

analysis), partitioning methods, and model checking for CTL Real-Time CTL, and LTL extended with past operators. Moreover, Bounded Model Checking for LTL and for checking invariants is available.
- **Rabbit** [36] - a tool for BDD-based verification of real-time systems, developed for an extension of TA, called Cottbus Timed Automata, and providing reachability analysis.
- **Red** [167] is a fully symbolic model checker based on data structures called Clock Restriction Diagrams (CRD) [168]. It supports TCTL model checking and backward reachability analysis.
- **UppAal2k** [121] (a successor of UppAal [100]) is a tool for checking reachability and for verification of properties expressible in a subset of TCTL. Many optimizations are implemented, e.g. application of Clock Difference Diagrams (CDDs) to symbolically represent unions of clock regions [24]. Moreover, cost-optimal search based on *Priced Timed Automata*, and parameterized verification are available as well [17].
- **VerICS** [65] - implements SAT-based BMC and UMC for verifying TCTL and reachability for timed automata and Estelle programs.

Further Reading

Some symbolic approaches to verification of time Petri Nets and timed automata are discussed in [54]. A survey on most recent approaches to SAT-based formal verification, also combined with BDDs, can be found in [131].

An overview of model checking techniques and tools for verifying timed automata can be found in [28].

Concluding Remarks and Future Research Directions

Time-dependent computer systems are used for controlling traffic lights, aircraft navigations, power-stations, nuclear submarines as well as tomographs and other sophisticated medical equipment. Usually, we believe that computer systems are less prone to errors than a human being operating the same system or machine. Unfortunately, the number of mistakes in software is growing with the number of the code lines. Moreover, it is clear that potential errors could result in fatal consequences for both people and hardware.

Formal methods are used in order to enhance reliability of time-dependent computer systems. It is very well known that proving full correctness of a system is a difficult and frequently unachievable task. Therefore, the formal methods aim mostly at detecting and eliminating errors. In principle, one could use all the formal methods like testing, deductive verification, and model checking in the verification process. This would increase the probability of finding an error, or when no error is found, the probability that the system is really correct. It is very interesting and tempting to combine tools based on fully automated methods like model checking with methods based on theorem proving or/and deductive verification.

Since model checking methods are usually applied to finite models, which does not mean that infinite models cannot be verified, it is important to investigate methods for obtaining finite abstract models and deductive methods for proving that correctness can be lifted from finite abstractions to actual time-dependent systems. It seems that even verification over finite models can still be improved. There are several ways of attacking this problem for models generated for timed systems. First of all, several reduction techniques known for untimed system verification, e.g. partial order and symmetry reductions, can be applied on-the-fly in the process of verification. Secondly, the existing symbolic verication methods, based on variants of BDDs and on SAT, can be combined. Potentially, there is still room for new symbolic methods known for solving other PSPACE-hard problems known in complexity theory.

Another important issue concerns methods for formal description techniques. They include higher-order languages like SDL, LOTOS, ESTELLE, and Timed-UML. The languages usually provide visual and textual representation, and quite recently several methods for translating from these formalisms to networks of timed automata and time Petri nets have been developed. It is more than necessary to combine these translations with different reduction techniques, e.g. slicing, in order to generate timed systems, automated verification of which would be feasible. It is obviously a trade-off between the expressiveness of higher-order languages and the efficiency of the translating methods.

Future research directions will include new areas of application of verification methods. Obviously, the most promising and exciting area of applications is connected with web-services and electronic commerce. Proving correctness of new internet protocols as well as of web-services will be soon unavoidable. Moreover, since security of internet protocols is frequently a source of fear of the suppliers and the customers in the e-commerce transactions, each party in a serious transaction has to be confident about several security properties. Methods developed for timed systems can be extended and applied to verify such properties. These will call for improving the existing verification tools as well as for the construction of new verification methods and tools.

References

1. W. van der Aalst. Interval timed coloured Petri nets and their analysis. In *Proc. of the 14th Int. Conf. on Applications and Theory of Petri Nets (ICATPN'93)*, volume 961 of *LNCS*, pages 452–472. Springer-Verlag, 1993.
2. M. Abadi and L. Lamport. An old-fashioned recipe for real time. In *REX workshop on Real-Time: Theory in Practice*, volume 600 of *LNCS*, pages 1–27. Springer-Verlag, 1991.
3. P. A. Abdulla and A. Nylén. Timed Petri nets and BQOs. In *Proc. of the 22nd Int. Conf. on Applications and Theory of Petri Nets (ICATPN'01)*, volume 2075 of *LNCS*, pages 53–70. Springer-Verlag, 2001.
4. R. Alur. Timed automata. NATO-ASI 1998 Summer School on Verification of Digital and Hybrid Systems; http://www.cis.upenn.edu/~alur/onlinepub.html, 1998.
5. R. Alur. Timed automata. In *Proc. of the 11th Int. Conf. on Computer Aided Verification (CAV'99)*, volume 1633 of *LNCS*, pages 8–22. Springer-Verlag, 1999.
6. R. Alur, C. Courcoubetis, and D. Dill. Model checking for real-time systems. In *Proc. of the 5th Symp. on Logic in Computer Science (LICS'90)*, pages 414–425. IEEE Computer Society, 1990.
7. R. Alur, C. Courcoubetis, and D. Dill. Model checking in dense real-time. *Information and Computation*, 104(1):2–34, 1993.
8. R. Alur, C. Courcoubetis, D. Dill, N. Halbwachs, and H. Wong-Toi. An implementation of three algorithms for timing verification based on automata emptiness. In *Proc. of the 13th IEEE Real-Time Systems Symposium (RTSS'92)*, pages 157–166. IEEE Computer Society, 1992.
9. R. Alur, C. Courcoubetis, D. Dill, N. Halbwachs, and H. Wong-Toi. Minimization of timed transition systems. In *Proc. of the 3rd Int. Conf. on Concurrency Theory (CONCUR'92)*, volume 630 of *LNCS*, pages 340–354. Springer-Verlag, 1992.
10. R. Alur and D. Dill. Automata for modelling real-time systems. In *Proc. of the 17th Int. Colloquium on Automata, Languages and Programming (ICALP'90)*, volume 443 of *LNCS*, pages 322–335. Springer-Verlag, 1990.
11. R. Alur and D. Dill. A theory of timed automata. *Theoretical Computer Science*, 126(2):183–235, 1994.

12. R. Alur and T. Henzinger. Logics and models of real time: A survey. In *Proc. of REX Workshop 'Real Time: Theory and Practice'*, volume 600 of *LNCS*, pages 74–106. Springer-Verlag, 1992.
13. R. Alur and T. Henzinger. Modularity for timed and hybrid systems. In *Proc. of the 8th Int. Conf. on Concurrency Theory (CONCUR'97)*, volume 1243 of *LNCS*, pages 74–88. Springer-Verlag, 1997.
14. R. Alur, T. Henzinger, and P. Ho. Automatic symbolic verification of embedded systems. *IEEE Trans. on Software Eng.*, 22(3):181–201, 1996.
15. R. Alur and R. Kurshan. Timing analysis in COSPAN. In *Hybrid Systems III*, volume 1066 of *LNCS*, pages 220–231. Springer-Verlag, 1996.
16. C. Amer-Yahia, N. Zerhouni, A. El Moundi, and M. Ferney. On finding deadlocks and traps in Petri nets. In *Proc. of System Analysis-Modelling-Simulation (SAMS'99)*, pages 495–507, 1999.
17. T. Amnell, G. Behrmann, J. Bengtsson, P. D'Argenio, A. David, A. Fehnker, T. Hune, B. Jeannet, K. G. Larsen, M. O. Möller, P. Pettersson, C. Weise, and W. Yi. UPPAAL - now, next, and future. In *Proc. of the 4th Summer School 'Modelling and Verification of Parallel Processes' (MOVEP'00)*, volume 2067 of *LNCS*, pages 99–124. Springer-Verlag, 2001.
18. E. Asarin, M. Bozga, A. Kerbrat, O. Maler, A. Pnueli, and A. Rasse. Data-structures for the verification of timed automata. In *Proc. of Int. Workshop on Hybrid and Real-Time Systems (HART'97)*, volume 1201 of *LNCS*, pages 346–360. Springer-Verlag, 1997.
19. G. Audemard, P. Bertoli, A. Cimatti, A. Kornilowicz, and R. Sebastiani. A SAT based approach for solving formulas over boolean and linear mathematical propositions. In *Proc. of the 18th Int. Conf. on Automated Deduction (CADE'02)*, volume 2392 of *LNCS*, pages 195–210. Springer-Verlag, 2002.
20. G. Audemard, A. Cimatti, A. Kornilowicz, and R. Sebastiani. Bounded model checking for timed systems. In *Proc. of the 22nd Int. Conf. on Formal Techniques for Networked and Distributed Systems (FORTE'02)*, volume 2529 of *LNCS*, pages 243–259. Springer-Verlag, 2002.
21. K. Barkaoui and J-F. Pradat-Peyre. Verification in concurrent programming with Petri nets structural techniques. In *Proc. of the 3rd IEEE Symp. on High-Assurance Systems (HASE'98)*, pages 124–133. IEEE Computer Society, November 1998.
22. G. Behrmann, P. Bouyer, E. Fleury, and K. G. Larsen. Static guard analysis in timed automata verification. In *Proc. of the 9th Int. Conf. on Tools and Algorithms for the Construction and Analysis of Systems (TACAS'03)*, volume 2619 of *LNCS*, pages 254–277. Springer-Verlag, 2003.
23. G. Behrmann, P. Bouyer, K. G. Larsen, and R. Pelánek. Lower and upper bounds in zone based abstractions of timed automata. In *Proc. of the 10th Int. Conf. on Tools and Algorithms for the Construction and Analysis of Systems (TACAS'04)*, volume 2988 of *LNCS*, pages 312–326. Springer-Verlag, 2004.
24. G. Behrmann, K. G. Larsen, J. Pearson, C. Weise, and W. Yi. Efficient timed reachability analysis using Clock Difference Diagrams. In *Proc. of the 11th Int. Conf. on Computer Aided Verification (CAV'99)*, volume 1633 of *LNCS*, pages 341–353. Springer-Verlag, 1999.
25. J. Bengtsson. *Clocks, DBMs and States in Timed Systems.* PhD thesis, Dept. of Information Technology, Uppsala University, 2002.
26. J. Bengtsson and W. Yi. On clock difference constraints and termination in reachability analysis in timed automata. In *Proc. of the 5th Int. Conf. on*

Formal Methods and Software Engineering (ICFEM'03), volume 2885 of *LNCS*, pages 491–503. Springer-Verlag, 2003.
27. J. Bengtsson and W. Yi. Timed automata: Semantics, algorithms and tools. In *Lectures on Concurrency and Petri Nets: Advances in Petri Nets*, volume 3098 of *LNCS*, pages 87–124. Springer-Verlag, 2004.
28. B. Bérard, M. Bidoit, A. Finkel, F. Laroussinie, A. Petit, L. Petrucci, P. Schnoebelen, and P. McKenzie. *Systems and Software Verification: Model-Checking Techniques and Tools.* Springer-Verlag, 2001.
29. B. Bérard, F. Cassez, S. Haddad, D. Lime, and O. H. Roux. Comparison of the expressiveness of timed automata and time Petri nets. In *Proc. of the 3rd Int. Workshop on Formal Analysis and Modeling of Timed Systems (FORMATS'05)*, volume 3829 of *LNCS*, pages 211–225. Springer-Verlag, 2005.
30. O. Bernholtz, M. Vardi, and P. Wolper. An automata-theoretic approach to branching-time model checking. In *Proc. of the 6th Int. Conf. on Computer Aided Verification (CAV'94)*, volume 818 of *LNCS*, pages 142–155. Springer-Verlag, 1994.
31. B. Berthomieu. Private communnication.
32. B. Berthomieu and M. Diaz. Modeling and verification of time dependent systems using time Petri nets. *IEEE Trans. on Software Eng.*, 17(3):259–273, 1991.
33. B. Berthomieu and M. Menasche. An enumerative approach for analyzing time Petri nets. In *Proc. of the 9th IFIP World Computer Congress*, volume 9 of *Information Processing*, pages 41–46. North Holland/ IFIP, September 1983.
34. B. Berthomieu, P-O. Ribet, and F. Vernadat. The tool TINA - construction of abstract state spaces for Petri nets and time Petri nets. *International Journal of Production Research*, 42(14), 2004.
35. B. Berthomieu and F. Vernadat. State class constructions for branching analysis of time Petri nets. In *Proc. of the 9th Int. Conf. on Tools and Algorithms for the Construction and Analysis of Systems (TACAS'03)*, volume 2619 of *LNCS*, pages 442–457. Springer-Verlag, 2003.
36. D. Beyer. Rabbit: Verification of real-time systems. In *Proc. of the Workshop on Real-Time Tools (RT-TOOLS'01)*, pages 13–21, 2001.
37. B. Bieber and H. Fleischhack. Model checking of time Petri nets based on partial order semantics. In *Proc. of the 10th Int. Conf. on Concurrency Theory (CONCUR'99)*, volume 1664 of *LNCS*, pages 210–225. Springer-Verlag, 1999.
38. A. Biere, A. Cimatti, E. Clarke, O. Strichman, and Y. Zhu. Bounded model checking. In *Highly Dependable Software*, volume 58 of *Advances in Computers*. Academic Press, 2003. Pre-print.
39. A. Bobbio and A. Horváth. Model checking time Petri nets using NuSMV. In *Proc. of the 5th Int. Workshop on Performability Modeling of Computer and Communication Systems (PMCCS5)*, pages 100–104, September 2001.
40. A. Bouajjani, J-C. Fernandez, N. Halbwachs, P. Raymond, and C. Ratel. Minimal state graph generation. *Science of Computer Programming*, 18:247–269, 1992.
41. A. Bouajjani, S. Tripakis, and S. Yovine. On-the-fly symbolic model checking for real-time systems. In *Proc. of the 18th IEEE Real-Time Systems Symposium (RTSS'97)*, pages 232–243. IEEE Computer Society, 1997.
42. A. Bouali, A. Ressouche, V. Roy, and R. de Simone. The FC2Tools set. In *Proc. of the 8th Int. Conf. on Computer Aided Verification (CAV'96)*, volume 1102 of *LNCS*, pages 441–445. Springer-Verlag, 1996.

43. H. Boucheneb and G. Berthelot. Towards a simplified building of time Petri nets reachability graph. In *Proc. of the 5th Int. Workshop on Petri Nets and Performance Models*, pages 46–55, October 1993.
44. H. Boucheneb and R. Hadjidj. CTL* model checking for time Petri nets. *Theoretical Computer Science*, 2006. http://www/sciencedirect.com.
45. P. Bouyer. Timed automata may cause some troubles. Technical Report LSV-02-9, ENS de Cachan, Cachan, France, July 2003.
46. F. D. J. Bowden. Modelling time in Petri nets. In *Proc. of the 2nd Australia-Japan Workshop on Stochastic Models (STOMOD'96)*, July 1996.
47. F. D. J. Bowden. A brief survey and synthesis of the roles of time in Petri nets. *Mathematical and Computer Modelling*, 31(10-12):55–68, 2000.
48. M. Bozga, O. Maler, A. Pnueli, and S. Yovine. Some progress in the symbolic verification of timed automata. In *Proc. of the 9th Int. Conf. on Computer Aided Verification (CAV'97)*, volume 1254 of *LNCS*, pages 179–190. Springer-Verlag, 1997.
49. M. C. Browne, E. Clarke, and O. Grumberg. Characterizing finite Kripke structures in propositional temporal logic. *Theoretical Computer Science*, 59(1/2):115–131, 1988.
50. R. Bryant. Graph-based algorithms for boolean function manipulation. *IEEE Transaction on Computers*, 35(8):677–691, 1986.
51. F. Cassez. Private communnication.
52. F. Cassez and O. H. Roux. Structural translation of time Petri nets to timed automata. In *Proc. of the 4th Int. Workshop on Automated Verification of Critical Systems (AVoCS'04)*, volume 128(6) of *ENTCS*, pages 145–160. Elsevier, 2005.
53. P. Chauhan, E. Clarke, and D. Kroening. Using SAT-based image computation for reachability analysis. Technical Report CMU-CS-03-151, Carnegie Mellon University, July 2003.
54. A. M. K. Cheng. *Real-Time Systems: Scheduling, Analysis, and Verification.* John Wiley & Sons, 2002.
55. S. Christensen, J. Jørgensen, and L. Kristensen. Design/CPN - a computer tool for coloured Petri nets. In *Proc. of the 3rd Int. Conf. on Tools and Algorithms for the Construction and Analysis of Systems (TACAS'97)*, volume 1217 of *LNCS*, pages 209–223. Springer-Verlag, 1997.
56. A. Cimatti, E. Clarke, E. Giunchiglia, F. Giunchiglia, M. Pistore, M. Roveri, R. Sebastiani, and A. Tacchella. NuSMV2: An open-source tool for symbolic model checking. In *Proc. of the 14th Int. Conf. on Computer Aided Verification (CAV'02)*, volume 2404 of *LNCS*, pages 359–364. Springer-Verlag, 2002.
57. E. Clarke, O. Grumberg, and D. Peled. *Model Checking.* MIT Press, 1999.
58. F. Commoner. Deadlocks in Petri nets. Technical Report CA-7206-2311, Massachusetts Computer Associates, Wakefield, Mass., June 1972.
59. J. Coolahan and N. Roussopoulos. Timing requirements for time-driven systems using augmented Petri nets. *IEEE Trans. on Software Eng.*, SE-9(5):603–616, 1983.
60. L. A. Cortés, P. Eles, and Z. Peng. Modeling and formal verification of embedded systems based on a Petri net representation. *Journal of Systems Architecture*, 49(12-15):571–598, 2003.
61. M. Davis, G. Logemann, and D. Loveland. A machine program for theorem proving. *Journal of the ACM*, 5(7):394–397, 1962.

62. M. Davis and H. Putnam. A computing procedure for quantification theory. *Journal of the ACM*, 7(3):201–215, 1960.
63. C. Daws and S. Tripakis. Model checking of real-time reachability properties using abstractions. In *Proc. of the 4th Int. Conf. on Tools and Algorithms for the Construction and Analysis of Systems (TACAS'98)*, volume 1384 of *LNCS*, pages 313–329. Springer-Verlag, 1998.
64. C. Daws and S. Yovine. Reducing the number of clock variables of timed automata. In *Proc. of the IEEE Real-Time Systems Symposium (RTSS'96)*, pages 73–81. IEEE Computer Society, 1996.
65. P. Dembiński, A. Janowska, P. Janowski, W. Penczek, A. Półrola, M. Szreter, B. Woźna, and A. Zbrzezny. VerICS: A tool for verifying timed automata and Estelle specifications. In *Proc. of the 9th Int. Conf. on Tools and Algorithms for the Construction and Analysis of Systems (TACAS'03)*, volume 2619 of *LNCS*, pages 278–283. Springer-Verlag, 2003.
66. P. Dembiński, W. Penczek, and A. Półrola. Verification of timed automata based on similarity. *Fundamenta Informaticae*, 51(1-2):59–89, 2002.
67. M. Dickhofer and T. Wilke. Timed alternating tree automata: The automata-theoretic solution to the TCTL model checking problem. In *Proc. of the 26th Int. Colloquium on Automata, Languages and Programming (ICALP'99)*, volume 1664 of *LNCS*, pages 281–290. Springer-Verlag, 1999.
68. D. Dill. Timing assumptions and verification of finite state concurrent systems. In *Automatic Verification Methods for Finite-State Systems*, volume 407 of *LNCS*, pages 197–212. Springer-Verlag, 1989.
69. E. A. Emerson. *Handbook of Theoretical Computer Science*, volume B: Formal Methods and Semantics, chapter Temporal and Modal Logic, pages 995–1067. Elsevier, 1990.
70. E. A. Emerson and C-L. Lei. Efficient model checking in fragments of the propositional mu-calculus. In *Proc. of the 1st Symp. on Logic in Computer Science (LICS'86)*, pages 267–278. IEEE Computer Society, 1986.
71. M. Fitting. *First-Order Logic and Automated Theorem Proving.* Springer-Verlag, 1990.
72. M. Ganai, A. Gupta, and P. Ashar. Efficient SAT-based unbounded symbolic model checking using circuit cofactoring. In *Proc. of the Int. Conf. on Computer-Aided Design (ICCAD'04)*, pages 510–517, 2004.
73. G. Gardey, O. H. Roux, and O. F. Roux. Using zone graph method for computing the state space of a time Petri net. In *Proc. of the 1st Int. Workshop on Formal Analysis and Modeling of Timed Systems (FORMATS'03)*, volume 2791 of *LNCS*, pages 246–259. Springer-Verlag, 2004.
74. R. Gawlick, R. Segala, J. Søgaard-Andersen, and N. Lynch. Liveness in timed and untimed systems. In *Proc. of the 21st Int. Colloquium on Automata, Languages and Programming (ICALP'94)*, volume 820 of *LNCS*, pages 166–177. Springer-Verlag, 1994.
75. R. Goldblatt. *Logics of Time and Computation.* CSLI Lecture Notes. CSLI Publications, Stanford University, 1992.
76. A. Göllü, A. Puri, and P. Varaiya. Discretization of timed automata. In *Proc. of the 33rd IEEE. Conf. on Decision and Control (CDC'94)*, pages 957–958, 1994.
77. Z. Gu and K. Shin. Analysis of event-driven real-time systems with time Petri nets. In *Proc. of the Int. Conf. on Design and Analysis of Distributed and

Embedded Systems (DIPES'02), volume 219 of *IFIP Conference Proceedings*, pages 31–40. Kluwer, 2002.
78. S. Haar, L. Kaiser, F. Simonot-Lion, and J. Toussaint. On equivalence between timed state machines and time Petri nets. Technical Report RR-4049, INRIA Rhône-Alpes, 655, avenue de l'Europe, 38330 Montbonnot-St-Martin, November 2000.
79. R. Hadjidj and H. Boucheneb. Much compact time Petri net state class spaces useful to restore CTL* properties. In *Proc. of the 5th Int. Conf. on Application of Concurrency to System Design (ACSD'05)*, pages 224–233. IEEE Computer Society, 2005.
80. H-M. Hanisch. Analysis of place/transition nets with timed arcs and its application to batch process control. In *Proc. of the 14th Int. Conf. on Applications and Theory of Petri Nets (ICATPN'93)*, volume 691 of *LNCS*, pages 282–299. Springer-Verlag, 1993.
81. T. Henzinger. It's about time: Real-time logics reviewed. In *Proc. of the 9th Int. Conf. on Concurrency Theory (CONCUR'98)*, volume 1466 of *LNCS*, pages 439–454. Springer-Verlag, 1998.
82. T. Henzinger and P. Ho. HyTech: The Cornell hybrid technology tool. In *Hybrid Systems II*, volume 999 of *LNCS*, pages 265–293. Springer-Verlag, 1995.
83. T. Henzinger, X. Nicollin, J. Sifakis, and S. Yovine. Symbolic model checking for real-time systems. In *Proc. of the 7th Symp. Logics in Computer Science (LICS'92)*, pages 394–406. IEEE Computer Society, 1992.
84. T. Henzinger, X. Nicollin, J. Sifakis, and S. Yovine. Symbolic model checking for real-time systems. *Information and Computation*, 111(2):193–224, 1994.
85. G. J. Holzmann. The model checker SPIN. *IEEE Trans. on Software Eng.*, 23(5):279–295, 1997.
86. G. E. Hughes and M. J. Cresswell. *An Introduction to Modal Logic*. Methuen, 1968.
87. G. E. Hughes and M. J. Cresswell. *A Companion to Modal Logic*. Methuen, 1984.
88. M. Huhn, P. Niebert, and F. Wallner. Verification based on local states. In *Proc. of the 4th Int. Conf. on Tools and Algorithms for the Construction and Analysis of Systems (TACAS'98)*, volume 1384 of *LNCS*, pages 36–51. Springer-Verlag, 1998.
89. H. Hulgaard and S. M. Burns. Efficient timing analysis of a class of Petri nets. In *Proc. of the 7th Int. Conf. on Computer Aided Verification (CAV'95)*, volume 939 of *LNCS*, pages 923–936. Springer-Verlag, 1995.
90. M. Huth and M. Ryan. *Logic in Computer Science: Modelling and Reasoning about Systems*. Cambridge University Press, 2004.
91. R. Janicki. Nets, sequential components and concurrency relations. *Theoretical Computer Science*, 29:87–121, 1984.
92. K. Jensen. *Coloured Petri Nets: Basic Concepts, Analysis Methods and Practical Use*. Monographs in Theoretical Computer Science. Springer-Verlag, 1995/96.
93. N. D. Jones, L. H. Landweber, and Y. E. Lien. Complexity of some problems in Petri nets. *Theoretical Computer Science*, 4(3):277–299, 1977.
94. I. Kang and I. Lee. An efficient state space generation for the analysis of real-time systems. In *Proc. of Int. Symposium on Software Testing and Analysis*, 1996.

95. D. Kozen. Results on the propositional mu-calculus. *Theoretical Computer Science*, 27:333–354, 1983.
96. O. Kupferman, T. Henzinger, and M. Vardi. A space-efficient on-the-fly algorithm for real-time model checking. In *Proc. of the 7th Int. Conf. on Concurrency Theory (CONCUR'96)*, volume 1119 of *LNCS*, pages 514–529. Springer-Verlag, 1996.
97. O. Kupferman, M. Vardi, and P. Wolper. An automata-theoretic approach to branching-time model checking. *Journal of the ACM*, 47(2):312–360, 2000.
98. K. G. Larsen, F. Larsson, P. Pettersson, and W. Yi. Efficient verification of real-time systems: Compact data structures and state-space reduction. In *Proc. of the 18th IEEE Real-Time System Symposium (RTSS'97)*, pages 14–24. IEEE Computer Society, 1997.
99. K. G. Larsen, J. Pearson, C. Weise, and W. Yi. Clock Difference Diagrams. *Nordic Journal of Computing*, 6(3):271–298, 1999.
100. K. G. Larsen, P. Pettersson, and W. Yi. UPPAAL in a nutshell. *International Journal of Software Tools for Technology Transfer*, 1(1/2):134–152, 1997.
101. D. Lee and M. Yannakakis. On-line minimization of transition systems. In *Proc. of the 24th ACM Symp. on the Theory of Computing*, pages 264–274, May 1992.
102. J. Lilius. Efficient state space search for time Petri nets. In *Proc. of MFCS Workshop on Concurrency, Brno'98*, volume 18 of *ENTCS*. Elsevier, 1999.
103. D. Lime and O. H. Roux. State class timed automaton of a time Petri net. In *Proc. of the 10th Int. Workshop on Petri Nets and Performance Models (PNPM'03)*. IEEE Computer Society, September 2003.
104. K. L. McMillan. The SMV system. Technical Report CMU-CS-92-131, Carnegie Mellon University, February 1992.
105. K. L. McMillan. Applying SAT methods in unbounded symbolic model checking. In *Proc. of the 14th Int. Conf. on Computer Aided Verification (CAV'02)*, volume 2404 of *LNCS*, pages 250–264. Springer-Verlag, 2002.
106. P. Merlin and D. J. Farber. Recoverability of communication protocols – implication of a theoretical study. *IEEE Trans. on Communications*, 24(9):1036–1043, 1976.
107. J. Møller, J. Lichtenberg, H. Andersen, and H. Hulgaard. Difference Decision Diagrams. In *Proc. of the 13th Int. Workshop Computer Science Logic (CSL'99)*, volume 1683 of *LNCS*, pages 111–125. Springer-Verlag, 1999.
108. J. Møller, J. Lichtenberg, H. Andersen, and H. Hulgaard. Fully symbolic model checking of timed systems using Difference Decision Diagrams. In *Proc. of the 2nd Federated Logic Conference (FLoC'99)*, volume 23(2) of *ENTCS*, 1999.
109. M. Moskewicz, C. Madigan, Y. Zhao, L. Zhang, and S. Malik. Chaff: Engineering an efficient SAT solver. In *Proc. of the 38th Design Automation Conference (DAC'01)*, pages 530–535, June 2001.
110. P. Niebert, M. Mahfoudh, E. Asarin, M. Bozga, O. Maler, and N. Jain. Verification of timed automata via satisfiability checking. In *Proc. of the 7th Int. Symp. on Formal Techniques in Real-Time and Fault Tolerant Systems (FTRTFT'02)*, volume 2469 of *LNCS*, pages 226–243. Springer-Verlag, 2002.
111. Y. Okawa and T. Yoneda. Symbolic CTL model checking of time Petri nets. *Electronics and Communications in Japan, Scripta Technica*, 80(4):11–20, 1997.
112. R. Paige and R. Tarjan. Three partition refinement algorithms. *SIAM Journal on Computing*, 16(6):973–989, 1987.

113. C. H. Papadimitriou. *Computational Complexity.* Addison Wesley, 1994.
114. D. Park. Concurrency and automata on infinite sequences. In *Proc. of the 5th GI Conf. on Theoretical Computer Science*, volume 104 of *LNCS*, pages 167–183. Springer-Verlag, 1981.
115. D. Peled. *Software Reliability Methods.* Springer-Verlag, 2001.
116. W. Penczek. Branching time and partial order in temporal logics. In L. Bolc and A. Szałas, editors, *Time and Logic: A Computational Approach*, pages 179–228. UCL Press Ltd., 1995.
117. W. Penczek and A. Półrola. Abstractions and partial order reductions for checking branching properties of time Petri nets. In *Proc. of the 22nd Int. Conf. on Applications and Theory of Petri Nets (ICATPN'01)*, volume 2075 of *LNCS*, pages 323–342. Springer-Verlag, 2001.
118. W. Penczek, A. Półrola, B. Woźna, and A. Zbrzezny. Bounded model checking for reachability testing in time Petri nets. In *Proc. of the Int. Workshop on Concurrency, Specification and Programming (CS&P'04)*, volume 170(1) of *Informatik-Berichte*, pages 124–135. Humboldt University, 2004.
119. W. Penczek, B. Woźna, and A. Zbrzezny. Bounded model checking for the universal fragment of CTL. *Fundamenta Informaticae*, 51(1-2):135–156, 2002.
120. W. Penczek, B. Woźna, and A. Zbrzezny. Towards bounded model checking for the universal fragment of TCTL. In *Proc. of the 7th Int. Symp. on Formal Techniques in Real-Time and Fault Tolerant Systems (FTRTFT'02)*, volume 2469 of *LNCS*, pages 265–288. Springer-Verlag, 2002.
121. P. Pettersson and K. G. Larsen. UPPAAL2k. *Bulletin of the European Association for Theoretical Computer Science*, 70:40–44, February 2000.
122. D. Plaisted, A. Biere, and Y. Zhu. A satisfiability procedure for quantified boolean formulae. *Discrete Applied Mathematics*, 130(2):291–328, 2003.
123. E. A. Pokozy. Toward verification of concurrent properties of time Petri nets. Preprint 61 of the A. P. Ershow Institute of Informatics Systems, Siberian Division of the Russian Academy of Sciences; http://www.iis.nsk.su/preprints/POKOZ/preprint/preprint_eng.html, 1999. In Russian.
124. A. Półrola and W. Penczek. Minimization algorithms for time Petri nets. *Fundamenta Informaticae*, 60(1-4):307–331, 2004.
125. A. Półrola, W. Penczek, and M. Szreter. Reachability analysis for timed automata using partitioning algorithms. *Fundamenta Informaticae*, 55(2):203–221, 2003.
126. A. Półrola, W. Penczek, and M. Szreter. Towards efficient partition refinement for checking reachability in timed automata. In *Proc. of the 1st Int. Workshop on Formal Analysis and Modeling of Timed Systems (FORMATS'03)*, volume 2791 of *LNCS*, pages 2–17. Springer-Verlag, 2004.
127. L. Popova. On time Petri nets. *Elektronische Informationsverarbeitung und Kybernetik*, 27(4):227–244, 1991.
128. L. Popova and S. Marek. TINA - a tool for analyzing paths in TPNs. In *Proc. of the Int. Workshop on Concurrency, Specification and Programming (CS&P'02)*, volume 110 of *Informatik-Berichte*, pages 195–196. Humboldt University, 1998.
129. L. Popova-Zeugmann. Essential states in time Petri nets. Informatik-Bericht 96, Humboldt University, 1998.
130. L. Popova-Zeugmann and D. Schlatter. Analyzing paths in time Petri nets. *Fundamenta Informaticae*, 37(3):311–327, 1999.

131. M. Prasad, A. Biere, and A. Gupta. A survey of recent advances in SAT-based formal verification. *International Journal of Software Tools for Technology Transfer*, 7(2):156–173, 2005.
132. V. R. Pratt. Two easy theories whose combination is hard. Memo sent to Nelson and Oppen concerning a preprint of their paper, available at http://boole.stanford.edu/abstracts.html, September 1977.
133. C. Ramchandani. Analysis of asynchronous concurrent systems by timed Petri nets. Technical Report MAC-TR-120, Massachusetts Institute of Technology, February 1974.
134. A. Ratzer, L. Wells, H. Lassen, M. Laursen, J. Qvortrup, M. Stissing, M. Westergaard, S. Christensen, and K. Jensen. CPN Tools for editing, simulating, and analyzing coloured Petri nets. In *Proc. of the 24th Int. Conf. on Applications and Theory of Petri Nets (ICATPN'03)*, volume 2679 of *LNCS*, pages 450–462. Springer-Verlag, 2003.
135. W. Reisig. *Petri Nets. An Introduction*, volume 4 of *EACTS Monographs on Theoretical Computer Science*. Springer-Verlag, 1985.
136. S. Roch and P. Starke. *INA: Integrated Net Analyser. Version 2.2*, 1999. Manual; http://www.informatik.hu-berlin.de/ ~starke/ina.html.
137. Romeo: A tool for time Petri net analysis. http://www.irccyn.ec-nantes.fr/irccyn/d/en/equipes/TempsReel/logs, 2000.
138. S. Samolej and T. Szmuc. Modelowanie systemów czasu rzeczywistego z zastosowaniem czasowych sieci Petriego. In *Mat. IX Konf. Systemy Czasu Rzeczywistego (SCR'02)*, pages 45–54. Instytut Informatyki Politechniki Śląskiej, 2002. In Polish.
139. P. Sénac, M. Diaz, and P. de Saqui Sannes. Toward a formal specification of multimedia scenarios. *Annals of Telecommunications*, 49(5-6):297–314, 1994.
140. S. Seshia and R. Bryant. Unbounded, fully symbolic model checking of timed automata using boolean methods. In *Proc. of the 15th Int. Conf. on Computer Aided Verification (CAV'03)*, volume 2725 of *LNCS*, pages 154–166. Springer-Verlag, 2003.
141. M. Sheeran, S. Singh, and G. Stålmarck. Checking safety properties using induction and a SAT-solver. In *Proc. of the Int. Conf. on Formal Methods in Computer-Aided Design (FMCAD'00)*, volume 1954 of *LNCS*, pages 108–125. Springer-Verlag, 2000.
142. N. V. Shilov and K. Yi. On expressive and model checking power of propositional program logics. In *Proc. of the 4th Int. Ershov Memorial Conf. 'Perspective of System Informatics' (PSI'01)*, volume 2244 of *LNCS*, pages 39–46. Springer-Verlag, 2001.
143. R. E. Shostak. Deciding linear inequalities by computing loop residues. *Journal of the ACM*, 28(4):769–779, 1981.
144. J. Sifakis and S. Yovine. Compositional specification of timed systems. In *Proc. of the 13th Annual Symposium on Theoretical Aspects of Computer Science (STACS'96)*, volume 1046 of *LNCS*, pages 347–359. Springer-Verlag, 1996.
145. M. Sorea. Bounded model checking for timed automata. In *Proc. of the 3rd Workshop on Models for Time-Critical Systems (MTCS'02)*, volume 68(5) of *ENTCS*. Elsevier, 2002.
146. R. L. Spelberg, H. Toetenel, and M. Ammerlaan. Partition refinement in real-time model checking. In *Proc. of the 5th Int. Conf. on Formal Techniques in Real-Time and Fault Tolerant Systems (FTRTFT'98)*, volume 1486 of *LNCS*, pages 143–157. Springer-Verlag, 1998.

147. J. Srba. Timed-arc Petri nets vs. networks of timed automata. In *Proc. of the 26th Int. Conf. on Applications and Theory of Petri Nets (ICATPN'05)*, volume 3536 of *LNCS*, pages 385–402. Springer-Verlag, 2005.
148. P. Starke. *Analyse von Petri-Netz-Modellen.* Teubner Verlag, 1990.
149. Ch. Stehno. PEP version 2.0. In *Tool Demonstrations on the 22nd Int. Conf. on Applications and Theory of Petri Nets (ICATPN'01)*, 2001.
150. C. Stirling. Comparing linear and branching time temporal logics. In *Proc. of the Int. Colloquium on Temporal Logic in Specification '87*, volume 398 of *LNCS*, pages 1–20. Springer-Verlag, 1989.
151. O. Strichman, S. Seshia, and R. Bryant. Deciding separation formulas with SAT. In *Proc. of the 14th Int. Conf. on Computer Aided Verification (CAV'02)*, volume 2404 of *LNCS*, pages 209–222. Springer-Verlag, 2002.
152. M. Szreter. Selective search in bounded model checking of reachability properties. In *Proc. of the 3rd Int. Symp. on Automated Technology for Verification and Analysis (ATVA'05)*, volume 3707 of *LNCS*, pages 159–173. Springer-Verlag, 2005.
153. M. Szreter. Generalized blocking clauses in unbounded model checking. In *Proc. of the 3rd Int. Workshop on Constraints in Formal Verification (CFV'05)*, 2006. To appear in ENTCS.
154. M. Talupur, N. Sinha, O. Strichman, and A. Pnueli. Range allocation for separation logic. In *Proc. of the 16th Int. Conf. on Computer Aided Verification (CAV'04)*, volume 3114 of *LNCS*, pages 148–161. Springer-Verlag, 2004.
155. A. Tarski. A lattice-theoretical fixpoint theorem and its applications. *Pacific Journal of Mathematics*, 5:285–309, 1955.
156. S. Tripakis. *L'Analyse Formelle des Systèmes Temporisés en Pratique.* PhD thesis, Joseph Fourier University, Grenoble, 1998.
157. S. Tripakis. Minimization of timed systems. http://verimag.imag.fr/~tripakis/dea.ps.gz, 1998.
158. S. Tripakis and S. Yovine. Analysis of timed systems based on time-abstracting bisimulations. In *Proc. of the 8th Int. Conf. on Computer Aided Verification (CAV'96)*, volume 1102 of *LNCS*, pages 232–243. Springer-Verlag, 1996.
159. S. Tripakis and S. Yovine. Analysis of timed systems using time-abstracting bisimulations. *Formal Methods in System Design*, 18(1):25–68, 2001.
160. S. Tripakis, S. Yovine, and A. Bouajjani. Checking timed Büchi automata emptiness efficiently. *Formal Methods in System Design*, 26(3):267–292, 2005.
161. J. Tsai, S. Yang, and Y. Chang. Timing constraint Petri nets and their application to schedulability analysis of real-time system specifications. *IEEE Trans. on Software Eng.*, 21(1):32–49, 1995.
162. K. Varpaaniemi. Efficient detection of deadlocks in Petri nets. Technical Report HUT-TCS-A26, Helsinki University of Technology, Digital Systems Laboratory, Espoo, Finland, October 1993.
163. I. B. Virbitskaite and E. A. Pokozy. A partial order method for the verification of time Petri nets. In *Fundamental of Computation Theory*, volume 1684 of *LNCS*, pages 547–558. Springer-Verlag, 1999.
164. W. Visser. *Efficient CTL* Model Checking Using Games and Automata.* PhD thesis, Faculty of Science and Engineering, University of Manchester, 1998.
165. B. Walter. Timed Petri nets for modelling and analysing protocols with real-time characteristics. In *Proc. of the 3rd IFIP Workshop on Protocol Specification, Testing, and Verification*, pages 149–159. North Holland, 1983.

166. F. Wang. Region Encoding Diagram for fully symbolic verification of real-time systems. In *Proc. of the 24th Int. Computer Software and Applications Conf. (COMPSAC'00)*, pages 509–515. IEEE Computer Society, October 2000.
167. F. Wang. RED: Model checker for timed automata with clock-restriction diagram. In *Proc. of the Int. Workshop on Real-Time Tools (RT-TOOLS'01)*, 2001.
168. F. Wang. Verification of timed automata with BDD-like data structures. In *Proc. of the 4th Int. Conf. on Verification, Model Checking, and Abstract Interpretation (VMCAI'03)*, volume 2575 of *LNCS*, pages 189–205. Springer-Verlag, 2003.
169. J. Wang. *Timed Petri Nets: Theory and Applications.* Kluwer Academic Publishers, 1998.
170. B. Woźna, A. Lomuscio, and W. Penczek. Bounded model checking for knowledge and real time. Submitted to Journal of Artificial Intelligence Research.
171. B. Woźna, A. Zbrzezny, and W. Penczek. Checking reachability properties for timed automata via SAT. *Fundamenta Informaticae*, 55(2):223–241, 2003.
172. M. Yannakakis and D. Lee. An efficient algorithm for minimizing real-time transition systems. In *Proc. of the 5th Int. Conf. on Computer Aided Verification (CAV'93)*, volume 697 of *LNCS*, pages 210–224. Springer-Verlag, 1993.
173. W. Yi, P. Pettersson, and M. Daniels. Automatic verification of real-time communicating systems by constraint-solving. In *Proc. of the 7th IFIP WG6.1 Int. Conf. on Formal Description Techniques (FORTE'94)*, volume 6 of *IFIP Conference Proceedings*, pages 243–258. Chapman & Hall, 1994.
174. T. Yoneda and H. Ryuba. CTL model checking of time Petri nets using geometric regions. *IEICE Trans. Inf. and Syst.*, 3:1–10, 1998.
175. S. Yovine. KRONOS: A verification tool for real-time systems. *International Journal of Software Tools for Technology Transfer*, 1(1/2):123–133, 1997.
176. S. Yovine. Model checking timed automata. In *Embedded Systems*, volume 1494 of *LNCS*, pages 114–152. Springer-Verlag, 1997.
177. A. Zbrzezny. Improvements in SAT-based reachability analysis for timed automata. *Fundamenta Informaticae*, 60(1-4):417–434, 2004.
178. A. Zbrzezny. SAT-based reachability checking for timed automata with diagonal constraints. *Fundamenta Informaticae*, 67(1-3):303–322, 2005.
179. L. Zhang, C. Madigan, M. Moskewicz, and S. Malik. Efficient conflict driven learning in a boolean satisfiability solver. In *Proc. of Int. Conf. on Computer-Aided Design (ICCAD'01)*, pages 279–285, 2001.

Index

Author Index